Mot

Pleasure, Technology and Stars

Christoph Schiller

MOTION MOUNTAIN

The Adventure of Physics
Volume V

Motion Inside Matter –
Pleasure, Technology
and Stars

Edition 25.62, free colour pdf with films
available at www.motionmountain.net

To Britta, Esther and Justus Aaron

τῷ ἐμοὶ δαίμονι

Die Menschen stärken, die Sachen klären.

PREFACE

This book is written for anybody who is curious about nature and motion. Curiosity about how bodies, images and empty space move leads to many adventures. This volume presents the best adventures about the motion inside people, inside animals, and inside any other type of matter – from the largest stars to the smallest nuclei.

Motion *inside* bodies – dead or alive – is described by quantum theory. Quantum theory describes all motion with the quantum of action $\hbar$, the smallest change observed in nature. Building on this basic idea, the text first shows how to describe life, death and pleasure. Then, the text explains the observations of chemistry, materials science, astrophysics and particle physics. In the structure of physics, these topics correspond to the three 'quantum' points in Figure 1. The story of motion inside living and non-living matter, from the coldest gases to the hottest stars, is told here in a way that is simple, up to date and captivating.

In order to be *simple*, the text focuses on concepts, while keeping mathematics to the necessary minimum. Understanding the concepts of physics is given precedence over using formulae in calculations. The whole text is within the reach of an undergraduate.

In order to be *up to date*, the text is enriched by the many gems – both theoretical and empirical – that are scattered throughout the scientific literature.

In order to be *captivating*, the text tries to startle the reader as much as possible. Reading a book on general physics should be like going to a magic show. We watch, we are astonished, we do not believe our eyes, we think, and finally we understand the trick. When we look at nature, we often have the same experience. Indeed, every page presents at least one surprise or provocation for the reader to think about. Numerous interesting challenges are proposed.

The motto of the text, *die Menschen stärken, die Sachen klären*, a famous statement by Hartmut von Hentig on pedagogy, translates as: 'To fortify people, to clarify things.' Clarifying things – and adhering only to the truth – requires courage, as changing the habits of thought produces fear, often hidden by anger. But by overcoming our fears we grow in strength. And we experience intense and beautiful emotions. All great adventures in life allow this, and exploring motion is one of them. Enjoy it!

Munich, 1 December 2013.

---

* 'First move, then teach.' In modern languages, the mentioned type of *moving* (the heart) is called *motivating*; both terms go back to the same Latin root.

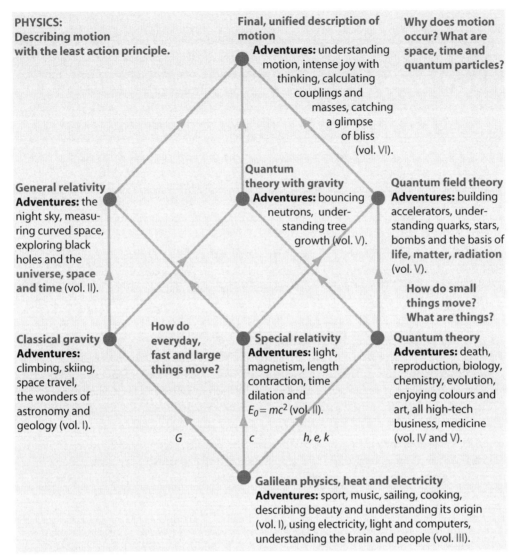

FIGURE 1 A complete map of physics: the connections are defined by the speed of light c, the gravitational constant G, the Planck constant h, the Boltzmann constant k and the elementary charge e.

ADVICE FOR LEARNERS

Learning widens knowledge, improves intelligence and generates proudness. And learning allows us to discover what kind of person we can be.

Learning from a book, especially one about nature, should be fast and fun. The most inefficient and most tedious learning method is to use a marker to underline text: it wastes time, provides false comfort and makes the text unreadable. Nobody marking text is learning efficiently or is enjoying it.

In my experience as a student and teacher, one learning method never failed to transform unsuccessful pupils into successful ones: if you read a text for study, summarize every section you read, *in your own words and images, aloud*. If you are unable to do so, read the section again. Repeat this until you can clearly summarize what you read in your own words and images, aloud. You can do this alone or with friends, in a room or while walking. If you do this with everything you read, you will reduce your learning and reading time significantly, enjoy learning from good texts much more and hate bad texts much less. Masters of the method can use it even while listening to a lecture, in a low voice, thus avoiding to ever take notes.

Advice for teachers

A teacher likes pupils and likes to lead them into exploring the field he chose. His or her enthusiasm for the job is the key to job satisfaction. If you are a teacher, before the start of a lesson, picture, feel and tell yourself how you enjoy the topic of the lesson; then picture, feel and tell yourself how you will lead each of your pupils into enjoying that topic as much as you do. Do this exercise consciously, every time. You will minimize trouble in your class and maximize your success.

This book is not written with exams in mind; it is written to make teachers and students *understand* and *enjoy* physics, the science of motion.

Using this book

Solutions and hints for *challenges* are given in the appendix. Challenges are classified as research level (r), difficult (d), standard student level (s) and easy (e). Challenges for which no solution has yet been included in the text are marked (ny).

The films mentioned in the text can be watched by downloading the pdf file of the book from www.motionmountain.net and then opening the file in the free Adobe Reader. The download is free and anonymous. It also contains the full colour version of all figures.

Feedback and support

This text is and will remain free to download from the internet. I would be delighted to receive an email from you at fb@motionmountain.net, especially on the following issues:

— What was unclear and should be improved?
— What story, topic, riddle, picture or movie did you miss?
— What should be corrected?

Challenge 1 s

In order to simplify annotations, the pdf file allows adding yellow sticker notes in Adobe Reader. Help on the specific points listed on the www.motionmountain.net/help.html web page would be particularly welcome. All feedback will be used to improve the next edition. On behalf of all readers, thank you in advance for your input. For a particularly useful contribution you will be mentioned – if you want – in the acknowledgements, receive a reward, or both.

Your donation to the charitable, tax-exempt non-profit organisation that produces, translates and publishes this book series is welcome! For details, see the web page www. motionmountain.net/donation.html. If you want, your name will be included in the sponsor list. Thank you in advance for your help, on behalf of all readers across the world.

The present paper edition of this book is printed on demand and delivered by
mail to any address of your choice. It can be ordered at www.lulu.com/spotlight/
motionmountain. But above all, enjoy the reading!

Contents

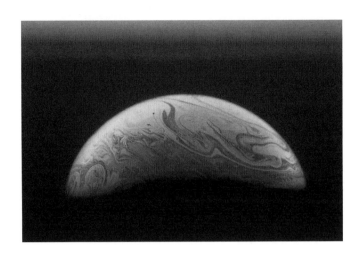

MOTION INSIDE MATTER –
PLEASURE, TECHNOLOGY AND STARS

In our quest to understand how things move
as a result of a smallest change value in nature, we discover
how pleasure appears,
why the floor does not fall but keeps on carrying us,
that interactions are exchanges of radiation particles,
that matter is not permanent,
how quantum effects increase human wealth and health,
why empty space pulls mirrors together,
why the stars shine,
where the atoms inside us come from,
how quantum particles make up the world,
and why swimming and flying is not so easy.

Chapter 1

MOTION FOR ENJOYING LIFE

> " Homo sum, humani nil a me alienum puto.* "
> Terence

SINCE we have explored quantum effects in the previous volume, let us now have some serious fun with *applied* quantum physics. The quantum of action $\hbar$ has significant consequences for medicine, biology, chemistry, materials science, engineering and the light emitted by stars. Also art, the colours and materials it uses, and the creative process in the artist, are based on the quantum of action. From a physics standpoint, all these domains study *motion inside matter*.

Inside matter, we observe, above all, small motions of quantum particles.** Thus the understanding and the precise description of matter requires quantum physics. In the following, we will only explore a cross-section of these topics, but it will be worth it. We start our exploration of motion inside matter with three special forms that are of special importance to us: life, reproduction and death. We mentioned at the start of quantum physics that none of these forms of motion can be described by classical physics. Indeed, Vol. IV, page 15 life, reproduction and death are quantum effects. In addition, every perception, every sense, and thus every kind of pleasure, are quantum effects. The same is true for all our actions. Let us find out why.

FROM QUANTUM PHYSICS TO BIOLOGICAL MACHINES AND MINIATURIZATION

We know that all of quantum theory can be resumed in one sentence:

> ▷ *In nature, actions or changes smaller than $\hbar = 1.1 \cdot 10^{-34}$ Js are not observed.*

In the following, we want to understand how this observation explains life pleasure and death. An important consequence of the quantum of action is well-known.

---

* 'I am a man and nothing human is alien to me.' Terence is Publius Terentius Afer (*c.* 190–159 BCE), the important roman poet. He writes this in his play *Heauton Timorumenos*, verse 77.
** The photograph on page 14 shows a soap bubble, the motion of the fluid in it, and the interference colours; it was taken and is copyright by Jason Tozer for Creative Review/Sony.

Starting size: the dot on the letter i

Final size:

http://swfsc.nmfs.noaa.gov/PRD/

FIGURE 2 Metabolic growth can lead from single cells, about 0.1 mm in diameter, to living beings of 25 m in size, such as the baobab or the blue whale (© Ferdinand Reus, NOAA).

▷ *If it moves, it is made of quantons, or quantum particles.*

Step by step we will discover how these statements are reflected in the behaviour of living beings. But what are living beings?

Living beings are physical systems that show metabolism, information processing, information exchange, reproduction and motion. Obviously, all these properties follow from a single one, to which the others are enabling means:

▷ *A living being is an object able to self-reproduce.*

By self-reproduction, we mean that an object uses its own metabolism to reproduce. From your biology lessons in secondary school you might remember the main properties of reproduction.* Reproduction is characterized by random changes, called *mutations*, that distinguish one generation from the next. The statistics of mutations, for example Mendel's 'laws' of heredity, and the lack of intermediate states, are direct consequences of quantum theory. In other words, reproduction and heredity are quantum effects.

Reproduction requires *growth*, and growth needs metabolism. *Metabolism* is a chemical process, and thus a quantum process, to harness energy, harness materials, realize growth, heal injuries and realize reproduction.

Page 54

In order to reproduce, living beings must be able to move in self-directed ways. An object able to perform self-directed motion is called a *machine*. All self-reproducing beings are machines. Even machines that do not grow still need fuel, and thus need a metabolism. All machines are based on quantum effects.

Reproduction, growth and motion are simpler the *smaller* the adult system is. Therefore, most living beings are extremely small machines for the tasks they perform. This is especially clear when they are compared to human-made machines. The smallness of living beings is often astonishing, because the design of human-made machines has considerably fewer requirements: human-made machines do not need to be able to reproduce; as a result, they do not need to be made of a single piece of matter, as all living beings have to. But despite all the strong engineering restrictions that life is subjected to, living beings hold many miniaturization world records for machines:

— The brain has the highest processing power per volume of any calculating device so far. Just look at the size of chess champion Gary Kasparov and the size of the computer against which he played and lost. Or look at the size of any computer that attempts to speak.
— The brain has the densest and fastest memory of any device so far. The set of compact discs (CDs) or digital versatile discs (DVDs) that compare with the brain is many thousand times larger in volume.

Vol. III, page 221

— Motors in living beings are several orders of magnitude smaller than human-built ones. Just think about the muscles in the legs of an ant.
— The motion of living beings beats the acceleration of any human-built machine by orders of magnitude. No machine moves like a grasshopper.

---

* There are examples of objects which reproduce and which nobody would call living. Can you find some examples? To avoid misunderstandings, whenever we say 'reproduction' in the following, we always mean 'self-reproduction'. Challenge 2 s

— Living being's sensor performance, such as that of the eye or the ear, has been surpassed by human machines only recently. For the nose, this feat is still far in the future. Nevertheless, the sensor *sizes* developed by evolution – think also about the ears or eyes of a common fly – are still unbeaten.
— Living beings that fly, swim or crawl – such as fruit flies, plankton or amoebas – are still thousands of times smaller than anything comparable that is built by humans. In particular, already the navigation systems built by nature are far smaller than anything built by human technology.

Challenge 3 s   — Can you spot more examples?

The superior miniaturization of living beings – compared to human-built machines – is due to their continuous strife for efficient construction. The efficiency has three main aspects. First of all, in the structure of living beings, *everything is connected to everything*. Each part influences many others. Indeed, the four basic processes in life, namely metabolic, mechanical, hormonal and electric, are intertwined in space and time. For example, in humans, breathing helps digestion; head movements pump liquid through the spine; a single hormone influences many chemical processes. Secondly, *all parts in living systems have more than one function*. For example, bones provide structure and produce blood; fingernails are tools and shed chemical waste. Living systems use many

Challenge 4 e   such optimizations. Thirdly, living machine are *well* miniaturized because they make efficient use of quantum effects. Indeed, every single function in living beings relies on the quantum of action. And every such function is extremely well miniaturized. We explore a few important cases.

REPRODUCTION

All the astonishing complexity of life is geared towards reproduction. *Reproduction* is the ability of an object to build other objects similar to itself. Quantum theory told us

Vol. IV, page 113   that it is only possible to build a *similar* object, since an *exact* copy would contradict the quantum of action. But this limitation is not a disadvantage: a similar, thus imperfect copy is all that is required for life; indeed, a similar, thus imperfect copy is essential for biological evolution, and thus for change and specialization.

Since reproduction requires an increase in mass, as shown in Figure 2, all reproducing objects show both metabolism and growth. In order that growth can lead to an object similar to the original, a construction *plan* is necessary. This plan must be similar to the plan used by the previous generation. Organizing growth with a construction plan is only possible if nature is made of smallest entities which can be assembled following that plan.

We thus deduce that reproduction and growth implies that matter is made of smallest entities. If matter were not made of smallest entities, there would be no way to realize reproduction. The observation of reproduction thus implies the existence of atoms and the necessity of quantum theory! Indeed, without the quantum of action there would be no DNA molecules and there would be no way to inherit our own properties – our own construction plan – to children.

Passing on a plan from generation to generation requires that living beings have ways to store information. Living beings must have some built-in memory storage. We know

Vol. I, page 343   already that a system with memory must be made of *many* particles: there is no other

FIGURE 3 A quantum machine (© Elmar Bartel).

way to store information and secure its stability over time. The large number of particles is necessary to protect the information from the influences of the outside world.

Our own construction plan is stored in DNA molecules in the nucleus and the mitochondria of each of the millions of cells inside our body. We will explore some details below. The plan is thus indeed stored and secured with the help of *many* particles. Therefore, reproduction is first of all a transfer of parent's DNA to the next generation. The transfer is an example of motion. It turns out that this and all other examples of motion in our bodies occur in the same way, namely with the help of molecular machines. Page 58

Quantum machines

Living beings move. Living beings, such as the one of Figure 3, are machines. How do these living machines work? From a fundamental point of view, we need only a few sections of our walk so far to describe them: we need QED and sometimes universal gravity. Simply stated, life is an electromagnetic process taking place in weak gravity.* But the details of this statement are tricky and interesting.

Table 1 gives an overview of processes in living beings. Above all, the table shows that all processes in living beings are due to *molecular machines*.

> ▷ *A living being is a huge collection of specialized molecular machines.*

Molecular machines are among the most fascinating devices found in nature. Table 1 also shows that nature only needs a few such devices to realize all the motion types used

---

* In fact, also the nuclear interactions play some role for life: cosmic radiation is one source for random mutations, which are so important in evolution. Plant growers often use radioactive sources to increase mutation rates. Radioactivity can also terminate life or be of use in medicine. Ref. 1

The nuclear interactions are also implicitly involved in life in several other ways. The nuclear interactions were necessary to form the atoms – carbon, oxygen, etc. – required for life. Nuclear interactions are behind the main mechanism for the burning of the Sun, which provides the energy for plants, for humans and for all other living beings (except a few bacteria in inaccessible places).

Summing up, the nuclear interactions play a role in the appearance and in the destruction of life; but they usually play no role for the actions or functioning of particular living beings.

TABLE 1 Motion and molecular machines found in living beings.

MOTION TYPE	EXAMPLES	INVOLVED MOTORS
Growth	collective molecular processes in cell growth, cell shape change, cell motility	linear molecular motors, ion pumps
	gene turn-on and turn-off	linear molecular motors
	ageing	linear molecular motors
Construction	material transport (polysaccharides, lipids, proteins, nucleic acids, others)	muscles, linear molecular motors
	forces and interactions between biomolecules and cells	pumps in cell membranes
Functioning	metabolism (respiration, digestion)	muscles, ATP synthase, ion pumps
	muscle working	linear molecular motors, ion pumps
	thermodynamics of whole living system and of its parts	muscles
	nerve signalling	ion motion, ion pumps
	brain working, thinking	ion motion, ion pumps
	memory: long-term potentiation	chemical pumps
	memory: synapse growth	linear molecular motors
	hormone production	chemical pumps
	illnesses	cell motility, chemical pumps
	viral infection of a cell	rotational molecular motors for RNA transport
Defence	the immune system	cell motility, linear molecular motors
	blood clotting	chemical pumps
	bronchial cleaning	cilial motors
Sensing	eye	chemical pumps, ion pumps
	ear	hair motion sensors, ion pumps, rotary molecular motors
	smell	ion pumps
	touch	ion pumps
Reproduction	information storage and retrieval	linear and rotational molecular motors inside nuclei
	cell division	linear molecular motors
	sperm motion	linear molecular motors
	courting, using brain and muscles	linear molecular motors, ion pumps

by humans and by all other living beings: *pumps* and *motors*. Given the long time that living systems have been around, these devices are extremely efficient. They are found in every cell, including those of Figure 5. The specialized molecular machines in living beings are ion pumps, chemical pumps and rotational and linear molecular motors. Ion and chemical pumps are found *in* membranes and transport matter *across* membranes. Rotational and linear motors move structures *against* membranes. Even though there is still a lot to be learned about molecular machines, the little that is known is already spectacular.

How do we move? – Molecular motors

How do our muscles work? What is the underlying motor? One of the beautiful results of modern biology is the elucidation of this issue.

> ▷ *Muscles work because they contain molecules which change shape when sup-plied with energy.*

This shape change is *repeatable*. A clever combination and repetition of these molecular shape changes is then used to generate macroscopic motion.

> ▷ *Each shape-changing molecule is a molecular motor.*

There are three basic classes of molecular motors: linear motors, rotational motors and pumps.

1. *Linear molecular motors* are at the basis of muscle motion; an example is given in Figure 4. Other linear motors separate genes during cell division. Linear motors also move organelles inside cells and displace cells through the body during embryo growth, when wounds heal, and in all other cases of cell motility. A typical molecular motor consumes around 100 to 1000 ATP molecules per second, thus about 10 to 100 aW. The numbers are small; however, we have to take into account that the power due to the white noise of the surrounding water is 10 nW. In other words, in every molecular motor, the power of the environmental noise is eight to nine orders of magnitude higher than the power consumed by the motor! The ratio shows what a fantastic piece of machinery such a molecular motor is. At our scale, this would correspond to a car that drives, all the time, through an ongoing storm and earthquake. Challenge 5 s

2. We encountered *rotational motors* already earlier on; nature uses them to rotate the cilia of many bacteria as well as sperm tails. Researchers have also discovered that evolution produced molecular motors which turn around DNA helices like a motorized bolt would turn around a screw. Such motors are attached at the end of some viruses and insert the DNA into virus bodies when they are being built by infected cells, or extract the DNA from the virus after it has infected a cell. The most important rotational motor, and the smallest known so far – 10 nm across and 8 nm high – is ATP synthase, a protein that synthesizes most ATP in cells. Vol. I, page 87
Ref. 2

Ref. 3
Page 26

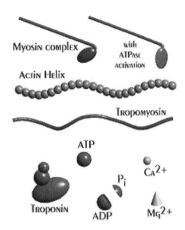 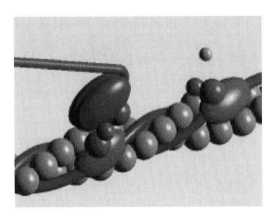

FIGURE 4 Left: myosin and actin are the two protein molecules that realize the most important linear molecular motor in living beings, including the motion in muscles. Right: the resulting motion step is 5.5 nm long; it has been slowed down by about a factor of ten (image and QuickTime film © San Diego State University, Jeff Sale and Roger Sabbadini).

3. *Molecular pumps* are equally essential to life. They pump chemicals, such as ions or specific molecules, into every cell or out of it, using energy, even if the concentration gradient tries to do the opposite. Molecular pumps are thus essential in ensuring that life is a process *far* from equilibrium. Malfunctioning molecular pumps are responsible for many health problems, for example for the water loss in cholera infection.

Ref. 4 In the following, we explore a few specific molecular motors found in cells. How molecules produce movement in linear motors was uncovered during the 1980s. The results started a wave of research on all other molecular motors found in nature. The research showed that molecular motors differ from most everyday motors: molecular motors do *not* involve temperature gradients, as car engines do, they do *not* involve electrical currents, as electrical motors do, and they do *not* rely on concentration gradients, as chemically induced motion, such as the rising of a cake, does.

LINEAR MOLECULAR MOTORS

The central element of the most important linear molecular motor is a combination of two protein molecules, namely myosin and actin. Myosin changes between two shapes and literally *walks* along actin. It moves in regular small steps, as shown in Figure 4. The motion step size has been measured, with the help of some beautiful experiments, to al-
Ref. 4 ways be an integer multiple of 5.5 nm. A step, usually forward, but sometimes backwards, results whenever an ATP (adenosine triphosphate) molecule, the standard biological fuel, hydrolyses to ADP (adenosine diphosphate) and releases the energy contained in the chemical bond. The force generated is about 3 to 4 pN; the steps can be repeated several times a second. Muscle motion is the result of thousand of millions of such elementary steps taking place in concert.

Why does this molecular motor work? The molecular motor is so small that the noise due to the Brownian motion of the molecules of the liquid around it is extremely intense.

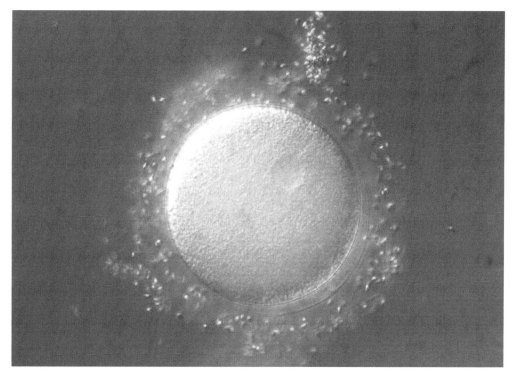

FIGURE 5 A sea urchin egg surrounded by sperm, or molecular motors in action: molecular motors make sperm move, make fecundation happen, and make cell division occur (photo by Kristina Yu, © Exploratorium www.exploratorium.edu).

Indeed, the transformation of disordered molecular motion into ordered macroscopic motion is one of the great wonders of nature.

Evolution is smart: with three tricks it takes advantage of Brownian motion and transforms it into macroscopic motion. (Molecular motors are therefore also called *Brownian motors*.) The first trick of evolution is the use of an asymmetric, but periodic potential, a so-called *ratchet*.* The second trick of evolution is a temporal variation of the potential of the ratchet, together with an energy input to make it happen. The two most important realizations are shown in Figure 6. Molecular motors thus work away from thermal equilibrium. The third trick is to take a large number of these molecular motors and to add their effects.

The periodic potential variation in a molecular motor ensures that for a short, recurring time interval the free Brownian motion of the moving molecule – typically 1 μm/s – affects its position. Subsequently, the molecule is fixed again. In most of the short time intervals of free Brownian motion, the position will not change. But if the position does change, the intrinsic asymmetry of the ratchet shape ensures that with high probability the molecule advances in the preferred direction. (The animation of Figure 4 lacks this irregularity.) Then the molecule is fixed again, waiting for the next potential change. On average, the myosin molecule will thus move in one direction. Nowadays the motion of single molecules can be followed in special experimental set-ups. These experiments

Ref. 5

---

* It was named after Ratchet Gearloose, the famous inventor from Duckburg.

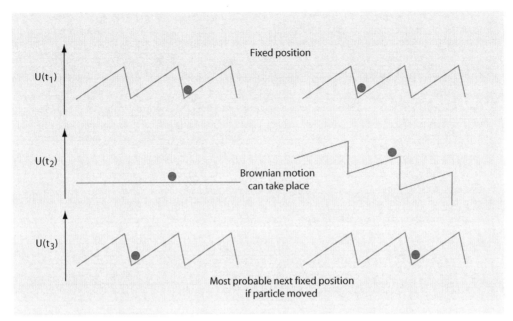

FIGURE 6 Two types of Brownian motors: switching potential (left) and tilting potential (right).

FIGURE 7 A classical ratchet, here of the piezoelectric kind, moves like a linear molecular motor (© PiezoMotor).

confirm that muscles use such a ratchet mechanism. The ATP molecule adds energy to the system and triggers the potential variation through the shape change it induces in the myosin molecule. Nature then takes millions of these ratchets together: that is how our muscles work.

Engineering and evolution took different choices. A moped contains one motor. An expensive car contains about 100 motors. A human contains at least 10^{16} motors.

Challenge 6 e

Another well-studied linear molecular motor is the kinesin–microtubule system that carries organelles from one place to the other within a cell. Like in the previous example, also in this case chemical energy is converted into unidirectional motion. Researchers were able to attach small silica beads to single molecules and to follow their motion. Using laser beams, they could even apply forces to these single molecules. Kinesin was found to move with around 800 nm/s, in steps lengths which are multiples of 8 nm, using one ATP molecule at a time, and exerting a force of about 6 pN.

Quantum ratchet motors do not exist only in living systems; they also exist as human-built systems. Examples are electrical ratchets that move single electrons and optical ratchets that drive small particles. These applications are pursued in various experimental research programmes.

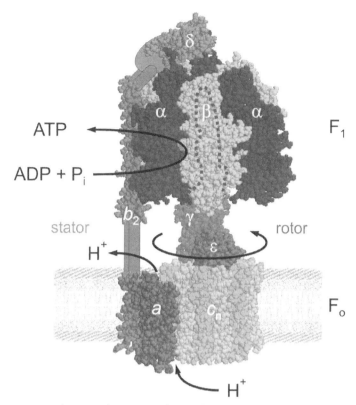

FIGURE 8 The structure of ATP synthase (© Joachim Weber).

Also *classical* ratchets exist. One example is found in every mechanical clock; another is shown in Figure 7. Many piezoelectric actuators work as ratchets and the internet is full of videos showing how they work. Piezoelectric ratchet, also called *ultrasound motors*, are found precision stages for probe motion and inside certain automatic zoom objectives in expensive photographic cameras. Also most atomic force microscopes and most scanning electron microscopes use ratchet actuators.

Molecular motors are essential for the growth and the working of nerves and the brain. A nerve contains large numbers of dozens of molecular motors types from all three main families: dynein, myosin and kinesin. These motors transport chemicals, called 'cargos', along axons and all have loading and unloading mechanisms at their ends. They are necessary to realize the growth of nerves, for example from the spine to the tip of the toes. Other motors control the growth of synapses, and thus ensure that we have long-term memory. Malfunctioning molecular motors are responsible for Alzheimer disease, Huntington disease, multiple sclerosis, certain cancers and many other diseases due either to genetic defects or to environmental poisons.

In short, without molecular motors, we could neither move nor think.

Ref. 6

A ROTATIONAL MOLECULAR MOTOR: ATP SYNTHASE

In cells, the usual fuel for most chemical reactions is adenosine triphosphate, or ATP. In plants, most ATP is produced on the membranes of cell organelles called *chloroplasts*, and in animal cells, in the so-called *mitochondria*. These are the power plants in most cells. ATP also powers most bacteria. It turns out that ATP is synthesized by a protein located in membranes. The protein is itself powered by protons, H^+, which form the basic fuel of the human body, whereas ATP is the high-level fuel. For example, most other molecular motors are powered by ATP. ATP releases its energy by being changed into to adenosine diphosphate, or ADP. The importance of ATP is simple to illustrate: every human synthesizes, during a typical day, an amount of ATP that is roughly equal to his or her body mass.

The protein that synthesizes ATP is simply called *ATP synthase*. In fact, ATP synthase differs slightly from organism to organism; however, the differences are so small that they can be neglected in most cases. (An important variation are those pumps where Na^+ ions replace protons.) Even though ATP synthase is a highly complex protein, its function it easy to describe: it works like a *paddle wheel* that is powered by a proton gradient across the membrane. Figure 8 gives an illustration of the structure and the process. The research that led to these discoveries was rewarded with the 1997 Nobel Prize in Chemistry.

Ref. 7

In fact, ATP synthase also works in the reverse: if there is a large ATP gradient, it pumps protons out of the cell. In short, ATP synthase is a rotational motor and molecular pump at the same time. It resembles the electric starting motor, powered by the battery, found in the older cars; in those cars, during driving, the electric motor worked as a dynamo charging the battery. ATP synthase has been studied in great detail. For example, it is known that it produces three ATP molecules per rotation, that it produces a torque of around $20kT/2\pi$. There are at least 10^{16} such motors in an adult body. The ATP synthase paddle wheel is one of the central building blocks of life.

ROTATIONAL MOTORS AND PARITY BREAKING

Why is our heart on the left side and our liver on the right side? The answer of this old question is known only since a few years. The left-right asymmetry, or chirality, of human bodies must be connected to the chirality of the molecules that make up life. In all living beings, sugars, proteins and DNS are chiral molecules, and only one types of molecules is actually found in living beings. But how does nature translate the chrality of molecules into a chirality of a body? Surprisingly, the answer, deduced only recently by Hirokawa Nobutaka and his team, depends on rotational molecular motors.

Ref. 8

The position of the internal organs is fixed during the early development of the embryo. At an early stage, a central part of the embryo, the so-called *node*, is covered with rotational cilia, i.e., rotating little hairs, as shown in Figure 9. (All vertebrates have a node at some stage of embryo development.) The nodal cilia are powered by a molecular motor; they all rotate in the *same* (clockwise) direction about ten times per second. The rotation direction is a consequence of the chirality of the molecules that are contained in the motors. However, since the cilia are inclined with respect to the surface – towards the tail end of the embryo – the rotating cilia effectively move the fluid above the node towards the left body side of the embryo. The fluid of this newly discovered *nodal flow*

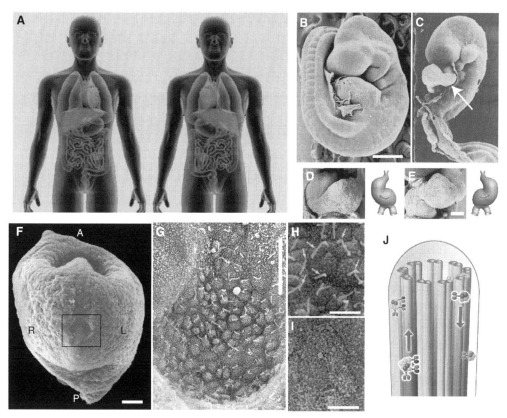

FIGURE 9 A: The asymmetric arrangement of internal organs in the human body: Normal arrangement, situs solitus, as common in most humans, and the mirrored arrangement, situs inversus. Images B to E are scanning electron micrographs of mouse embryos. B: Healthy embryos at this stage already show a right-sided tail. C: In contrast, mutant embryos with defect cilial motors remain unturned, and the heart loop is inverted, as shown by the arrow. D: Higher-magnification images and schematic representations of a normal heart loop. E: Similar image showing an inverted loop in a mutant embryo. Images F to I are scanning electron micrographs of a mouse node. F: A low-magnification view of a 7.5 day-old mouse embryo observed from the ventral side, with the black rectangle indicating the node. The orientation is indicated with the letters A for anterior, P for posterior, L for left and R for right; the scale bar is 100 *muunit*m. G: A higher-magnification image of the mouse node; the scale bar is 20 *muunit*m. H: A still higher-magnification view of healthy nodal cilia, indicated by arrows, and of the nodal pit cells; the scale bar is 5 *muunit*m. I: The nodal pit cells of mutant embryos lacking cilia. J: Illustration of the molecular transport inside a healthy flagellum (© Hirokawa Nobutaka).

contains substances – nodal vesicular parcels – that accumulate on the left side of the node and subsequently trigger processes that determine the position of the heart. If the cilia do not rotate, or if the flow is reversed by external means, the heart and other organs get misplaced. This connection also explains the other consequences of such genetic defects.

In other words, through the rotation of the cilia and the mentioned mechanisms, the chirality of molecules is mapped to the chirality of the whole vertebrate organism. It might even be that similar processes occur also elsewhere in nature, for example in the development of the brain asymmetry. This is still a field of intense research. Molecular

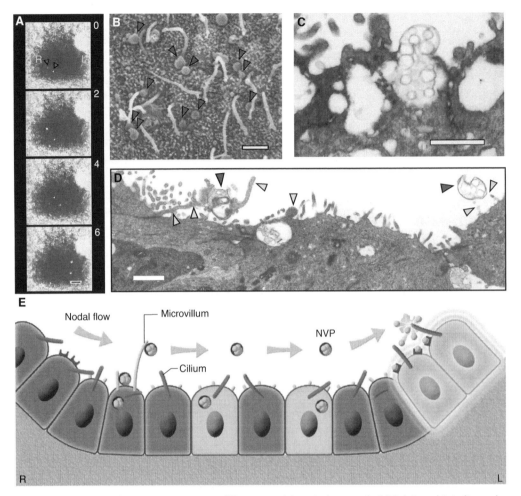

FIGURE 10 A: Optical microscope images of flowing nodal vesicular parcels (NVPs). L and R indicate the orientation. The NVPs, indicated by arrowheads, are transported to the left side by the nodal flow. The scale bar is 10 µm. B: A scanning electron micrograph of the ventral surface of nodal pit cells. The red arrowheads indicate NVP precursors. The scale bar is 2 µm. C and D: Transmission electron micrographs of nodal pit cells. The scale bar is 1 µm. E: A schematic illustration of NVP flow induced by the cilia. The NVPs are released from dynamic microvilli, are transported to the left side by the nodal flow due to the cilia, and finally are fragmented with the aid of cilia at the left periphery of the node. The green halos indicate high calcium concentration – a sign of cell activation that susequently starts organ formation (© Hirokawa Nobutaka).

motors are truly central to our well-being and life.

CURIOSITIES AND FUN CHALLENGES ABOUT BIOLOGY

> " Una pelliccia è una pelle che ha cambiato
> bestia.*
>   Girolamo Borgogelli Avveduti "

---

* 'A *fur* is a skin that has changed beast.'

Discuss the following argument: If nature were classical instead of quantum, there would not be just *two* sexes – nor any other *discrete* number of them, as in some lower animals – but there would be a *continuous range* of them. In a sense, there would be an infinite number of sexes. True?

Challenge 7 s

* *

Biological evolution can be summarized in three principles:

1. All living beings are different – also in a species.
2. All living beings have a tough life – due to competition.
3. Living beings with an advantage will survive and reproduce.

As a result of these three principles, with each generation, *species and living beings can change*. The result of accumulated generational change is called *biological evolution*. The last principle is often called the 'survival of the fittest'.

These three principles explain, among others, the change from unicellular to multicellular life, from fish to land animals, and from animals to people.

We note that quantum physics enters in all three principles that lead to evolution. For example, the differences mentioned in the first principle are due to quantum physics: perfect copies of macroscopic systems are impossible. And of course, life and metabolism are quantum effects. The second principle mentions competition; that is a type of measurement, which, as we saw, is only possible due to the existence of a quantum of action. The third principle mentions reproduction: that is again a quantum effect, based on the copying of genes, which are quantum structures. In short, biological evolution is a process due to the quantum of action.

Vol. IV, page 113

Vol. IV, page 19
Page 18

* *

How would you determine which of two identical twins is the father of a baby?

Challenge 8 d

* *

Can you give at least five arguments to show that a human clone, if there will ever be one, is a completely different person than the original?

Challenge 9 s

It is well known that the first ever cloned cat, *copycat*, born in 2002, looked completely different from the 'original' (in fact, its mother). The fur colour and its patch pattern were completely different from that of the mother. Analogously, identical human twins have different finger prints, iris scans, blood vessel networks and intrauterine experiences, among others.

Man properties of a mammal are not determined by genes, but by the environment of the pregnancy, in particular by the womb and the birth experience. Womb influences include fur patches, skin fold shapes, but also character traits. The influence of the birth experience on the character is well-known and has been studied by many psychologists.

* *

A famous unanswered question on evolution: how did the first *kefir grains* form? Kefir grains produce the kefir drink when covered with milk for about 8 to 12 hours. The grains consist of a balanced mixture of about 40 types of bacteria and yeasts. All kefir grains in the world are related. But how did the first ones form, about 1000 years ago?

Challenge 10 r

* *

Molecular motors are quite capable. The molecular motors in the sooty shearwater (*Puffinus griseus*), a 45 cm long bird, allow it to fly 74 000 km in a year, with a measured record of 1094 km a day.

* *

When the ciliary motors that clear the nose are overwhelmed and cannot work any more, they send a distress signal. When enough such signals are sent, the human body triggers the sneezing reaction. The sneeze is a reaction to blocked molecular motors.

* *

The growth of human embryos is one of the wonders of the world. The website embryo. soad.umich.edu provides extensive data, photos, animations and magnetic resonance images on the growth process.

* *

Many molecules found in living beings, such as sugar, have mirror molecules. However, in all living beings only one of the two sorts is found. Life is intrinsically asymmetric. Challenge 11 s   How can this be?

* *

How is it possible that the genetic difference between man and chimpanzee is regularly given as about 1 %, whereas the difference between man and woman is one chromosome Challenge 12 s   in 46, in other words, about 2.2 %?

* *

What is the longest time a single bacterium has survived? It is more than the 5000 years of the bacteria found in Egyptian mummies. For many years, the time was estimated to Ref. 9   lie at over 25 million years, a value claimed for the bacteria spores resurrected from the intestines in insects enclosed in amber. Then it was claimed to lie at over 250 million years, the time estimated that certain bacteria discovered in the 1960s by Heinz Dombrowski in (low-radioactivity) salt deposits in Fulda, in Germany, have hibernated there before being brought back to life in the laboratory. A similar result has been recently claimed by the discovery of another bacterium in a North-American salt deposit in the Ref. 10   Salado formation.

However, these values are now disputed, as DNA sequencing has shown that these bacteria were probably due to sample contamination in the laboratory, and were not part Ref. 11   of the original sample. So the question is still open.

* *

In 1967, a TV camera was deposited on the Moon. Unknown to everybody, it contained a small patch of *Streptococcus mitis*. Three years later, the camera was brought back to Earth. The bacteria were still alive. They had survived for three years without food, water or air. Life can be resilient indeed. This widely quoted story is so unbelievable that it was checked again in 2011. The conclusion: the story is false; the bacteria were added by mistake in the laboratory after the return of the camera.

TABLE 2 Approximate number of living species.

LIFE GROUP	DESCRIBED SPECIES	ESTIMATED SPECIES	
		MIN.	MAX.
Viruses	$4 \cdot 10^3$	$50 \cdot 10^3$	$1 \cdot 10^6$
Prokaryotes ('bacteria')	$4 \cdot 10^3$	$50 \cdot 10^3$	$3 \cdot 10^6$
Fungi	$72 \cdot 10^3$	$200 \cdot 10^3$	$2.7 \cdot 10^6$
Protozoa	$40 \cdot 10^3$	$60 \cdot 10^3$	$200 \cdot 10^3$
Algae	$40 \cdot 10^3$	$150 \cdot 10^3$	$1 \cdot 10^6$
Plants	$270 \cdot 10^3$	$300 \cdot 10^3$	$500 \cdot 10^3$
Nematodes	$25 \cdot 10^3$	$100 \cdot 10^3$	$1 \cdot 10^6$
Crustaceans	$40 \cdot 10^3$	$75 \cdot 10^3$	$200 \cdot 10^3$
Arachnids	$75 \cdot 10^3$	$300 \cdot 10^3$	$1 \cdot 10^6$
Insects	$950 \cdot 10^3$	$2 \cdot 10^6$	$100 \cdot 10^6$
Molluscs	$70 \cdot 10^3$	$100 \cdot 10^3$	$200 \cdot 10^3$
Vertebrates	$45 \cdot 10^3$	$50 \cdot 10^3$	$55 \cdot 10^3$
Others	$115 \cdot 10^3$	$200 \cdot 10^3$	$800 \cdot 10^3$
Total	$\mathbf{1.75 \cdot 10^6}$	$\mathbf{3.6 \cdot 10^6}$	$\mathbf{112 \cdot 10^6}$

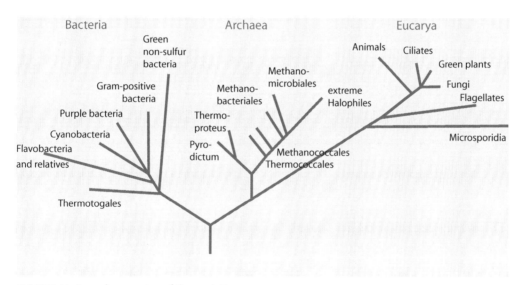

FIGURE 11 A modern version of the evolutionary tree.

* *

In biology, classifications are extremely useful. (This is similar to the situation in astrophysics, but in full contrast to the situation in physics.) Table 2 gives an overview of the magnitude of the task. This wealth of material can be summarized in one graph, shown Ref. 13 in Figure 11. Newer research seems to suggest some slight changes to the picture. So far however, there still is only a single root to the tree.

* *

Muscles produce motion through electrical stimulation. Can technical systems do the same? Candidate are appearing: so-called *electroactive polymers* change shape when they are activated with electrical current or with chemicals. They are lightweight, quiet and simple to manufacture. However, the first arm wrestling contest between human and artificial muscles, held in 2005, was won by a teenage girl. The race to do better is ongoing.

* *

Life is not a clearly defined concept. The definition used above, the ability to self-reproduce, has its limits. Can it be applied to old animals, to a hand cut off by mistake, to sperm, to ovules or to the first embryonal stages of a mammal? The definition of life also gives problems when trying to apply it to single cells. Can you find a better definition? Is Challenge 13 e   the definition of living beings as 'what is made of cells' useful?

* *

Every example of growth is a type of motion. Some examples are extremely complex. Take the growth of acne. It requires a lack of zinc, a weak immune system, several bacteria, as well as the help of *Demodex brevis*, a mite (a small insect) that lives in skin pores. With a size of 0.3 mm, somewhat smaller than the full stop at the end of this sentence, this and other animals living on the human face can be observed with the help of a strong magnifying glass.

* *

Humans have many living beings on board. For example, humans need bacteria to live. It is estimated that 90% of the bacteria in the human mouth alone are not known yet; only about 1000 species have been isolated so far.

     Bacteria are essential for our life: they help us to digest and they defend us against Ref. 14   illnesses due to dangerous bacteria. In fact, the number of bacteria in a human body is estimated to be 10^{14}, whereas the number of cells in a human body is estimated to lie between 10^{13} and $5 \cdot 10^{13}$. In short, a human body contains about ten times more bacteria than own cells! Nevertheless, the combined mass of all bacteria in a human body is estimated to be only about 1 kg, because bacteria are much smaller than human cells, on average.

     Of the around 100 groups of bacteria in nature, the human body mainly contains species from four of them: actinobacteria, bacteroidetes, firmicutes and proteobacteria. They play a role in obesity, malnutrition, heart disease, diabetes, multiple sclerosis, autism and many more. The connections are an important domain of research.

* *

How do trees grow? When a tree – biologically speaking, a *monopodal phanerophyte* – grows and produces leaves, between 40% and 60% of the mass it consists of, namely the water and the minerals, has to be lifted upwards from the ground. (The rest of the mass comes from the CO_2 in the air.) How does this happen? The materials are pulled upwards by the water columns inside the tree; the pull is due to the negative pressure that is created when the top of the column evaporates. This is called the *transpiration-cohesion-tension model*. (This summary is the result of many experiments.) In other words, no energy is needed for the tree to pump its materials upwards.

Trees do not need energy to transport water. As a consequence, a tree grows purely by adding material to its surface. This implies that when a tree grows, a branch that is formed at a given height is also found at that *same* height during the rest of the life of that tree. Just check this observation with the trees in your garden.   Challenge 14 e

* *

Mammals have a narrow operating temperature. In contrast to machines, humans function only if the internal temperature is within a narrow range. Why? And does this requirement also apply to extraterrestrials – provided they exist?   Challenge 15 d

* *

How did the first cell arise? This question is still open. As a possible step towards the answer, researchers have found several substances that spontaneously form closed membranes in water. Such substances also form foams. It might well be that life formed in foam. Other options discussed are that life formed underwater, at the places where magma rises into the ocean. Elucidating the question is one of the great open riddles of biology.   Challenge 16 r

* *

Could life have arrived to Earth from outer space?   Challenge 17 s

* *

Is there life elsewhere in the universe? The answer is clear. First of all, there *might* be life elsewhere, though the probability is extremely small, due to the long times involved and the requirements for a stable stellar system, a stable planetary system, and a stable geological system. In addition, so far, all statements that claim to have detected an example were *lies*. Not mistakes, but actual lies. The fantasy of extraterrestrial life poses an interesting challenge to everybody: Why would an extraterrestrial being be of interest to you? If you can answer, realize the motivation in some other way, *now*, without waiting. If you cannot answer, do something else.   Challenge 18 e

* *

What could *holistic medicine* mean to a scientist, i.e., avoiding nonsense and false beliefs? Holistic medicine means treating illness with view on the whole person. That translates to four domains:
— *physical* support, to aid mechanical or thermal healing processes in the body;
— *chemical* support, with nutrients or vitamins;
— *signalling* support, with electrical or chemical means, to support the signalling system of the body;
— *psychological* support, to help all above processes.
When all theses aspects are taken care of, healing is as rapid and complete as possible. However, one main rule remains: *medicus curat, natura sanat.**

---

* 'The physician helps, but nature heals.'

* *

Life is, above all, beautiful. For example, go to www.thedeepbook.org to enjoy the beauty of life deep in the ocean.

* *

What are the effects of *environmental pollution* on life? Answering this question is an intense field of modern research. Here are some famous stories.

— Herbicides and many genetically altered organisms kill bees. For this reason, bees are dying (since 2007) in the United States; as a result, many crops – such as almonds and oranges – are endangered there. In countries where the worst herbicides and genetically modified crops have been banned, bees have no problems. An example is France, where the lack of bees posed a threat to the wine industry.
— Chemical pollution leads to malformed babies. In mainland China, one out of 16 children is malformed for this reason (in 2007). In Japan, malformations have been much reduced – though not completely – since strict anti-pollution laws have been passed.
— Radioactive pollution kills. In Russia, the famous Lake Karachay had to be covered by concrete because its high radioactivity killed anybody that walked along it for an hour.
— Smoking kills – though slowly. Countries that have lower smoking rates or that have curbed smoking have reduced rates for cancer and several other illnesses.
— Eating tuna is dangerous for your health, because of the heavy metals it contains.
— Cork trees are disappearing. The wine industry has started large research programs to cope with this problem.
— Even arctic and antarctic animals have livers full of human-produced chemical poisons.
— Burning fuels rises the CO_2 level of the atmosphere. This leads to many effects for the Earth's climate, including a slow rise of average temperature and sea level.

Page 179

Ecological research is uncovering many additional connections. Let us hope that the awareness for these issues increases across the world.

* *

Some researchers prefer to define living beings as *self-reproducing* systems, others prefer to define them as *metabolic* systems. Among the latter, Mike Russell and Eric Smith propose the following definition of life: '*The purpose of life is to hydrogenate carbon dioxide.*' In other terms, the aim of life is to realize the reaction

$$CO_2 + 4\,H_2 \rightarrow CH_4 + 2\,H_2O \ . \tag{1}$$

This beautifully dry description is worth pondering – and numerous researchers are indeed exploring the consequences of this view.

THE PHYSICS OF PLEASURE

> "What is mind but motion in the intellectual sphere?
> Oscar Wilde (1854–1900) *The Critic as Artist.*

Pleasure is a quantum effect. The reason is simple. Pleasure comes from the senses. All senses measure. And all measurements rely on quantum theory.

The human body, like an expensive car, is full of sensors. Evolution has build these sensors in such a way that they trigger pleasure sensations whenever we do with our body what we are made for. Of course, no researcher will admit that he studies pleasure. Therefore the researcher will say that he or she studies the senses, and that he or she is doing *perception research*. But pleasure and all human sensors exist to let life continue. Pleasure is highest when life is made to continue. In the distant past, the appearance of new sensors in living systems has always had important effects of evolution, for example during the Cambrian explosion.

Research into pleasure and biological sensors is a fascinating field that is still evolving; here we can only have a quick tour of the present knowledge.

The *ear* is so sensitive and at the same time so robust against large signals that the experts are still studying how it works. No known sound sensor can cover an energy range of 10^{13}; indeed, the detected sound intensities range from $1\,\mathrm{pW/m^2}$ (some say $50\,\mathrm{pW/m^2}$) to $10\,\mathrm{W/m^2}$, the corresponding air pressures vary from $20\,\mu\mathrm{Pa}$ to $60\,\mathrm{Pa}$. The lowest intensity that can be heard is that of a $20\,\mathrm{W}$ sound source heard at a distance of $10\,000\,\mathrm{km}$, if no sound is lost in between. Audible sound wavelengths span from $17\,\mathrm{m}$ (for $20\,\mathrm{Hz}$) to $17\,\mathrm{mm}$ (for $20\,\mathrm{kHz}$). In this range, the ear, with its $16\,000$ to $20\,000$ hair cells, is able to distinguish at least 1500 pitches. But the ear is also able to distinguish nearby frequencies, such as 400 and 401 Hz, using a special pitch sharpening mechanism.

The *eye* is a position dependent photon detector. Each eye contains around 126 million separate detectors on the retina. Their spatial density is the highest possible that makes sense, given the diameter of the lens of the eye. They give the eye a resolving power of $1'$, or $0.29\,\mathrm{mrad}$, and the capacity to consciously detect down to 60 *incident* photons in $0.15\,\mathrm{s}$, or 4 *absorbed* photons in the same time interval.

Each eye contains 120 million highly sensitive general light intensity detectors, the *rods*. They are responsible for the mentioned high sensitivity. Rods cannot distinguish colours. Before the late twentieth century, human built light sensors with the same sensitivity as rods had to be helium cooled, because technology was not able to build sensors at room temperature that were as sensitive as the human eye.

The human eye contains about 6 million not so sensitive colour detectors, the *cones*, Vol. III, page 166 whose distribution we have seen earlier on. The different chemicals in the three cone types (red, green, blue) lead to different sensor speeds; this can be checked with the simple test shown in Figure 12. The sensitivity difference between the colour-detecting cones Ref. 15 and the colour-blind rods is the reason that at night all cats are grey.

The images of the eye are only sharp if the eye constantly moves in small random motions. If this motion is stopped, for example with chemicals, the images produced by the eye become unsharp.

FIGURE 12 The different speed of the eye's colour sensors, the cones, lead to a strange effect when this picture (in colour version) is shaken right to left in *weak* light.

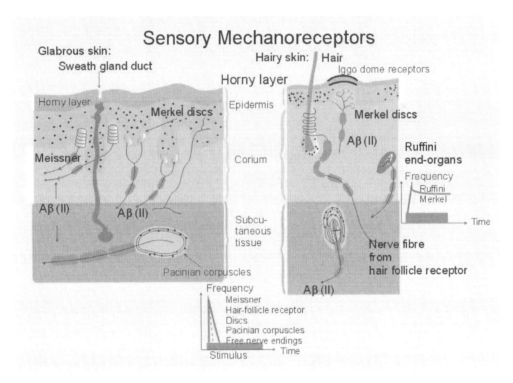

FIGURE 13 The five sensors of touch in humans: hair receptors, Meissner's corpuscules, Merkel cells, Ruffini corpuscules, and Pacinian corpuscules.

Human *touch sensors* are distributed over the skin, with a surface density which varies from one region to the other. The density is lowest on the back and highest in the face and on the tongue. There are separate sensors for pressure, for deformation, for vibration, and for tickling; there are additional separate sensors for heat, for coldness,* and

---

* There are *four* sensors for heat; one is triggered above 27°C, one above 31°C, one above 42°C, and one above 52°C. The sensor for temperatures above 42°C, TRPV1, is also triggered by capsaicin, the sharp chemical in chilli peppers.

for pain. Some of the sensors, whose general appearance is shown in Figure 13, react proportionally to the stimulus intensity, some differentially, giving signals only when the stimulus changes. Many of these sensors are also found inside the body – for example on the tongue. The sensors are triggered when external pressure deforms them; this leads to release of Na^+ and K^+ ions through their membranes, which then leads to an electric signal that is sent via nerves to the brain.

The human body also contains *orientation sensors* in the ear, *extension sensors* in each muscle, and *pain sensors* distributed with varying density over the skin and inside the body.

The *taste sensor* mechanisms of tongue are only partially known. The tongue is known to produce six taste signals* – sweet, salty, bitter, sour, proteic and fatty – and the mechanisms are just being unravelled. The sense for proteic, also called *umami*, has been discovered in 1907, by Ikeda Kikunae; the sense for 'fat' has been discovered only in 2005. Ref. 16

In ancient Greece, Democritus imagined that taste depends on the shape of atoms. Today it is known that sweet taste is connected with certain shape of molecules. Modern research is still unravelling the various taste receptors in the tongue. At least three different sweetness receptors, dozens of bitterness receptors, and one proteic and one fattiness receptor are known. In contrast, the sour and salty taste sensation are known to be due to ion channels. Despite all this knowledge, no sensor with a distinguishing ability of the same degree as the tongue has yet been built by humans. A good taste sensor would have great commercial value for the food industry. Research is also ongoing to find substances to block taste receptors; one aim is to reduce the bitterness of medicines or of food.

The *nose* has about 350 different smell receptors; through combinations it is estimated that the nose can detect about 10 000 different smells.** Together with the six signals that the sense of taste can produce, the nose also produces a vast range of taste sensations. It protects against chemical poisons, such as smoke, and against biological poisons, such as faecal matter. In contrast, artificial gas sensors exist only for a small range of gases. Good artificial taste and smell sensors would allow checking wine or cheese during their production, thus making their inventor extremely rich. At the moment, humans, with all Challenge 20 ny their technology at their disposal, are not even capable of producing sensors as good as those of a bacterium; it is known that *Escherichia coli* can sense at least 30 substances in its environment.

Other animals feature additional types of sensors. Sharks can *feel electrical fields*. Many Vol. III, page 31 snakes have *sensors for infrared light*, such as the pit viper or vampire bats. These sensors are used to locate prey or food sources. Some beetles, such as *Melanophila acuminata*,

---

There seems to be only *one* sensor for coldness, the ion channel TRPM8, triggered between 8 and 26 °C. It is also triggered by menthol, a chemical contained in mojito and mint. Coldness neurons, i.e., neurons with TRPM8 at their tips, can be seen with special techniques using fluorescence and are known to arrive into the teeth; they provide the sensation you get at the dentist when he applies his compressed air test.

* Taste sensitivity is *not* separated on the tongue into distinct regions; this is an incorrect idea that has been copied from book to book for over a hundred years. You can perform a falsification by yourself, using sugar or salt grains. Challenge 19 s

** Linda Buck and Richard Axel received the 2004 Nobel Prize for medicine and physiology for their unravelling of the working of the sense of smell.

can also detect infrared; they use this sense to locate the wildfires they need to make their eggs hatch. Also other insects have such organs. Pigeons, trout and sharks can *feel magnetic fields*, and use this sense for navigation. Many birds and certain insects can *see UV light*. Bats and dolphins are able to *hear ultrasound* up to 100 kHz and more. Whales and elephants can detect and localize *infrasound* signals.

Vol. I, page 281

In summary, the sensors with which nature provides us are state of the art; their sensitivity and ease of use is the highest possible. Since all sensors trigger pleasure or help to avoid pain, nature obviously wants us to enjoy life with the most intense pleasure possible. Studying physics is one way to do this.

Ref. 17

> There are two things that make life worth living: Mozart and quantum mechanics.
> Victor Weisskopf*

THE NERVES AND THE BRAIN

> There is no such thing as perpetual tranquillity of mind while we live here; because life itself is but motion, and can never be without desire, nor without fear, no more than without sense.
> Thomas Hobbes (1588–1679) *Leviathan.*

The main unit processing all the signals arriving from the sensors, the brain, is essential for all feelings of pleasure. The human brain has the highest complexity of all brains known.** In addition, the processing power and speed of the human brain is still larger than any device build by man.

Vol. I, page 272

Vol. III, page 218

We saw already earlier on how electrical signals from the sensors are transported into the brain. In the brain itself, the arriving signals are *classified* and *stored*, sometimes for a short time, sometimes for a long time. Most storage mechanisms take place in the structure and the connection strength between brain cells, the *synapses*, as we have seen. The process remaining to understand is the classification, a process we usually call *thinking*. For certain low level classifications, such as geometrical shapes for the eye or sound harmonies for the ear, the mechanisms are known. But for high-level classifications, such as the ones used in conceptual thinking, the aim is not yet achieved. It is not yet known how to describe the processes of reading or understanding in terms of signal motions. Research is still in full swing and will probably remain so for a large part of the twenty-first century.

---

* Victor Friedrich Weisskopf (b. 1908 Vienna, d. 2002 Cambridge), acclaimed theoretical physicist who worked with Einstein, Born, Bohr, Schrödinger and Pauli. He catalysed the development of quantum electrodynamics and nuclear physics. He worked on the Manhattan project but later in life intensely campaigned against the use of nuclear weapons. During the cold war he accepted the membership in the Soviet Academy of Sciences. He was professor at MIT and for many years director of CERN, in Geneva. He wrote several successful physics textbooks. The author heard him making the above statement in 1982, during one of his lectures.
** This is not in contrast with the fact that a few *whale* species have brains with a larger mass. The larger mass is due to the protection these brains require against the high pressures which appear when whales dive (some dive to depths of 1 km). The number of neurons in whale brains is considerably smaller than in human brains.

In the following we look at a few abilities of our brain, of our body and of other bodies that are important for the types of pleasure that we experience when we study motion.

LIVING CLOCKS

> « L'horloge fait de la réclame pour le temps.* »
> Georges Perros

We have given an overview of living clocks already at the beginning of our adventure. Vol. I, page 44 They are common in bacteria, plants and animals. And as Table 3 shows, without biological clocks, neither life nor pleasure would exist.

When we sing a musical note that we just heard we are able to reproduce the original frequency with high accuracy. We also know from everyday experience that humans are able to keep the beat to within a few per cent for a long time. When doing sport or when Ref. 18 dancing, we are able to keep the timing to high accuracy. (For shorter or longer times, the internal clocks are not so precise.) All these clocks are located in the brain.

Brains process information. Also computers do this, and like computers, all brains need a clock to work well. Every clock is made up of the same components. It needs an *oscillator* determining the rhythm and a mechanism to feed the oscillator with energy. In addition, every clock needs an oscillation *counter*, i.e., a mechanism that reads out the clock signal, and a means of *signal distribution* throughout the system is required, synchronizing the processes attached to it. Finally, a clock needs a *reset mechanism*. If the clock has to cover many time scales, it needs several oscillators with different oscillation frequencies and a way to reset their relative phases.

Even though physicists know fairly well how to build good clocks, we still do not know many aspects of biological clocks. Most biological oscillators are chemical systems; some, like the heart muscle or the timers in the brain, are electrical systems. The general eluci- Ref. 19 dation of chemical oscillators is due to Ilya Prigogine; it has earned him a Nobel Prize for chemistry in 1977. But not all the chemical oscillators in the human body are known yet, not to speak of the counter mechanisms. For example, a 24-minute cycle inside each human cell has been discovered only in 2003, and the oscillation mechanism is not yet fully clear. (It is known that a cell fed with heavy water ticks with 27–minute instead of 24–minute rhythm.) It might be that the daily rhythm, the circadian clock, is made up Ref. 20 of or reset by 60 of these 24–minute cycles, triggered by some master cells in the human body. The clock reset mechanism for the circadian clock is also known to be triggered by daylight; the cells in the eye who perform this resetting action have been pinpointed only in 2002. The light signal from these cells is processed by the superchiasmatic nuclei, two dedicated structures in the brain's hypothalamus. The various cells in the human body act differently depending on the phase of this clock.

The clocks with the longest cycle in the human body control *ageing*. One of the more famous ageing clock limits the number of divisions that a cell can undergo. Indeed, the number of cell divisions is finite for most cell types of the human body and typically lies between 50 and 200. (An exception are reproductory cells – we would not exist if they would not be able to divide endlessly.) The cell division counter has been identified; it is embodied in the *telomeres*, special structures of DNA and proteins found at both ends

---

* 'Clocks are ads for time.'

TABLE 3 Examples of biological rhythms and clocks.

LIVING BEING	OSCILLATING SYSTEM	PERIOD
Sand hopper (*Talitrus saltator*)	knows in which direction to flee from the position of the Sun or Moon	circadian
Human (*Homo sapiens*)	gamma waves in the brain	0.023 to 0.03 s
	alpha waves in the brain	0.08 to 0.13 s
	heart beat	0.3 to 1.5 s
	delta waves in the brain	0.3 to 10 s
	blood circulation	30 s
	cellular circahoral rhythms	1 to 2 ks
	rapid-eye-movement sleep period	5.4 ks
	nasal cycle	4 to 14 ks
	growth hormone cycle	11 ks
	suprachiasmatic nucleus (SCN), circadian hormone concentration, temperature, etc.; leads to jet lag	90 ks
	skin clock	circadian
	monthly period	2.4(4) Ms
	built-in aging	3.2(3) Gs
Common fly (*Musca domestica*)	wing beat	30 ms
Fruit fly (*Drosophila melanogaster*)	wing beat for courting	34 ms
Most insects (e.g. wasps, fruit flies)	winter approach detection (diapause) by length of day measurement; triggers metabolism changes	yearly
Algae (*Acetabularia*)	Adenosinetriphosphate (ATP) concentration	
Moulds (e.g. *Neurospora crassa*)	conidia formation	circadian
Many flowering plants	flower opening and closing	circadian
Tobacco plant	flower opening clock (photoperiodism); triggered by length of days, discovered in 1920 by Garner and Allard	annual
Arabidopsis	circumnutation	circadian
	growth	a few hours
Telegraph plant (*Desmodium gyrans*)	side leaf rotation	200 s
Forsythia europaea, F. suspensa, F. viridissima, F. spectabilis	Flower petal oscillation, discovered by Van Gooch in 2002	5.1 ks

of each chromosome. These structures are reduced by a small amount during each cell division. When the structures are too short, cell division stops. The purely theoretical prediction of this mechanism by Alexei Olovnikov in 1971 was later proven by a number of researchers. (Only the latter received the Nobel Prize in medicine, in 2009, for this confirmation.) Research into the mechanisms and the exceptions to this process, such as cancer and sexual cells, is ongoing.

Not all clocks in human bodies have been identified, and not all mechanisms are known. For example, basis of the monthly period in women is interesting, complex, and unclear.

Other fascinating clocks are those at the basis of *conscious* time. Of these, the brain's stopwatch or *interval timer* has been most intensely studied. Only recently was its mechanism uncovered by combining data on human illnesses, human lesions, magnetic resonance studies and effects of specific drugs. The basic interval timing mechanism takes place in the striatum in the basal ganglia of the brain. The striatum contains thousands of timer cells with different periods. They can be triggered by a 'start' signal. Due to their large number, for small times of the order of one second, every time interval has a different pattern across these cells. The brain can read these patterns and learn them. In this way we can time music or specific tasks to be performed, for example, one second after a signal. Ref. 21

Even though not all the clock mechanisms in humans are known, biological clocks share a property with all human-built and all non-living clocks: they are limited by quantum mechanics. Even the simple pendulum is limited by quantum theory. Let us explore the topic.

WHEN DO CLOCKS EXIST?

> Die Zukunft war früher auch besser.*
>
> Karl Valentin.

When we explored general relativity we found out that purely gravitational clocks do not exist, because there is no unit of time that can be formed using the constants c and G. Vol. II, page 263 Clocks, like any measurement standard, need matter and non-gravitational interactions to work. This is the domain of quantum theory. Let us see what the situation is in this case.

First of all, in quantum theory, the time is *not* an observable. Indeed, the time operator is not Hermitean. In other words, quantum theory states that there is no physical Ref. 22 observable whose value is proportional to time. On the other hand, clocks are quite common; for example, the Sun or Big Ben work to most people's satisfaction. Observations thus encourages us to look for an operator describing the position of the hands of a clock. However, if we look for such an operator we find a strange result. Any quantum system having a Hamiltonian bounded from below – having a lowest energy – lacks a Hermitean operator whose expectation value increases monotonically with time. This result can be proven rigorously. Challenge 21 ny

Take a mechanical pendulum clock. In all such clocks the weight has to stop when the chain end is reached. More generally, all clocks have to stop when the battery or the energy source is empty. In other words, in all real clocks the Hamiltonian is bounded from below. And the above theorem from quantum theory then states that such a clock cannot really work.

---

* 'Also the future used to be better in the past.' Karl Valentin (b. 1882 Munich, d. 1948 Planegg), German author and comedian.

In short, quantum theory shows that exact *clocks do not exist in nature*. Quantum theory states that any clock can only be *approximate*. Time cannot be measured exactly; time can only be measured approximately. Obviously, this result is of importance for high precision clocks. What happens if we try to increase the precision of a clock as much as possible?

High precision implies high sensitivity to fluctuations. Now, all clocks have an oscillator inside, e.g., a motor, that makes them work. A high precision clock thus needs a high precision oscillator. In all clocks, the position of this oscillator is read out and shown on the dial. Now, the quantum of action implies that even the most precise clock oscillator has a position indeterminacy. The precision of any clock is thus limited.

Worse, like any quantum system, any clock oscillator even has a small, but finite probability to stop or to run backwards for a while. You can check this conclusion yourself. Just have a look at a clock when its battery is almost empty, or when the weight driving the pendulum has almost reached the bottom position. The clock will start doing funny things, like going backwards a bit or jumping back and forward. When the clock works normally, this behaviour is strongly suppressed; however, it is still possible, though with Challenge 22 e   low probability. This is true even for a sundial.

In summary, clocks necessarily have to be *macroscopic* in order to work properly. A clock must be as large as possible, in order to average out its fluctuations. Astronomical systems are good examples. A good clock must also be *well-isolated* from the environment, such as a freely flying object whose coordinate is used as time variable. For example, this is regularly done in atomic optical clocks.

The precision of clocks

Given the limitations due to quantum theory, what is the ultimate accuracy τ of a clock? To start with, the indeterminacy relation provides the limit on the mass of a clock. The Challenge 23 ny   clock mass M must obey

$$M > \frac{\hbar}{c^2 \tau} \tag{2}$$

Challenge 24 e   which is obviously always fulfilled in everyday life. But we can do better. Like for a pendulum, we can relate the accuracy τ of the clock to its maximum reading time T. The Ref. 23   idea was first published by Salecker and Wigner. They argued that

$$M > \frac{\hbar}{c^2 \tau} \frac{T}{\tau} \tag{3}$$

where T is the time to be measured. You might check that this condition directly requires Challenge 25 e   that any clock must be *macroscopic*.

Let us play with the formula by Salecker and Wigner. It can be rephrased in the following way. For a clock that can measure a time t, the size l is connected to the mass m by

$$l > \sqrt{\frac{\hbar t}{m}} \ . \tag{4}$$

How close can this limit be achieved? It turns out that the smallest clocks known, as well as the clocks with most closely approach this limit, are bacteria. The smallest bacteria, the *mycoplasmas*, have a mass of about $8 \cdot 10^{-17}$ kg, and reproduce every 100 min, with a precision of about 1 min. The size predicted from expression (4) is between 0.09 μm and 0.009 μm. The observed size of the smallest mycoplasmas is 0.3 μm. The fact that bacteria can come so close to the clock limit shows us again what a good engineer evolution has been.

Ref. 24

Note that the requirement by Salecker and Wigner is not in contrast with the possibility to make the *oscillator* of the clock very small; researchers have built oscillators made of a single atom. In fact, such oscillations promise to be the most precise human built clocks. But the oscillator is only one part of any clock, as explained above.

Ref. 25
Page 39

In the real world, the clock limit can be tightened even more. The whole mass M cannot be used in the above limit. For clocks made of atoms, only the binding energy between atoms can be used. This leads to the so-called *standard quantum limit for clocks*; it limits the accuracy of their frequency ν by

$$\frac{\delta \nu}{\nu} = \sqrt{\frac{\Delta E}{E_{\text{tot}}}} \qquad (5)$$

where $\Delta E = \hbar/T$ is the energy indeterminacy stemming from the finite measuring time T and $E_{\text{tot}} = N E_{\text{bind}}$ is the total binding energy of the atoms in the metre bar. So far, the quantum limit has not yet been achieved for any clock, even though experiments are getting close to it.

In summary, clocks exist only in the limit of $\hbar$ being negligible. In practice, the errors made by using clocks and metre bars can be made as small as required; it suffices to make the clocks large enough. Clock built into human brains comply with this requirement. We can thus continue our investigation into the details of matter without much worry, at least for a while. Only in the last part of our mountain ascent, where the requirements for precision will be even higher and where general relativity will limit the size of physical systems, trouble will appear again: the impossibility to build precise clocks will then become a central issue.

Vol. VI, page 59

WHY ARE PREDICTIONS SO DIFFICULT, ESPECIALLY OF THE FUTURE?

> Future: that period of time in which our affairs
> prosper, our friends are true, and our happiness
> is assured.
>
> Ambrose Bierce

Nature limits predictions. We have seen that quantum theory, through the uncertainty relations, limits the precision of time measurements. Thus, the quantum of action makes it hard to determine initial states to full precision. We also have found in our adventure that predictions of the future are made difficult by nonlinearities and by the divergence from similar conditions. We have seen that many particles make it difficult to predict the future due to the statistical nature of their initial conditions. We have seen that a non-trivial space-time topology can limit predictability. Finally, we will discover that black hole and horizons can limit predictability due to their one-way inclusion of energy, mass and signals.

Predictability and measurements are thus limited. The main reason for this limit is the quantum of action. If, due to the quantum of action, perfect clocks do not exist, is determinism still the correct description of nature? And does time exist after all? The answer is clear: yes and no. We learned that all the mentioned limitations of clocks can be overcome for limited time intervals; in practice, these time intervals can be made so large that the limitations do *not* play a role in everyday life. As a result, in quantum systems both determinism and time remain applicable, and our ability to enjoy the pleasures due to the flow of time remains intact.

However, when extremely large momentum flows or extremely large dimensions need to be taken into account, quantum theory cannot be applied alone; in those cases, general relativity needs to be taken into account. The fascinating effects that occur in those
Vol. VI, page 52 situations will be explored later on.

Decay and the golden rule

> I prefer most of all to remember the future.
> Salvador Dalì

All pleasure only makes sense in the face of death. And death is a form of decay. Decay is any spontaneous change. Like the wave aspect of matter, decay is a process with no classical counterpart. Of course, any decay – including the emission of light by a lamp, the triggering of a camera sensor, radioactivity or the ageing of humans – can be observed classically; however, the *origin* of decay is a pure quantum effect.

In any decay of unstable systems or particles, the decoherence of superpositions of
Vol. IV, page 132 macroscopically distinct states plays an important role. Indeed, experiments confirm that the prediction of decay for a specific system, like a scattering of a particle, is only possible on *average*, for a large number of particles or systems, and never for a single one. These observations confirm the quantum origin of decay. In every decay process, the superposition of macroscopically distinct states – in this case those of a decayed and an undecayed particle – is made to decohere rapidly by the interaction with the environment. Usually the 'environment' vacuum, with its fluctuations of the electromagnetic, weak and strong fields, is sufficient to induce decoherence. As usual, the details of the involved environment states are unknown for a single system and make any prediction for a specific system impossible.

What is the *origin* of decay? Decay is always due to *tunnelling*. With the language of quantum electrodynamics, we can say that decay is motion induced by the vacuum fluctuations. Vacuum fluctuations are random. The experiment between the plates confirms the importance of the environment fluctuations for the decay process.

Quantum theory gives a simple description of decay. For a system consisting of a large number N of decaying identical particles, any *decay* is described by

$$\dot{N} = -\frac{N}{\tau} \quad \text{where} \quad \frac{1}{\tau} = \frac{2\pi}{\hbar} \left| \langle \psi_{\text{initial}} | H_{\text{int}} | \psi_{\text{final}} \rangle \right|^2 . \tag{6}$$

This result was named the *golden rule* by Fermi,* because it works so well despite being an approximation whose domain of applicability is not easy to specify. The golden rule leads to

Challenge 26 e

$$N(t) = N_0 \, e^{-t/\tau} \, . \qquad (7)$$

Decay is thus predicted to follow an exponential law, independently of the details of the physical process. In addition, the decay time τ depends on the interaction and on the square modulus of the transition matrix element. For almost a century, all experiments confirmed that quantum decay is *exponential*.

On the other hand, when quantum theory is used to derive the golden rule, it is found that decay is exponential only in certain special systems. A calculation that takes into account higher order terms predicts two deviations from exponential decay for completely isolated systems: for short times, the decay rate should *vanish*; for long times, the decay rate should follow an *algebraic* – not an exponential – dependence on time, in some cases even with superimposed oscillations. After an intense experimental search, deviations for short times have been observed. The observation of deviations at long times are rendered impossible by the ubiquity of thermal noise. In summary, it turns out that decay is exponential only when the environment is noisy, the system made of many weakly interacting particles, or both. Since this is usually the case, the mathematically exceptional exponential decrease becomes the (golden) rule in the description of decay.

Ref. 26

Ref. 27

Can you explain why human life, despite being a quantum effect, is not observed to follow an exponential decay?

Challenge 27 s

THE PRESENT IN QUANTUM THEORY

> *Utere temporibus.**
>
> Ovidius

Many sages advise to enjoy the present. As shown by perception research, what humans call 'present' has a duration of between 20 and 70 milliseconds. This result on the biological present leads us to ask whether the *physical* present might have a duration as well.

Ref. 28

In everyday life, we are used to imagine that shortening the time taken to measure the position of a point object as much as possible will approach the ideal of a particle fixed at a given point in space. When Zeno discussed the flight of an arrow, he assumed that this is possible. However, quantum theory changes the situation.

Can we really say that a moving system is at a given spot at a given time? In order to find an answer through experiment, we could use a photographic camera whose shutter time can be reduced at will. What would we find? When the shutter time approaches the oscillation period of light, the sharpness of the image would decrease; in addition, the colour of the light would be influenced by the shutter motion. We can increase the energy of the light used, but the smaller wavelengths only shift the problem, they do not solve it. Worse, at extremely small wavelengths, matter becomes transparent, and shutters cannot

---

* Originally, the golden rule is a statement from the christian bible, (Matthew 7,12) namely the sentence 'Do to others what you want them to do to you'.
** 'Use the occasions.' *Tristia* 4, 3, 83

be realized any more. All such investigations confirm: Whenever we reduce shutter times as much as possible, observations become unsharp. The lack of sharpness is due to the quantum of action. Quantum theory thus does not confirm the naive expectation that shorter shutter times lead to sharper images. In contrast, the quantum aspects of nature show us that there is no way in principle to approach the limit that Zeno was discussing.

In summary, the indeterminacy relation and the smallest action value prevent that moving objects are at a fixed position at a given time. Zeno's discussion was based on an extrapolation of classical concepts into domains where it is not valid any more. Every observation, like every photograph, implies a *time average*: observations average interactions over a given time.* For a photograph, the duration is given by the shutter time; for a measurement, the average is defined by the details of the set-up. Whatever this set-up might be, the averaging time is never zero. There is no 'point-like' instant of time that describes the present. The observed, physical *present* is always an average over a non-vanishing interval of time. In nature, the present has a finite duration. To give a rough value that guides our thought, in most situations the length of the present will be less than a yoctosecond, so that it can usually be neglected.

Why can we observe motion?

Zeno of Elea was thus wrong in assuming that motion is a sequence of specific positions in space. Quantum theory implies that motion is *only approximately* the change of position with time.

Why then can we observe and describe motion in quantum theory? Quantum theory shows that motion is the *low energy approximation* of quantum evolution. Quantum evolution *assumes* that space and time measurements of sufficient precision can be performed. We know that for any given observation energy, we can build clocks and metre bars with much higher accuracy than required, so that in practice, quantum evolution is applicable in all cases. As long as energy and time have no limits, all problems are avoided, and *motion is a time sequence of quantum states*.

In summary, we can observe motion because for any known observation energy we can find a still higher energy and a still longer averaging time that can be used by the measurement instruments to define space and time with higher precision than for the Vol. VI, page 42 system under observation. In the final part of our mountain ascent, we will discover that there is a maximum energy in nature, so that we will need to change our description in those situations. However, this energy value is so huge that it does not bother us at all at the present point of our exploration.

Rest and the quantum Zeno effect

The quantum of action implies that there is no rest in nature. Rest is thus always either an approximation or a time average. For example, if an electron is bound in an atom, not freely moving, the probability cloud, or density distribution, is stationary in time. But there is another apparent case of rest in quantum theory, the *quantum Zeno effect*. Usually, observation *changes* the state of a system. However, for certain systems, observation can have the opposite effect, and *fix* a system.

---

* Also the discussion of the quantum Zeno effect, below, does not change the conclusions of this section.

Quantum mechanics predicts that an unstable particle can prevented from decaying if it is continuously observed. The reason is that an observation, i.e., the interaction with the observing device, yields a non-zero probability that the system does not evolve. If the frequency of observations is increased, the probability that the system does not decay at all approaches 1. Three research groups – Alan Turing by himself in 1954, the group of A. Degasperis, L. Fonda and G.C. Ghirardi in 1974, and George Sudarshan and Baidyanath Misra in 1977 – have independently predicted this effect, today called the *quantum Zeno effect*. In sloppy words, the quantum Zeno effect states: if you look at a system all the time, nothing happens.

The quantum Zeno effect is a natural consequence of quantum theory; nevertheless, its strange circumstances make it especially fascinating. After the prediction, the race for the first observation began. The effect was partially observed by David Wineland and his group in 1990, and definitively observed by Mark Raizen and his group in 2001. In the meantime, other groups have confirmed the measurements. Thus, quantum theory has been confirmed also in this surprising aspect. Ref. 29

The quantum Zeno effect is also connected to the deviations from exponential decay – due to the golden rule – that are predicted by quantum theory. Indeed, quantum theory predicts that every decay is exponential only for intermediate times, and quadratic for short times and polynomial for extremely long times. These issues are research topics to this day.

In a fascinating twist, in 2002, Saverio Pascazio and his team have predicted that the quantum Zeno effect can be used to realize X-ray tomography of objects with the lowest radiation levels imaginable. Ref. 30

In summary, the quantum Zeno effect does not contradict the statement that there is no rest in nature; in situations showing the effect, there is an non-negligible interaction between the system and its environment. The details of the interaction are important: in certain cases, frequent observation can actually accelerate the decay or evolution. Quantum physics still remains a rich source of fascinating effects. Ref. 31

Consciousness – a result of the quantum of action

In the pleasures of life, consciousness plays an essential role. *Consciousness* is our ability to observe what is going on in our mind. This activity, like any type of change, can itself be observed and studied. Obviously, consciousness takes place in the brain. If it were not, there would be no way to keep it connected with a given person. Simply said, we know that each brain located on Earth moves with over one million kilometres per hour through the cosmic background radiation; we also observe that consciousness moves along with it. Vol. III, page 284

The brain is a quantum system: it is based on molecules and electrical currents. The changes in consciousness that appear when matter is taken away from the brain – in operations or accidents – or when currents are injected into the brain – in accidents, experiments or misguided treatments – have been described in great detail by the medical profession. Also the observed influence of chemicals on the brain – from alcohol to hard drugs – makes the same point. The brain is a quantum system.

Modern imaging machines can detect which parts of the brain work when sensing, remembering or thinking. Not only is sight, noise and thought processed in the brain; we

Page 146 can follow these processes with measurement apparatus. The best imaging machines are based on magnetic resonance, as described below. Another, more questionable imaging technique, positron tomography, works by letting people swallow radioactive sugar. Both techniques confirm the findings on the location of thought and on its dependence on chemical fuel. In addition, we already know that memory depends on the particle nature of matter. All these observations depend on the quantum of action.

Today, we are thus in the same situation today as material scientists were a century ago: they knew that matter is made of charged particles, but they could not say *how* matter is built up. Similarly, we know today that consciousness is made from the signal propagation and signal processing in the brain; we know that consciousness is an electrochemical process. But we do not know yet the details of *how* the signals make up consciousness. Unravelling the workings of this fascinating quantum system is the aim of neurological science. This is one of the great challenges of twenty-first century science.

Can you add a few arguments to the ones given here, showing that consciousness is a physical process? Can you show in particular that not only the consciousness of others, but also your own consciousness is a quantum process? Can you show, in addition, that Challenge 28 s despite being a quantum process, coherence plays no essential role in consciousness?

In short, our consciousness is a consequence of the matter that makes us up. Consciousness and pleasure depend on matter, it interactions, and the quantum of action.

WHY CAN WE OBSERVE MOTION? – AGAIN

Studying nature can be one of the most intense pleasures of life. All pleasures are based on our ability to observe or detect motion. And our human condition is central to this ability. In particular, in our adventure so far we found that we experience motion

— only because we are of finite size,
— only because we are made of a large but finite number of atoms (to produce memory and enable observations),
— only because we have a finite but moderate temperature (finite so that we have a lifetime, not zero so that we can be working machines),
— only because we are a mixture of liquids and solids (enabling us to move and thus to experiment),
— only because we are approximately electrically neutral (thus avoiding that our sensors get swamped),
— only because we are large compared to a black hole of our same mass (so that we have useful interactions with our environment),
— only because we are large compared to our quantum mechanical wavelength (so that we do not experience wave effects in everyday life),
— only because we have a limited memory (so that we can clear it),
— only because our brain forces us to approximate space and time by continuous entities (otherwise we would not form these concepts),
— only because our brain cannot avoid describing nature as made of different parts (otherwise we would not be able to talk or think),
— only because our ancestors reproduced,
— only because we are animals (and thus have a brain),
— only because life evolved here on Earth,

— only because we live in a relatively quiet region of our galaxy (which allowed evolution), and

— only because the human species evolved long after the big bang (when the conditions were more friendly to life).

If any of these conditions – and many others – were not fulfilled we would not observe motion; we would have no fun studying physics. In fact, we can also say: if any of these conditions were not fulfilled, motion would not exist. In many ways motion is thus an illusion, as Zeno of Elea had claimed a long time ago. Of course, it is a *inevitable* illusion, Vol. I, page 15 one that is shared by many other animals and machines. To say the least, the observation and the concept of motion is a result of the properties and limitations of the human condition. A complete description of motion and nature must take this connection into account. Before we attempt that in the last volume of this adventure, we explore a few additional details.

CURIOSITIES AND FUN CHALLENGES ABOUT QUANTUM EXPERIENCE

Most clocks used in everyday life, those built inside the human body and those made by humans, are electromagnetic. Any clock on the wall, be it mechanical, quartz controlled, radio or solar controlled, is based on electromagnetic effects. Do you know an exception? Challenge 29 s

* *

The sense of smell is quite complex. For example, the substance that smells most badly to humans is *skatole*, also called, with his other name, *3-methylindole*. This is the molecule to which the human nose is most sensitive. Skatole makes faeces smell bad; it is a result of haemoglobin entering the digestive tract through the bile. Skatole does not smell bad to all animals; in contrast to humans, flies are attracted by its smell. Skatole is also produced by some plants for this reason.

On the other hand, *small* levels of skatole do not smell bad to humans. Skatole is also used by the food industry in small quantities to give smell and taste to vanilla ice cream – though under the other name.

* *

It is worth noting that human senses detect energies of quite different magnitudes. The eyes can detect light energies of about 1 aJ, whereas the sense of touch can detect only energies as small as about 10 μJ. Is one of the two systems relativistic? Challenge 30 s

* *

The human construction plan is stored in the DNA. The DNA is structured into 20 000 genes, which make up about 2% of the DNA, and 98% non-coding DNA, once called "junk DNA". Humans have as many genes as worms; plants have many more. Only around 2010 it became definitely clear, through the international 'Encode' project, what the additional 98% of the DNA do: they switch genes on and off. They form the *administration* of the genes and work mostly by binding to specific proteins.

Research suggests that most genetic defects, and thus of genetic diseases, are not due to errors in genes, but in errors of the control switches. All this is an ongoing research field.

* *

Even at perfect darkness, the eye does not yield a black impression, but a slightly brighter one, called *eigengrau*. This is a result of noise created inside the eye, probably triggered by spontaneous decay of rhodopsin, or alternatively, by spontaneous release of neurotransmitters.

* *

The high sensitivity of the ear can be used to *hear* light. To do this, take an empty 750 ml jam glass. Keeping its axis horizontal, blacken the upper half of the inside with a candle. The lower half should remain transparent. After doing this, close the jam glass with its lid, and drill a 2 to 3 mm hole into it. If you now hold the closed jam glass with the hole to your ear, keeping the black side up, and shining into it from below with a 50 W light bulb, something strange happens: you hear a 100 Hz sound. Why?

Challenge 31 s

* *

Most senses work already before birth. It is well-known since many centuries that playing the violin to a pregnant mother every day during the pregnancy has an interesting effect. Even if nothing is told about it to the child, it will become a violin player later on. In fact, most musicians are 'made' in this way.

* *

There is ample evidence that not using the senses is damaging. People have studied what happens when in the first years of life the vestibular sense – the one used for motion detection and balance restoration – is not used enough. Lack of rocking is extremely hard to compensate later in life. Equally dangerous is the lack of use of the sense of touch. Babies, like all small mammals, that are generally and systematically deprived of these experiences tend to violent behaviour during the rest of their life.

Ref. 33

* *

The importance of ion channels in the human body can not be overstressed. Ion channels malfunctions are responsible for many infections, for certain types of diabetes and for many effects of poisons. But above all, ion channels, and electricity in general, are essential for life.

Ref. 34

* *

It is still unknown why people *yawn*. This is still a topic of research.

* *

Nature has invented the senses to increase pleasure and avoid pain. But neurologists have found out that nature has gone even further; there is a dedicated *pleasure system* in the brain, shown in Figure 15, whose function is to decide which experiences are pleasurable and which not. The main parts of the pleasure system are the *ventral tegmental area* in the midbrain and the *nucleus accumbens* in the forebrain. The two parts regulate each other mainly through *dopamine* and *GABA*, two important neurotransmitters. Research has shown that dopamine is produced whenever pleasure exceeds expectations. Nature has thus developed a special signal for this situation.

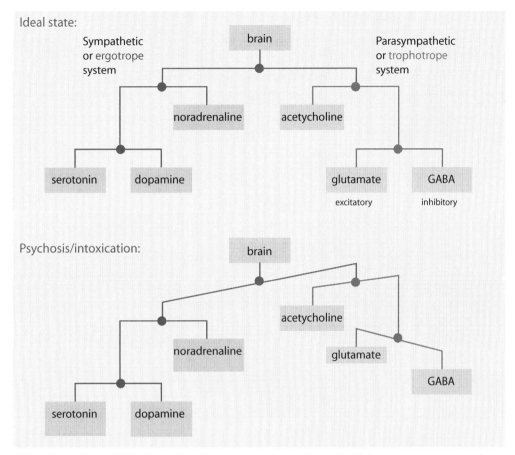

FIGURE 14 The 'neurochemical mobile' model of well-being, with one of the way it can get out of balance.

In fact, well-being and pleasure are controlled by a large number of neurotransmitters and by many additional regulation circuits. Researchers are trying to model the pleasure system with hundreds of coupled differential equations, with the distant aim being to understand addiction and depression, for example. On the other side, also simple models of the pleasure system are possible. One, shown in Figure 14, is the 'neurochemical mobile' model of the brain. In this model, well-being is achieved whenever the six most important neurotransmitters are in relative equilibrium. The different possible departures from equilibrium, at each joint of the mobile, can be used to describe depression, schizophrenia, psychosis, the effect of nicotine or alcohol intake, alcohol dependency, delirium, drug addiction, detoxication, epilepsy and more.

Ref. 35

* *

The pleasure system in the brain is not only responsible for addiction. It is also responsible, as Helen Fisher showed through MRI brain scans, for romantic love. Romantic love, directed to one single other person, is a state that is created in the *ventral tegmental area* and in the *nucleus accumbens*. Romantic love is thus a part of the reptilian brain; indeed,

Ref. 36

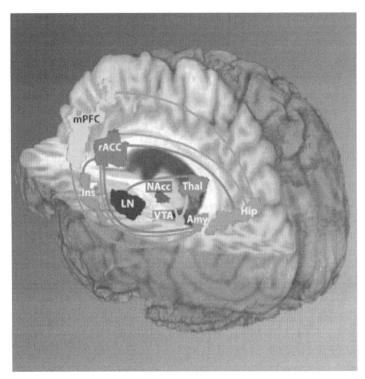

FIGURE 15 The location of the ventral tegmental area (VTA) and of the nucleus accumbens (NAcc) in the brain. The other regions involved in pleasure and addiction are Amy, the amygdala, Hip, the hippocampus, Thal, the thalamus, rACC, the rostral anterior cingulate cortex, mPFC, the medial prefrontal cortex, Ins, the insula, and LN, the lentiform nucleus (courtesy NIH).

romantic love is found in many animal species. Romantic love is a kind of positive addiction, and works like cocaine. In short, in life, we can all chose between addiction and love.

* *

An important aspect of life is death. When we die, conserved quantities like our energy, momentum, angular momentum and several other quantum numbers are redistributed. They are redistributed because conservation means that nothing is lost. What does all this imply for what happens after death?

Challenge 32 s

* *

Many plants have built-in sensors and clocks that measure the length of the day. For example, *spinach* does not grow in the tropics, because in order to flower, spinach must sense for at least fourteen days in a row that the day is at least 14 hours long – and this never happens in the tropics. This property of plants, called *photoperiodism*, was discovered in 1920 by Wightman Garner and Harry Allard while walking across tobacco fields. They discovered – and proved experimentally – that tobacco plants and soja plants only florish when the length of the day gets sufficiently short, thus around September. Garner and Allard found that plants could be divided into species that flower when days are short – such as chrysanthemum or coffee – others that flower when days are long – such as carnation or clover – and still others that do not care about the length of the day at all – such as roses or tomatoes. The measurement precision for the length of the day is around 10 min. The day length sensor itself was discovered only much later; it is located in the leaves of the plant, is called the *phytochrome system* and is based on specialised proteins. The proteins are able to measure the ratio between bright red and dark red light and control the moment of flower opening in such plants.

SUMMARY ON BIOLOGY AND PLEASURE

To ensure the successful reproduction of living being, evolution has pursued miniaturization as much as possible. Molecular motors, including molecular pumps, are the smallest motors known so far; they are found in huge numbers in every living cell. Molecular motors work as quantum ratchets.

To increase pleasure and avoid pain, evolution has supplied the human body with numerous sensors, sensor mechanisms, and a pleasure system deep inside the brain. In short, nature has invented pleasure as a guide for behaviour. Neurologists have thus proven what Epicurus* said 25 centuries ago and Sigmund Freud said one century ago: pleasure controls human life. All biological pleasure sensors and all pleasure systems are based on quantum motion, especially on chemistry and materials science. We therefore explore both fields in the following.

---

* Though it is often forgotten that Epicurus also said: "It is impossible to live a pleasant life without living wisely and honourably and justly, and it is impossible to live wisely and honourably and justly without living pleasantly."

CHAPTER 2

CHANGING THE WORLD WITH QUANTUM EFFECTS

T HE discovery of quantum effects has changed everyday life. It has allowed the distribution of speech, music and films. The numerous possibilities of telecommunications and of the internet, the progress in chemistry, materials science, electronics and medicine would not have been possible without quantum effects. Many other improvements of our everyday life are due to quantum physics, and many are still expected. In the following, we give a short overview of this vast field.

CHEMISTRY – FROM ATOMS TO DNA

> " Bier macht dumm.*
>
> Albert Einstein "

It is an old truth that Schrödinger's equation contains all of chemistry. With quantum theory, for the first time people were able to calculate the strengths of chemical bonds, and what is more important, the angle between them. Quantum theory thus explains the *shape* of molecules and thus indirectly, the shape of all matter. In fact, the correct statement is: the *Dirac* equation contains all of chemistry. The relativistic effects that distinguish the two equations are necessary, for example, to understand why gold is yellow and does not rust or why mercury is liquid.

To understand molecules and everyday matter, the first step is to understand atoms. The early quantum theorists, lead by Niels Bohr, dedicated their life to understanding their atoms and their detailed structure. The main result of their efforts is what you learn in secondary school: in atoms with more than one electron, the various electron clouds form spherical layers around the nucleus. The electron layers can be grouped into groups of related clouds that are called *shells*. For electrons outside the last fully occupied shell, the nucleus and the inner shells, the atomic *core*, can often be approximated as a single charged entity.

Shells are numbered from the inside out. This *principal quantum number*, usually written n, is deduced and related to the quantum number that identifies the states in the hydrogen atom. The relation is shown in Figure 16.

Quantum theory shows that the first atomic shell has room for two electrons, the second for 8, the third for 18, and the general n-th shell for $2n^2$ electrons. The (neutral)

Ref. 37

---

* 'Beer makes stupid.'

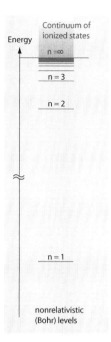

FIGURE 16 The principal quantum numbers in hydrogen.

atom with one electron is hydrogen, the atom with two electrons is called helium. Every chemical element has a specific number of electrons (and the same number of protons, as we will see). A way to picture this connection is shown in Figure 17. It is called the *periodic table of the elements*. The standard way to show the table is shown on page 320 and, more vividly, in Figure 18. (For a periodic table with a video about each element, see www.periodicvideos.com.)

Ref. 38

Experiments show that different atoms that share the *same* number of electrons in their outermost shell show *similar* chemical behaviour. Chemists know that the chemical behaviour of an element is decided by the ability of its atoms to from bonds. For example, the elements with one electron in their outer s shell are the *alkali metals* lithium, sodium, potassium, rubidium, caesium and francium; hydrogen, the exception, is conjectured to be metallic at high pressures. The elements with filled outermost shells are the *noble gases* helium, neon, argon, krypton, xenon, radon and ununoctium.

ATOMIC BONDS

When two atoms approach each other, their electron clouds are deformed and mixed. The reason for these changes is the combined influence of the two nuclei. These cloud changes are highest for the outermost electrons: they form chemical *bonds*.

Bonds can be pictured, in the simplest approximation, as cloud overlaps that *fill* the outermost shell of both atoms. These overlaps lead to a gain in energy. The energy gain is the reason that fire is hot. In wood fire, chemical reactions between carbon and oxygen atoms lead to a large release of energy. After the energy has been released, the atomic bond produces a fixed distance between the atoms, as shown in Figure 19. This distance is due to an energy minimum: a lower distance would lead to electrostatic repulsion

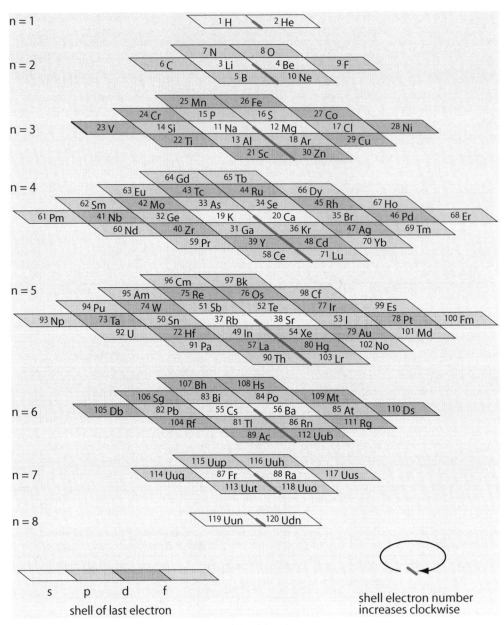

FIGURE 17 An unusual form of the periodic table of the elements.

between the atomic cores, a higher distance would increase the electron cloud energy.

Many atoms can bind to more than one neighbours. In this case, energy minimization also leads to specific bond angles, as shown in Figure 20. Maybe you remember those funny pictures of school chemistry about orbitals and dangling bonds. Such dangling bonds can now be measured and observed. Several groups were able to image them using scanning force or scanning tunnelling microscopes, as shown in Figure 21.

The repulsion between the clouds of each bond explains why angle values near that of tetrahedral skeletons ($2 \arctan \sqrt{2} = 109.47°$) are so common in molecules. For example,

Ref. 39

Ref. 40

Challenge 33 e

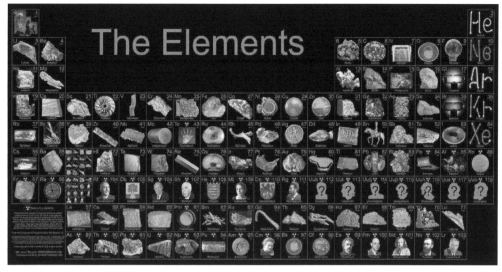

FIGURE 18 A modern periodic table of the elements (© Theodore Gray, for sale at www.theodoregray. com).

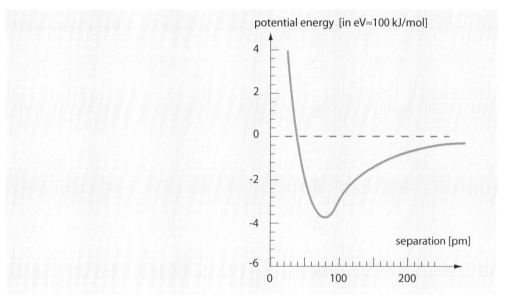

FIGURE 19 The forming of a chemical bond between two atoms, and the related energy minimum.

the H–O–H angle in water molecules is 107°.

Atoms can also be connected by *multiple* bonds. Double bonds appear in carbon dioxide, or CO_2, which is therefore often written as O = C = O, triple bonds appear in carbon monoxide, CO, which is often written as C ≡ O. Both double and triple bonds are common in organic compounds. (In addition, the well-known hexagonal benzene ring molecule C_6H_6, like many other compounds, has a one-and-a-half-fold bond.) Higher bonds are rare but do exist; quadruple bonds occur among transition metals atomes such as rhenium or tungsten. Research also confirmed that the uranium U_2 molecule, among

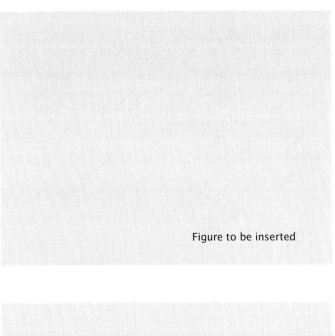

Figure to be inserted

FIGURE 20 An illustration of chemical bond angles when several atoms are involved.

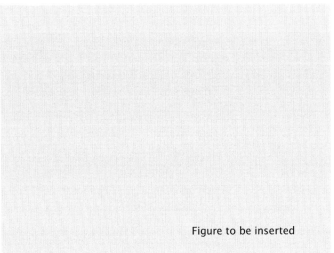

Figure to be inserted

FIGURE 21 Measured chemical bonds.

Ref. 41 others, has a quintuple bond, and that the tungsten W_2 molecule has a hextuple bond.

RIBONUCLEIC ACID AND DEOXYRIBONUCLEIC ACID

Probably the most fascinating molecule of all is human deoxyribonucleic acid, better known with its abbreviation DNA. The nucleic acids where discovered in 1869 by the Swiss physician Friedrich Miescher (1844–1895) in white blood cells. He also found it in cell nuclei, and thus called the substance 'Nuklein'. In 1874 he published an important study showing that the molecule is contained in spermatozoa, and discussed the question if this substance could be related to heredity. With his work, Miescher paved the way to a research field that earned many colleagues Nobel Prizes (though not for himself, as he died before they were established). They changed the name to 'nucleic acid'.

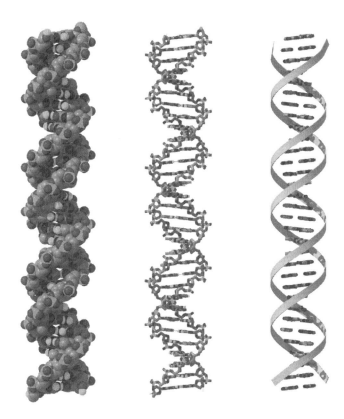

FIGURE 22 Several ways to picture B-DNA, all in false colours (© David Deerfield).

DNA is, as shown in Figure 22, a polymer. A polymer is a molecule built of many similar units. In fact, DNA is among the longest molecules known. Human DNA molecules, for example, can be up to 5 cm in length. Inside each human cell there are 46 chromosomes. In other words, inside each human cell there are molecules with a total length of 2 m. The way nature keeps them without tangling up and knotting is a fascinating topic in itself. All DNA molecules consist of a double helix of sugar derivates, to which four nuclei acids are attached in irregular order. Nowadays, it is possible to make images of single DNA molecules; an example is shown in Figure 23.   Ref. 42

At the start of the twentieth century it became clear that Desoxyribonukleinsäure (DNS) – translated as deoxyribonucleic acid (DNA) into English – was precisely what Erwin Schrödinger had predicted to exist in his book *What Is Life?* As central part of the chromosomes contained the cell nuclei, DNA is responsible for the storage and reproduction of the information on the construction and functioning of Eukaryotes. The information is coded in the ordering of the four nucleic acids. DNA is the carrier of hereditary information. DNA determines in great part how the single cell we all once have been grows into the complex human machine we are as adults. For example, DNA determines the hair colour, predisposes for certain illnesses, determines the maximum size one can grow to, and much more. Of all known molecules, human DNA is thus most intimately related to human existence. The large size of the molecules is the reason that understanding its full structure and its full contents is a task that will occupy scientists for several

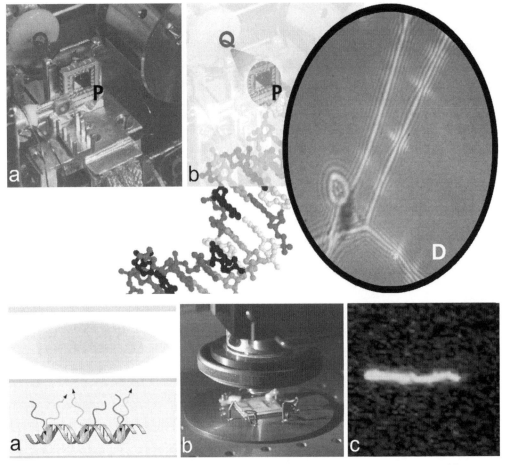

FIGURE 23 Two ways to image single DNA molecules: by holography with electrons emitted from atomically sharp tips (top) and by fluorescence microscopy, with a commercial optical microscope (bottom) (© Hans-Werner Fink/Wiley VCH).

generations to come.

To experience the wonders of DNA, have a look at the animations of DNA copying and of other molecular processes at the unique website www.wehi.edu.au/education/wehitv.

Curiosities and fun challenges about chemistry

Among the fascinating topics of chemistry are the studies of substances that influence humans: toxicology explores *poisons*, pharmacology explores *medicines* (pharmaceutical drugs) and endocrinology explores *hormones*.

Over 50 000 poisons are known, starting with water (usually kills when drunk in amounts larger than about 10 l) and table salt (can kill when 100 g are ingested) up to polonium 210 (kills in doses as low as 5 ng, much less than a spec of dust). Most countries have publicly accessible poison databases; see for example www.gsbl.de.

Can you imagine why 'toxicology', the science of poisons, actually means 'bow science' in Greek? In fact, not all poisons are chemical. Paraffin and oil for lamps, for example,

Challenge 34 e

regularly kill children who taste it because some oil enters the lung and forms a thin film over the alveoles, preventing oxygen intake. This so-called *lipoid pneumonia* can be deadly even when only a *single drop* of oil is in the mouth and then inhaled by a child. Paraffin should never be present in homes with children.

* *

Hormones are signalling substances produced by the human body. Chemically, they can be peptides, lipids or monoamines. Hormones induce mood swings, organize the fight, flight or freeze responses, stimulate growth, start puberty, control cell death and ageing, activate or inhibit the immune system, regulate the reproductive cycle and activate thirst, hunger and sexual arousal.

* *

When one mixes 50 ml of distilled water and 50 ml of ethanol (alcohol), the volume of the mixture has less than 100 ml. Why?

Challenge 35 s

* *

Why do *organic* materials, i.e., materials that contain several carbon atoms, usually burn at much lower temperature than *inorganic* materials, such as aluminium or magnesium?

Challenge 36 ny

* *

A cube of sugar does not burn. However, if you put some cigarette ash on top of it, it burns. Why?

Challenge 37 ny

* *

Sugars are essential for life. One of the simplest sugars is *glucose*, also called *dextrose* or *grape sugar*. Glucose is a so-called *monosaccharide*, in contrast to cane sugar, which is a disaccharide, or starch, which is a polysaccharide.

The digestion of glucose and the burning, or combustion, of glucose follow the same chemical reaction:

$$C_6H_{12}O_6 + 6\,O_2 \rightarrow 6\,CO_2 + 6\,H_2O + 2808\,kJ \tag{8}$$

This is the simplest and main reaction that fuels the muscle and brain activities in our body. The reaction is the reason we have to eat even if we do not grow in size any more. The required oxygen, O_2, is the reason that we breathe in, and the resulting carbon dioxide, CO_2, is the reason that we breathe out. Life, in contrast to fire, is thus able to 'burn' sugar at 37°. That is one of the great wonders of nature. Inside cells, the energy gained from the digestion of sugars is converted into adenosinetriphosphate (ATP) and then converted into motion of molecules.

* *

Chemical reactions can be slow but still dangerous. Spilling mercury on aluminium will lead to an amalgam that reduces the strength of the aluminium part after some time. That is the reason that bringing mercury thermometers on aeroplanes is strictly forbidden.

* *

Challenge 38 s What happens if you take the white power potassium iodide – KJ – and the white power lead nitrate – $Pb(NO_3)_2$ – and mix them with a masher? (This needs to be done with proper protection and supervision.)

* *

Writing on paper with a pen filled with lemon juice instead of ink produces invisible writing. Later on, the secret writing can be made visible by carefully heating the paper on top of a candle flame.

* *

In 2008, it was shown that perispinal infusion of a single substance, *etanercept*, reduced Alzheimer's symptoms in a patient with late-onset Alzheimer's disease, within a few minutes. Curing Alzheimer's disease is one of the great open challenges for modern medicine.

* *

Fireworks fascinate many. A great challenge of firework technology is to produce forest green and greenish blue colours. Producers are still seeking to solve the problem. For more information about fireworks, see the cc.oulu.fi/~kempmp website.

MATERIALS SCIENCE

> " Did you know that one cannot use a boiled egg
> as a toothpick?
>
> Karl Valentin "

We mentioned several times that the quantum of action explains all properties of matter. Many researchers in physics, chemistry, metallurgy, engineering, mathematics and biology have cooperated in the proof of this statement. In our mountain ascent we have only a little time to explore this vast but fascinating topic. Let us walk through a selection.

WHY DOES THE FLOOR NOT FALL?

We do not fall through the mountain we are walking on. Some interaction keeps us from falling through. In turn, the continents keep the mountains from falling through them. Also the liquid magma in the Earth's interior keeps the continents from sinking. All these statements can be summarized in two ideas: First, atoms do not penetrate each other: despite being mostly empty clouds, atoms keep a distance. Secondly, atoms cannot be compressed in size. Both properties are due to Pauli's exclusion principle between electrons. Vol. IV, page 125 The fermion character of electrons avoids that atoms shrink or interpenetrate – on Earth.

In fact, not all floors keep up due to the fermion character of electrons. Atoms are not impenetrable at all pressures. At sufficiently large pressures, atoms can collapse, and form new types of floors. Such floors do not exist on Earth. These floors are so exciting to study that people have spent their whole life to understand why they do not fall, or when they do, how it happens: the *surfaces of stars*.

In usual *stars*, such as in the Sun, the gas pressure takes the role which the incompressibility of solids and liquids has for planets. The pressure is due to the heat produced by the nuclear reactions.

In most stars, the radiation pressure of the light plays only a minor role. Light pressure does play a role in determining the size of *red giants*, such as Betelgeuse; but for average stars, light pressure is negligible.

The next star type appears whenever light pressure, gas pressure and the electronic Pauli pressure cannot keep atoms from interpenetrating. In that case, atoms are compressed until all electrons are pushed into the protons. Protons then become neutrons, and the whole star has the same mass density of atomic nuclei, namely about $2.3 \cdot 10^{17}$ kg/m^3. A drop weighs about 200 000 tons. In these so-called *neutron stars*, the floor – or better, the size – is also determined by Pauli pressure; however, it is the Pauli pressure between neutrons, triggered by the nuclear interactions. These neutron stars are Page 170 all around 10 km in radius.

If the pressure increases still further, the star becomes a *black hole*, and never stops collapsing. Black holes have no floor at all; they still have a constant size though, deter- Vol. II, page 243 mined by the horizon curvature.

The question whether other star types exist in nature, with other floor forming mechanisms – such as *quark stars* – is still a topic of research.

Rocks and stones

If a geologist takes a stone in his hands, he is usually able to give, within an error of a few per cent, the age of the stone simply by looking at it. The full story behind this ability forms a large part of geology, but the general lines should be known to every physicist.

Every stone arrives in your hand through the *rock cycle(s)*. The main rock cycle is a process that transforms magma from the interior of the Earth into *igneous (or magmatic) rocks* through cooling and crystallization. Igneous rocks, such as basalt, can transform through erosion, transport and deposition into *sedimentary rocks*. Either of these two rock types can be transformed through high pressures or temperatures into *metamorphic rocks*, such as marble. Finally, most rocks are generally – but not always – transformed back into magma.

The main rock cycle takes around 110 to 170 million years. For this reason, rocks that are older than this age are much less common on Earth. Any stone is the product of erosion of one of the rock types. A geologist can usually tell, simply by looking at it, the type of rock it belongs to; if he sees the original environment, he can also give the age, without any laboratory.

For a physicist, most rocks are mixtures of crystals. *Crystals* are solids with a regular arrangement of atoms. They form a fascinating topic by themselves.

Some interesting crystals

Have you ever admired a quartz crystal or some other crystalline material? The beautiful shape and atomic arrangement has formed spontaneously, as a result of the motion of atoms under high temperature and pressure, during the time that the material was deep under the Earth's surface. The details of crystal formation are complex and interesting.

For example, are regular crystal lattices *energetically* optimal? This simple question

TABLE 4 The types of rocks and stones.

TYPE	PROPERTIES	SUBTYPE	EXAMPLE
Igneous rocks (magmatites)	formed from magma, 95% of all rocks	volcanic or extrusive	basalt (ocean floors, Giant's Causeway), andesite, obsidian
		plutonic or intrusive	granite, gabbro
Sedimentary rocks (sedimentites)	often with fossils	clastic	shale, siltstone, sandstone
		biogenic	limestone, chalk, dolostone
		precipitate	halite, gypsum
Metamorphic rocks (metamorphites)	transformed by heat and pressure	foliated	slate, schist, gneiss (Himalayas)
		non-foliated (grandoblastic or hornfelsic)	marble, skarn, quartzite
Meteorites	from the solar system	rock meteorites	
		iron meteorites	

leads to a wealth of problems. We might start with the much simpler question whether a regular dense packing of spheres is the most *dense* possible. Its density is $\pi/\sqrt{18}$, i.e., a bit over 74 %. Even though this was conjectured to be the maximum possible value already in 1609 by Johannes Kepler, the statement was proven only in 1998 by Tom Hales. The proof is difficult because in small volumes it is possible to pack spheres up to almost 78 %. To show that over large volumes the lower value is correct is a tricky business.

Challenge 39 s

Ref. 93

Next, does a regular crystal of solid spheres, in which the spheres do *not* touch, have the highest possible entropy? This simple problem has been the subject of research only in the 1990s. Interestingly, *for low temperatures*, regular sphere arrangements indeed show the largest possible entropy. At low temperatures, spheres in a crystal can oscillate around their average position and be thus more disordered than if they were in a liquid; in the liquid state the spheres would block each other's motion and would not allow reaching the entropy values of a solid.

This many similar results deduced from the research into these so-called *entropic forces* show that the transition from solid to liquid is – at least in part – simply a geometrical effect. For the same reason, one gets the surprising result that even slightly repulsing spheres (or atoms) can form crystals and melt at higher temperatures. These are beautiful examples of how classical thinking can explain certain material properties, using from quantum theory only the particle model of matter.

Ref. 94

But the energetic side of crystal formation provides other interesting questions. Quantum theory shows that it is possible that *two* atoms repel each other, while *three* attract each other. This beautiful effect was discovered and explained by Hans-Werner Fink in 1984. He studied rhenium atoms on tungsten surfaces and showed, as observed, that they

Ref. 95

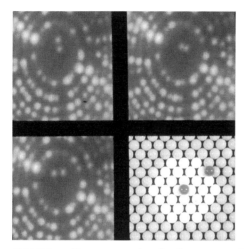 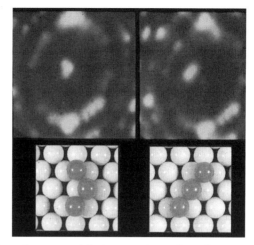

FIGURE 24 On tungsten tips, rhenium atoms, visible at the centre of the images, do not form dimers (left) but do form trimers (right) (© Hans-Werner Fink, APS, AIP from Ref. 95).

FIGURE 25 Some snow flakes (© Furukawa Yoshinori).

cannot form dimers – two atoms moving together – but readily form trimers. This is an example contradicting classical physics; the effect is impossible if one pictures atoms as immutable spheres, but becomes possible when one remembers that the electron clouds around the atoms rearrange depending on their environment.

For an exact study of crystal energy, the interactions between all atoms have to be included. The simplest question is to determine whether a regular array of alternatively charged spheres has lower energy than some irregular collection. Already such simple questions are still topic of research; the answer is still open.

Another question is the mechanism of face formation in crystals. Can you confirm that crystal faces are those planes with the *slowest* growth speed, because all fast growing planes are eliminated? The finer details of the process form a complete research field in itself.

However, not always the slowest growing planes win out. Figure 25 shows some well-known exceptions. Explaining such shapes is possible today, and Furukawa Yoshinori is one of the experts in the field, heading a dedicated research team. Indeed, there remains the question of symmetry: why are crystals often symmetric, such as snowflakes, instead of asymmetric? This issue is a topic of self-organization, as mentioned already in the section of classical physics. It turns out that the symmetry is an automatic result of the way molecular systems grow under the combined influence of diffusion and non-linear

Challenge 40 s

Ref. 96

Ref. 97

Vol. I, page 354

processes. The details are still a topic of research.

SOME INTERESTING CRYSTALS

Every crystal, like every structure in nature, is the result of growth. Every crystal is thus the result of motion. To form a crystal whose regularity is as high as possible and whose shape is as symmetric as possible, the required motion is a *slow* growth of facets from the liquid (or gaseous) basic ingredients. The growth requires a certain pressure, temperature and temperature gradient for a certain time. For the most impressive crystals, the *gemstones*, the conditions are usually quite extreme; this is the reason for their durability. The conditions are realized in specific rocks deep inside the Earth, where the growth process can take thousands of years. Mineral crystals can form in all three types of rocks:

Page 63 igneous (magmatic), metamorphic, and sedimentary. Other crystals can be made in the laboratory in minutes, hours or days and have led to a dedicated industry. Only a few crystals grow from liquids at standard conditions; examples are gypsum and several other sulfates, which can be crystallized at home, potassium bitartrate, which appears in the making of wine, and the crystals grown inside plants or animals, such as teeth, bones or magnetosensitive crystallites.

Growing, cutting, treating and polishing crystals is an important industry. Especially the growth of crystals is a science in itself. Can you show with pencil and paper that

Challenge 41 e only the *slowest* growing facets are found in crystals? In the following, a few important crystals are presented.

* *

Quartz, amethyst (whose colour is due to radiation and iron Fe^{4+} impurities), *citrine* (whose colour is due to Fe^{3+} impurities), *smoky quartz* (with colour centres induced by radioactivity), *agate* and *onyx* are all forms of crystalline silicon dioxide or SiO_2. Quartz forms in igneous and in magmatic rocks; crystals are also found in many sedimentary rocks. Quartz crystals can sometimes be larger than humans. By the way, most amethysts lose their colour with time, so do not waste money buying them.

Quartz is the most common crystal on Earth's crust and is also grown synthetically for many high-purity applications. The structure is rombohedral, and the ideal shape is a six-sided prism with six-sided pyramids at its ends. Quartz melts at 1986 K and is piezo- and pyroelectric. Its piezoelectricity makes it useful as electric oscillator and filter. A film

Vol. I, page 254 of an oscillating clock quartz is found in the first volume. Quartz is also used for glass production, in communication fibres, for coating of polymers, in gas lighters, as source of silicon and for many other applications.

* *

Corundum, ruby and *sapphire* are crystalline variations of alumina, or Al_2O_3. Corundum is pure and colourless crystalline alumina, ruby is Cr doped and blue sapphire is Ti or Fe doped. They all have trigonal crystal structure and melt at 2320 K. Natural gems are formed in metamorphic rocks. Yellow, green, purple, pink, brown, grey and salmon-coloured sapphires also exist, when doped with other impurities. The colours of natural sapphires, like that of many other gemstones, are often changed by baking and other treatments.

FIGURE 26 Quartz found at St. Gotthard, Switzerland, picture size 12 cm (© Rob Lavinsky).

FIGURE 27 Citrine found on Magaliesberg, South Africa, crystal height 9 cm (© Rob Lavinsky).

FIGURE 28 Amethystine and orange quartz found in the Orange River, Namibia, picture size 6 cm (© Rob Lavinsky).

Corundum, ruby and sapphire are used in jewellery, as heat sink and growth substrate, and for lasers. Corundum is also used as scratch-resistant 'glass' in watches. Ruby was the first gemstone that was grown synthetically in gem quality, in 1892 by Auguste Verneuil (1856-1913), who made his fortune in this way.

FIGURE 29 Corundum found in Laacher See, Germany, picture size 4 mm (© Stephan Wolfsried).

FIGURE 30 Ruby found in Jagdalak, Afghanistan, picture height 2 cm (© Rob Lavinsky).

FIGURE 31 Sapphire found in Ratnapura, Sri Lanka, crystal size 1.6 cm (© Rob Lavinsky).

* *

Tourmaline is a frequently found mineral and can be red, blue, green, orange, yellow,

FIGURE 32 Raw crystals, or boules, of synthetic corundum, picture size c. 50 cm (© Morion Company).

pink or black, depending on its composition. The chemical formula is astonishingly complex and varies from type to type. Tourmaline has trigonal structure and usually forms columnar crystals that have triangular cross-section. It is only used in jewellery. Paraiba tourmalines, a very rare type of green or blue tourmaline, are among the most beautiful gemstones and can be, if natural and untreated, more expensive than diamonds.

FIGURE 33 Natural bicoloured tourmaline found in Paprok, Afghanistan, picture size 9 cm (© Rob Lavinsky).

FIGURE 34 Cut Paraiba tourmaline from Brazil, picture size 3 cm (© Manfred Fuchs).

* *

Garnets are a family of compounds of the type $X_2Y_3(SiO_4)_3$. They have cubic crystal structure. They can have any colour, depending on composition. They show no cleavage and their common shape is a rhombic dodecahedron. Some rare garnets differ in colour when looked at in daylight or in incandescent light. Natural garnets form in metamorphic rocks and are used in jewellery, as abrasive and for water filtration. Synthetic garnets are used in many important laser types.

FIGURE 35 Red garnet with smoky quartz found in Lechang, China, picture size 9 cm (© Rob Lavinsky).

FIGURE 36 Green demantoid, a garnet owing its colour to chromium doping, found in Tubussis, Namibia, picture size 5 cm (© Rob Lavinsky).

FIGURE 37 Synthetic Cr,Tm,Ho:YAG, a doped yttrium aluminium garnet, picture size 25 cm (© Northrop Grumman).

* *

Alexandrite, a chromium-doped variety of *chrysoberyl*, is used in jewellery and in lasers. Its composition is $BeAl_2O_4$; the crystal structure is orthorhombic. Chrysoberyl melts at 2140 K. Alexandrite is famous for its colour-changing property: it is green in daylight or fluorescent light but amethystine in incandescent light, as shown in Figure 35. The effect is due to its chromium content: the ligand field is just between that of chromium in red ruby and that in green emerald. A few other gems also show this effect, in particular the rare blue garnet and some Paraiba tourmalines.

FIGURE 38 Alexandrite found in the Setubal river, Brazil, crystal height 1.4 cm, illuminated with daylight (left) and with incandescent light (right) (© Trinity Mineral).

FIGURE 39 Synthetic alexandrite, picture size 20 cm (© Northrop Grumman).

* *

Perovskites are a large class of cubic crystals used in jewellery and in tunable lasers. Their general composition is XYO_3, XYF_3 or $XYCl_3$.

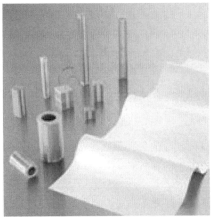

FIGURE 40 Perovskite found in Hillesheim, Germany. Picture width 3 mm (© Stephan Wolfsried).

FIGURE 41 Synthetic PZT, or lead zirconium titanate, is a perovskite used in numerous products. Picture width 20 cm (© Ceramtec).

* *

Diamond is a metastable variety of *graphite*, thus pure carbon. Theory says that graphite is the stable form; practice says that diamond is still more expensive. In contrast to graphite, diamond has face-centred cubic structure, is a large band gap semiconductor and typically has octahedral shape. Diamond burns at 1070 K; in the absence of oxygen it converts to graphite at around 1950 K. Diamond can be formed in magmatic and in metamorphic rocks. Diamonds can be synthesized in reasonable quality, though gemstones of large size and highest quality are not yet possible. Diamond can be coloured and be doped to achieve electrical conductivity in a variety of ways. Diamond is mainly used in jewellery, for hardness measurements and as abrasive.

* *

Silicon, Si, is not found in nature in pure form; all crystals are synthetic. The structure is face-centred cubic, thus diamond-like. It is moderately brittle, and can be cut in thin wafers which can be further thinned by grinding or chemical etching, even down to a thickness of 10 μm. Being a semiconductor, the band structure determines its black colour, its metallic shine and its brittleness. Silicon is widely used for silicon chips and electronic semiconductors. Today, human-sized silicon crystals can be grown free of dislocations and other line defects. (They will still contain some point defects.)

* *

Teeth are the structures that allowed animals to be so successful in populating the Earth. They are composed of several materials; the outer layer, the *enamel*, is 97% hydroxylapatite, mixed with a small percentage of two proteins groups, the amelogenins and the enamelins. The growth of teeth is still not fully understood; neither the molecular level nor the shape-forming mechanisms are completely clarified. Hydroxylapatite is soluble

FIGURE 42 Natural diamond from Saha republic, Russia, picture size 4 cm (© Rob Lavinsky).

FIGURE 43 Synthetic diamond, picture size 20 cm (© Diamond Materials GmbH).

FIGURE 44 Ophtalmic diamond knife, picture size 1 cm (© Diamatrix Ltd.).

in acids; addition of fluorine ions changes the hydroxylapatite to fluorapatite and greatly reduces the solubility. This is the reason for the use of fluorine in tooth paste.

Hydroxylapatite (or hydroxyapatite) has the chemical formula $Ca_{10}(PO_4)_6(OH)_2$, possesses hexagonal crystal structure, is hard (more than steel) but relatively brittle. It occurs as mineral in sedimentary rocks (see Figure 46), in bones, renal stones, bladders tones, bile stones, atheromatic plaque, cartilage arthritis and teeth. Hydroxylapatite is mined as a phosphorus ore for the chemical industry, is used in genetics to separate single and double-stranded DNA, and is used to coat implants in bones.

* *

Pure metals, such as *gold*, *silver* and even *copper*, are found in nature, usually in magmatic rocks. But only a few metallic compounds form crystals, such as *pyrite*. Monocrystalline pure metal crystals are all synthetic. Monocrystalline metals, for example iron, aluminium, gold or copper, are extremely soft and ductile. Either bending them repeatedly – a process called cold working – or adding impurities, or forming alloys makes them hard and strong. Stainless steel, a carbon-rich iron alloy, is an example that uses all three processes.

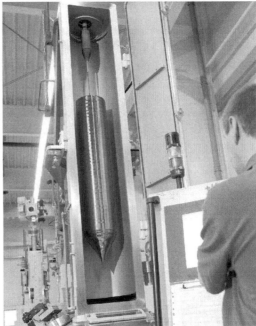

FIGURE 45 A silicon crystal growing machine and two resulting crystals, with a length of c. 2 m (© www.PVATePla.com).

* *

Ref. 43 In 2009, Luca Bindi of the Museum of Natural History in Florence, Italy, made headlines across the world with his discovery of the first natural *quasicrystal*. Quasicrystals are materials that show non-crystallographic symmetries. Until 2009, only synthetic materials were known. Then, in 2009, after years of searching, Bindi discovered a specimen in his collection whose grains clearly show fivefold symmetry.

* *

FIGURE 46 Hydroxylapatite found in Snarum, Norway, picture size 5 cm (© Rob Lavinsky).

FIGURE 47 The main and the reserve teeth on the jaw bone of a shark, all covered in hydroxylapatite, picture size 15 cm (© Peter Doe).

FIGURE 48 Pyrite, found in Navajún, Spain, picture width 5.7 cm (© Rob Lavinsky).

FIGURE 49 Silver from Colquechaca, Bolivia, picture width 2.5 cm (© Rob Lavinsky).

FIGURE 50 Synthetic copper single crystal, picture width 30 cm (© Lachlan Cranswick).

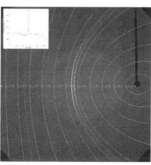

FIGURE 51 The specimen, found in the Koryak Mountains in Russia, is part of a triassic mineral, about 220 million years old; the black material is mostly khatyrkite ($CuAl_2$) and cupalite ($CuAl_2$) but also contains quasicrystal grains with composition $Al_{63}Cu_{24}Fe_{13}$ that have fivefold symmetry, as clearly shown in the X-ray diffraction pattern and in the transmission electron image. (© Luca Bindi).

There are about 4000 known mineral types. On the other hand, there are ten times as many obsolete mineral names, namely around 40 000. An official list can be found in various places on the internet, including www.mindat.org or www.minieralienatlas.de. To explore the world of crystal shapes, see the www.smorf.nl website. Around new 40 minerals are discovered each year. Searching for minerals and collecting them is a fascinating pastime.

How can we look through matter?

Vol. IV, page 27 The quantum of action tells us that all obstacles have only finite potential heights. The quantum of action implies that matter is penetrable. That leads to a question: Is it possible to look through solid matter? For example, can we see what is hidden inside a mountain? To achieve this, we need a signal which fulfils two conditions: the signal must be able to *penetrate* the mountain, and it must be scattered in a *material-dependent* way. Indeed, such signals exist, and various techniques use them. Table 5 gives an overview of the possibilities.

TABLE 5 Signals penetrating mountains and other matter.

Signal	Penetration depth in stone	Achieved resolution	Material dependence	Use
Fluid signals				
Diffusion of gases, such as helium	*c.* 5 km	*c.* 100 m	medium	exploring vacuum systems and tube systems
Diffusion of water or liquid chemicals	*c.* 5 km	*c.* 100 m	medium	mapping hydrosystems
Sound signals				
Infrasound and earthquakes	100 000 km	100 km	high	mapping of Earth crust and mantle
Sound, explosions, short seismic waves	0.1 – 10 m	*c.* $\lambda/100$	high	oil and ore search, structure mapping in rocks, searching for underwater treasures in sunken ship with sub-bottom-profilers
Ultrasound		1 mm	high	medical imaging, acoustic microscopy, sonar, echo systems
Acousto-optic or photoacoustic imaging		1 mm	medium	blood stream imaging, mouse imaging
Electromagnetic signals				
Static magnetic field variations			medium	cable search, cable fault localization, search for structures and metal inside soil, rocks and the seabed
Electrical currents				soil and rock investigations, search for tooth decay
Electromagnetic sounding, 0.2 – 5 Hz				soil and rock investigations in deep water and on land
Radio waves	10 m	30 m to 1 mm	small	soil radar (up to 10 MW), magnetic resonance imaging, research into solar interior

TABLE 5 (Continued) Signals penetrating mountains and other matter.

SIGNAL	PENE-TRATION DEPTH IN STONE	ACHIE-VED RESOLU-TION	MATE-RIAL DEPEN-DENCE	USE
Ultra-wide band radio	10 cm	1 mm	sufficient	searching for wires and tubes in walls, breast cancer detection
THz and mm waves	below 1 mm	1 mm		see through clothes, envelopes and teeth Ref. 44
Infrared	$c.$ 1 m	0.1 m	medium	mapping of soil over 100 m
Visible light	$c.$ 1 cm	0.1 µm	medium	imaging of many sorts, including breast tumor screening
X-rays	a few metres	5 µm	high	medicine, material analysis, airports, food production check
γ-rays	a few metres	1 mm	high	medicine
Matter particle signals				
Neutrons from a reactor	up to $c.$ 1 m	1 mm	medium	tomography of metal structures, e.g., archaeologic statues or car engines
Muons created by cosmic radiation or technical sources	up to $c.$ 300 m	0.1 m	small	finding caves in pyramids, imaging interior of trucks
Positrons	up to $c.$ 1 m	2 mm	high	brain tomography
Electrons	up to $c.$ 1 µm	10 nm	small	transmission electron microscopes
Neutrino beams	light years	none	very weak	studies of Sun
Radioactivity mapping	1 mm to 1 m			airport security checks
Gravitation				
Variation of g		50 m	low	oil and ore search

We see that many signals are able to penetrate a mountain, and ven more signals are able to penetrate other condensed matter. To distinguish different materials, or to distinguish solids from liquids and from air, sound and radio waves are the best choice. In addition, any useful method requires a large number of signal sources and of signal receptors, and thus a large amount of money. Will there ever be a simple method that allows looking into mountains as precisely as X-rays allow looking into human bodies? For example, is it possible to map the interior of the pyramids? A motion expert should be able to give a definite answer.

Challenge 42 s

One of the high points of twentieth century physics was the development of the best method so far to look into matter with dimensions of about a metre or less: magnetic resonance imaging. We will discuss it later on.

Page 146

Various modern imaging techniques, such as X-rays, ultrasound imaging and several future ones, are useful in medicine. As mentioned before, the use of ultrasound imaging

Vol. I, page 270

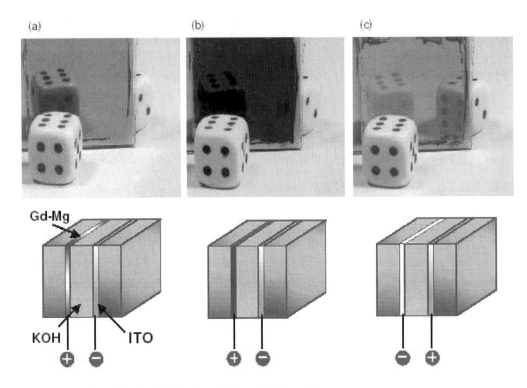

FIGURE 52 A switchable Mg-Gd mirror (© Ronald Griessen).

Vol. I, page 270
for prenatal diagnostics of embryos is not recommended. Studies have found that ultrasound produces extremely high levels of audible sound to the embryo, especially when the ultrasound is repeatedly switched on and off, and that babies react negatively to this loud noise.

WHAT IS NECESSARY TO MAKE MATTER INVISIBLE?

You might have already imagined what adventures would be possible if you could be *invisible* for a while. In 1996, a team of Dutch scientists found a material, yttrium hydride or YH_3, that can be switched from mirror mode to transparent mode using an electrical Ref. 45 signal. A number of other materials were also discovered. An example of the effect for Mg-Gd layers is shown in Figure 52.

Switchable mirrors might seem a first step to realize the dream to become invisible and visible at will. In 2006, and repeatedly since then, researchers made the headlines in the popular press by claiming that they could build a *cloak of invisibility*. This is a blatant lie. This lie is frequently used to get funding from gullible people, such as buyers of bad science fiction books or the military. For example, it is often claimed that objects can be made invisible by covering them with metamaterials. The impossibility of this aim has Vol. III, page 146 been already shown earlier on. But we now can say more.

Nature shows us how to realize invisibility. An object is invisible if it has no surface, no absorption and small size. In short, invisible objects are either small clouds or composed of them. Most atoms and molecules are examples. Homogeneous non-absorbing gases also realize these conditions. That is the reason that air is (usually) invisible. When air is not homogeneous, it can be visible, e.g. above hot surfaces.

In contrast to gases, solids or liquids do have surfaces. Surfaces are usually visible, even if the body is transparent, because the refractive index changes there. For example, quartz can be made so transparent that one can look through 1 000 km of it; pure quartz is thus *more* transparent than usual air. Still, objects made of pure quartz are visible to the eye, due to their refractive index. Quartz can be invisible only when submerged in liquids with the same refractive index.

In short, anything that has a shape cannot be invisible. If we want to become invisible, we must transform ourselves into a diffuse gas cloud of non-absorbing atoms. On the way to become invisible, we would lose all memory and all genes, in short, we would lose all our individuality, because an individual cannot be made of gas. An individual is defined through its boundary. There is no way that we can be invisible and alive at the same time; a way to switch back to visibility is even less likely. In summary, quantum theory shows that only the dead can be invisible. Quantum theory has a reassuring side: we already found that quantum theory forbids ghosts; we now find that it also forbids Vol. IV, page 126 any invisible beings.

WHAT MOVES INSIDE MATTER?

All matter properties are due to the motion of the components of matter. Therefore, we can correctyl argue that understanding the motion of electrons and nuclei implies understanding all properties of matter. Sometimes, however, it is more practical to explore the motion of collections of electrons or nuclei as a whole. Here is a selection of such collective motions.

In crystalline solids, sound waves can be described as the motion of *phonons*. For example, transverse phonons and longitudinal phonons describe many processes in semiconductors, in solid state lasers and in ultrasound systems. Phonon approximatively behave as bosons.

In metals, the motion of crystal defects, the so-called *dislocations* and *disclinations*, is Page 273 central to understand their hardening and their breaking.

Also in metals, the charge waves of the conductive electron plasma, can be seen as composed of so-called *plasmons*. Plasmons are also important in the behaviour of high-speed electronics.

In magnetic materials, the motion of spin orientation is often best described with the help of magnons. Understanding the motion of magnons and that of magnetic domain walls is useful to understand the magnetic properties of magnetic material, e.g., in permanent magnets, magnetic storage devices, or electric motors. Magnons behave approximately like bosons.

In semiconductors and insulators, the motion of *conduction electrons* and *electron holes*, is central for the description and design of most electronic devices. They behave as fermions with spin 1/2, elementary electric charge, and a mass that depends on the material, on the specific conduction band and on the spicific direction of motion. The bound system of a conduction electron and a hole is called an exciton. It can have spin 0 or spin 1.

In polar materials, the motion of light through the material is often best described in terms of *polaritons*, i.e., the coupled motion of photons and diople caryying material excitations. Polaritons are approximate bosons.

In dielectric crystals, such as in many anorganic ionic crystals, the motion of an electron is often best described in terms of *polarons*, the coupled motion of the electron with the coupled polarzation region that surrounds it. Polarons are fermions.

In fluids, the motion of vortices is central in understanding turbulence or air hoses. Especially in superfluids, vortex motion is quantized in terms of *rotons* which determines flow properties. Also in fluids, bubble motion is often useful to describe mixing processes.

In superconductors, not only the motion of *Cooper pairs*, but also the motion of *magnetic flux tubes* determines the temperature behaviour. Especially in thin and flat superconductors – so-called 'two-dimensional' systems – such tubes have particle-like properties.

In all condensed matter systems, the motion of *surface states* – such as surface magnons, surface phonons, surface plasmons, surface vortices – has also to be taken into account.

Many other, more exotic quasiparticles exist in matter. Each is in itself is an important research field where quantum physics and material science come together. To clarify the concepts, we mention that a *soliton* is not, in general, a quasiparticle. 'Soliton' is a mathematical concept; it applies to macroscopic waves with only one crest that remain unaltered after collisions. Many domain walls can be seen as solitons. But quasiparticles are concepts that describe physical observations similar to quantum particles.

Vol. I, page 273

In summary, all the mentioned examples of collective motion inside matter, both macroscopic and quantized, are of importance in electronics, photonics, engineering and medical applications. Many are quantized and their motion can be studied like the motion of real quantum particles.

CURIOSITIES AND FUN CHALLENGES ABOUT MATERIALS SCIENCE

Ref. 46

What is the maximum height of a mountain? This question is of course of interest to all climbers. Many effects limit the height. The most important is that under heavy pressure, solids become liquid. For example, on Earth this happens for a mountain with a height of about 27 km. This is quite a bit more than the highest mountain known, which is the volcano Mauna Kea in Hawaii, whose top is about 9.45 km above the base. On Mars gravity is weaker, so that mountains can be higher. Indeed the highest mountain on Mars, Olympus mons, is 80 km high. Can you find a few other effects limiting mountain height?

Challenge 43 ny

Challenge 44 s

* *

Challenge 45 r

Do you want to become rich? Just invent something that can be produced in the factory, is cheap and can fully substitute duck feathers in bed covers, sleeping bags or in badminton shuttlecocks. Another industrial challenge is to find an artificial substitute for latex, and a third one is to find a substitute for a material that is rapidly disappearing due to pollution: cork.

* *

How much does the Eiffel tower change in height over a year due to thermal expansion

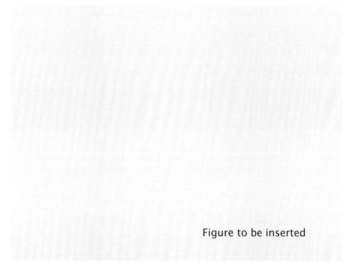

FIGURE 53 A Gekko sticks to glass and other surfaces using the van der Waals force.

and contraction?

Challenge 46 s

* *

What is the difference between the makers of bronze age knifes and the builders of the Eiffel tower? Only their control of defect distributions. The main defects in metals are *disclinations* and *dislocations*. Disclinations are crystal defects in form of surfaces; they are the microscopic aspect of grain boundaries. Dislocations are crystal defects in form of curved lines; above all, their distribution and their motion in a metal determines the stiffness. For a picture of dislocations, see below.

Page 274

* *

What is the difference between solids, liquids and gases?

Challenge 47 e

* *

Quantum theory shows that tight walls do not exist. Every material is penetrable. Why?

Challenge 48 s

* *

Quantum theory shows that even if tight walls would exist, the lid of a box made of such walls can never be tightly shut. Can you provide the argument?

Challenge 49 s

* *

Heat can flow. Like for all flows, quantum theory predicts that heat transport quantized. This implied that thermal conductance is quantized. And indeed, in the year 2000, experiments have confirmed the prediction. Can you guess the smallest unit of thermal conductance?

Ref. 47

Challenge 50 s

* *

Robert Full has shown that van der Waals forces are responsible for the way that geckos

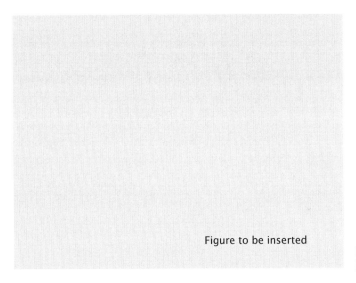

Figure to be inserted

FIGURE 54 The structure of bones.

Ref. 48   walk on walls and ceilings, as shown in Figure 53. The gekko, a small reptile with a mass of about 100 g, uses an elaborate structure on its feet to perform the trick. Each foot has 500 000 hairs each split in up to 1000 small spatulae, and each spatula uses the van der Waals force (or alternatively, capillary forces) to stick to the surface. As a result, the gekko can walk on vertical glass walls or even on glass ceilings; the sticking force can be as high as 100 N per foot.

The same mechanism is used by jumping spiders (*Salticidae*). For example, *Evarcha arcuata* have hairs at their feet which are covered by hundred of thousands of setules.

Ref. 49   Again, the van der Waals force in each setule helps the spider to stick on surfaces.

Researchers have copied these mechanisms for the first time in 2003, using microlithography on polyimide, and hope to make durable sticky materials in the near future.

* *

One of the most fascinating material in nature are bones. Bones are light, stiff, and can

Ref. 50   heal after fractures. If you are interested in composite materials, read more about bones: their structure, shown in Figure 54, and their material properties are fascinating and complex.

* *

Millimetre waves or terahertz waves are emitted by all bodies at room temperature. Modern camera systems allow producing images with them. In this way, it is possible to see through clothes, as shown by Figure 55. (Caution is needed; there is the widespread suspicion that the image is a fake produced to receive more development funding.) This ability could be used in future to detect hidden weapons in airports. But the development of a practical and affordable detector which can be handled as easily as a binocular is still under way. The waves can also be used to see through paper, thus making it unnecessary to open letters in order to read them. Secret services are exploiting this technique. A

FIGURE 55 An alleged image acquired with terahertz waves. Can you explain why it is a fake? (© Jefferson Lab)

Figure to be inserted

FIGURE 56 Some plasma machines.

third application of terahertz waves might be in medical diagnostic, for example for the search of tooth decay. Terahertz waves are almost without side effects, and thus superior to X-rays. The lack of low-priced quality sources is still an obstacle to their application.

* *

Does the melting point of water depend on the magnetic field? This surprising claim was made in 2004 by Inaba Hideaki and colleagues. They found a change of 0.9 mK/T. It is known that the refractive index and the near infrared spectrum of water is affected by magnetic fields. Indeed, not everything about water might be known yet.   Ref. 51

* *

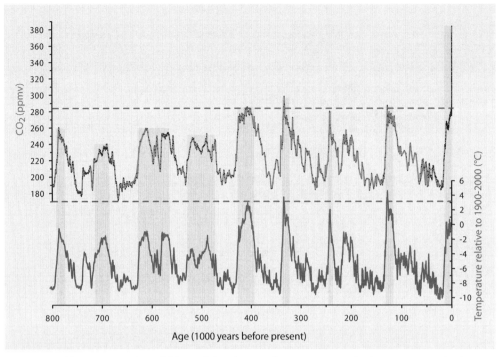

FIGURE 57 The concentration of CO_2 and the change of average atmospheric temperature in the past 0.8 million years (© Dieter Lüthi).

Plasmas, or ionized gases, are useful for many applications. A few are shown in Figure 56. Not only can plasmas be used for heating or cooking and generated by chemical means (such plasmas are variously called fire or flames) but they can also be generated electrically and used for lighting or deposition of materials. Electrically generated plasmas are even being studied for the disinfection of dental cavities.

* *

It is known that the concentration of CO_2 in the atmosphere between 1800 and 2005 has increased from 280 to 380 parts per million, as shown in Figure 57. (How would you measure this?) It is known without doubt that this increase is due to human burning of fossil fuels, and not to natural sources such as the oceans or volcanoes. There are three arguments. First of all, there was a parallel decline of the $^{14}C/^{12}C$ ratio. Second, there was a parallel decline of the $^{13}C/^{12}C$ ratio. Finally, there was a parallel decline of the oxygen concentration. All three measurements independently imply that the CO_2 increase is due to the burning of fuels, which are low in ^{14}C and in ^{13}C, and at the same time decrease the oxygen ratio. Natural sources do not have these three effects. Since CO_2 is a major greenhouse gas, the data implies that humans are also responsible for a large part of the temperature increase during the same period. Global warming exists and is mainly due to humans. The size of the effect, however, is still a matter of heated dispute. On average, the Earth has cooled over the past 10 million years; since a few thousand years, the temperature is, however, rising and at the same level as 3 million years ago. How do the decade trend global warming, the thousand year warming trend and the million year

Ref. 52
Challenge 51 s

Ref. 53

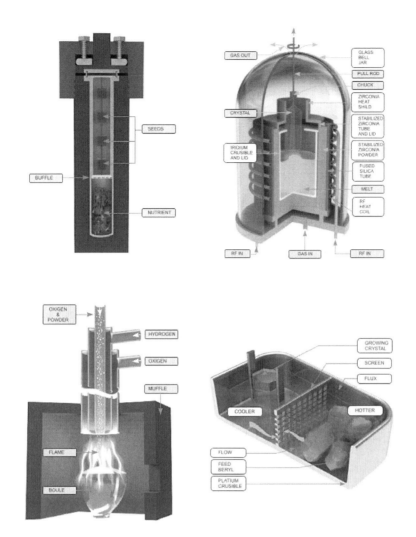

FIGURE 58 Four crystal and synthetic gemstone growing methods. Top: the hydrothermal technique, used to grow emeralds, quartz, rock crystal and amethyst, and Czochralski's pulling technique, used for growing ruby, sapphire, spinel, yttrium-aluminum-garnet, gadolinium-gallium-garnet and alexandrite. Bottom: Verneuil's flame fusion technique, used for growing corundum, sapphire, ruby and spinel boules, and the flux process, used for chrysoberyl (© Ivan Golota).

cooling trend interact? This is a topic of intense research and discussions.

∗ ∗

Making crystals can make one rich. The first man who did so, the Frenchman Auguste Verneuil (1856–1913), sold rubies grown in his laboratory for many years without telling anybody. Many companies now produce synthetic gems with machines that are kept secret. An example is given in Figure 58.

Synthetic diamonds have now displaced natural diamonds in almost all applications.

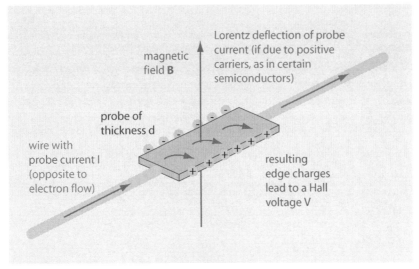

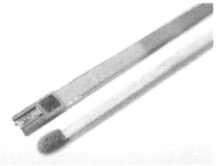

FIGURE 59 The (classical) Hall effect and a modern miniature Hall probe using it (© Metrolab).

In the last years, methods to produce large, white, jewel-quality diamonds of ten carats and more are being developed. These advances will lead to a big change in all the domains that depend on these stones, such as the production of the special surgical knives used in eye lens operation.

Ref. 54

* *

The technologies to produce *perfect* crystals, without grain boundaries or dislocations, are an important part of modern industry. Perfectly regular crystals are at the basis of the integrated circuits used in electronic appliances, are central to many laser and telecommunication systems and are used to produce synthetic jewels.

* *

Challenge 52 s   How can a small plant pierce through tarmac?

* *

If you like abstract colour images, do not miss looking at liquid crystals through a microscope. You will discover a wonderful world. The best introduction is the text by Ingo Dierking.

Ref. 55

* *

The Lorentz force leads to an interesting effect inside materials. If a current flows along a conducting strip that is in a (non-parallel) magnetic field, a voltage builds up between two edges of the conductor, because the charge carriers are deflected in their flow. This effect is called the (classical) *Hall effect* after the US-American physicist Edwin Hall (1855–1938), who discovered it in 1879, during his PhD. The effect, shown in Figure 59, is regularly used, in so-called *Hall probes*, to measure magnetic fields; the effect is also used to read data from magnetic storage media or to measure electric currents (of the order of 1 A or more) in a wire without cutting it. Typical Hall probes have sizes of around 1 cm down to 1 μm and less. The Hall voltage V turns out to be given by

$$V = \frac{IB}{ned} \, , \qquad\qquad (9)$$

where n is the electron number density, e the electron charge, and d is the thickness of the probe, as shown in Figure 59. Deducing the equation is a secondary school exercise. The Hall effect is a material effect, and the material parameter n determines the Hall Challenge 53 e voltage. The sign of the voltage also tells whether the material has positive or negative charge carriers; indeed, for metal strips the voltage polarity is opposite to the one shown in the figure. Challenge 54 e

Many variations of the Hall effect have been studied. For example, the *quantum* Hall effect and the fractional quantum Hall effect will be explored below. Page 92

In 1998, Geert Rikken and his coworkers found that in certain materials photons can also be deflected by a magnetic field; this is the *photonic Hall effect*. Ref. 56

In 2005, again Geert Rikken and his coworkers found a material, a terbium gallium garnet, in which a flow of phonons in a magnetic field leads to temperature difference on the two sides. They called this the *phonon Hall effect*. Ref. 57

* *

Do magnetic fields influence the crystallization of calcium carbonate in water? This issue is topic of intense debates. It might be, or it might not be, that magnetic fields change the Ref. 58 crystallization seeds from calcite to aragonite, thus influencing whether water tubes are covered on the inside with carbonates or not. The industrial consequences of reduction in *scaling*, as this process is called, would be enormous. But the issue is still open, as are convincing data sets.

* *

It has recently become possible to make the thinnest possible sheets of graphite and Ref. 59 other materials (such as BN, MoS_2, $NbSe_2$, $Bi_2Sr_2CaCu_2O_x$): these crystal sheets are precisely one atom thick! The production of *graphene* – that is the name of a monoatomic graphite layer – is extremely complicated: you need graphite from a pencil and a roll of adhesive tape. That is probably why it was necessary to wait until 2004 for the development of the technique. (In fact, the stability of monoatomic sheets was questioned for many years before that. Some issues in physics cannot be decided with paper and pencil; sometimes you need adhesive tape as well.) Graphene and the other so-called *two-dimensional crystals* (this is, of course, a tabloid-style exaggeration) are studied for their

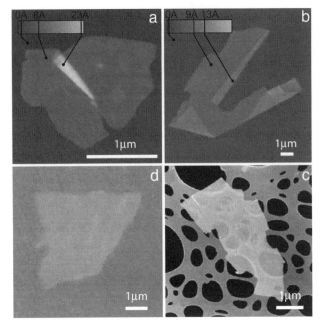

FIGURE 60 Single atom sheets, mapped by atomic force microscopy, of a: NbSe$_2$, b: of graphite or graphene, d: a single atom sheet of MoS$_2$ imaged by optical microscopy, and c: a single atom sheet of Bi$_2$Sr$_2$CaCu$_2$O$_x$ on a holey carbon film imaged by scanning electron microscopy (from Ref. 59, © 2005 National Academy of Sciences).

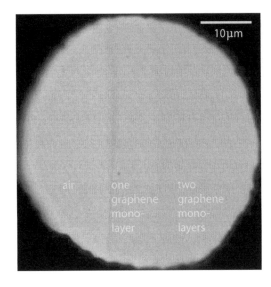

FIGURE 61 A microscope photograph shows the absorption of a single and of a double layer of graphene – and thus provides a way to see the fine structure constant. (© Andre Geim).

electronic and mechanical properties; in the future, they might even have applications in high-performance batteries.

* *

A monolayer of graphene has an astonishing optical property. Its optical absorption over the full optical spectrum is $\pi\alpha$, where α is the fine structure constant. (The exact expression for the absorption is $A = 1 - (1 + \pi\alpha/2)^{-2}$.) The expression for the absorption is the consequence of the electric conductivity $G = e^2/4\hbar$ for every monolayer of graphene.

FIGURE 62 The beauty of materials science: the surface of a lotus leaf leads to almost spherical water droplets; plasma-deposited PTFE, or teflon, on cotton leads to the same effect for the coloured water droplets on it (© tapperboy, Diener Electronics).

The numeric value of the absorption is about 2.3%. This value is visible by the naked eye, as shown in Figure 61. Graphene thus yields a way to 'see' the fine structure constant.   Ref. 60

* *

Gold absorbs light. Therefore it is used, in expensive books, to colour the edges of pages. Apart from protecting the book from dust, it also prevents that sunlight lets the pages turn yellow near the edges.

* *

Like trees, crystals can have *growth rings*. Smoke quartz is known for these so-called *phantoms*, but also fluorite and calcite.

* *

The science and art of surface treatment is still in full swing, as Figure 62 shows. Making hydrophobic surfaces is an important part of modern materials science, that copies what the lotus, *Nelumbo nucifera*, does in nature. Hydrophobic surfaces allow that water droplets bounce on them, like table tennis balls on a table. The lotus surface uses   Vol. I, page 39 this property to clean itself, hence the name *lotus effect*. This is also the reason that lotus plants have become a symbol of purity.

* *

Sometimes research produces bizarre materials. An example are the so-called *aerogels*, highly porous solids, shown in Figure 63. Aerogels have a density of a few g/l, thus a few hundred times lower than water and only a few times that of air. Like any porous material, aerogels are good insulators; however, they are easily destroyed and therefore have not found important applications up to now.

* *

Where do the minerals in the Amazonian rainforest come from? The Amazonas river washes many nutrient minerals into the Atlantic Ocean. How does the rainforest gets it minerals back? It was a long search until it became clear that the largest supply of

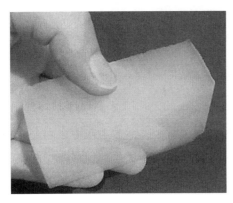

FIGURE 63 A piece of aerogel, a solid that is so porous that it is translucent (courtesy NASA).

minerals is airborne: from the Sahara. Winds blow dust from the Sahara desert to the Amazonas basin, across the Atlantic Ocean. It is estimated that 40 million tons of dust are moved from the Sahara to the Amazonas rainforest every year.

* *

Some materials undergo almost unbelievable transformations. What is the final state of *moss*? Large amounts of moss often become peat (turf). Old turf becomes lignite, or brown coal. Old lignite becomes black coal (bituminous coal). Old black coal can become *diamond*. In short, diamonds can be the final state of moss.

QUANTUM TECHNOLOGY

> I were better to be eaten to death with a rust than to be scoured to nothing with perpetual motion.
> William Shakespeare (1564–1616) *King Henry IV*.

Quantum effects do not appear only in microscopic systems or in material properties. Also *applied* quantum effects are important in modern life: technologies such as transistors, lasers, superconductivity and others effects and systems have deeply affected our civilisation.

TRANSISTORS

Transistors are found in almost all devices that improve health, as well as in almost all devices for telecommunication. A *transistor*, shown in Figure 64 is a device that allows to control a large electric current with the help of a small one; therefore it can play the role of an electrically controlled switch or of an amplifier. Transistors are made from silicon and can be as small as a 2 by 2 μm and as large as 10 by 10 cm. Transistors are used to control the signals in pacemakers for the heart and the current of electric train engines. Amplifying transistors are central to the transmitter in a mobile phone and switching transistors are central to the computer and display in it.

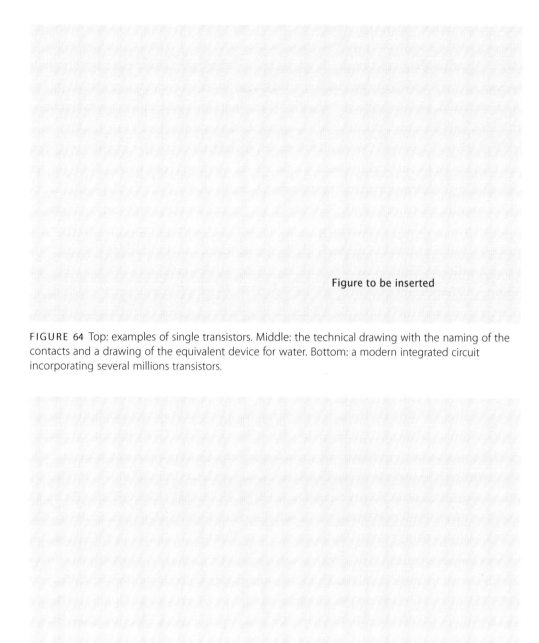

FIGURE 64 Top: examples of single transistors. Middle: the technical drawing with the naming of the contacts and a drawing of the equivalent device for water. Bottom: a modern integrated circuit incorporating several millions transistors.

FIGURE 65 Top: the layer structure of a transistor. Bottom: the corresponding band structure, with and without control voltage, and the resulting conductivity change.

Transistors are based on *semiconductors*, i.e., on materials where the electrons that are responsible for electric conductivity are *almost* free. The devices are built in such a way that applying an electric signal changes the conductivity. Every explanation of a transistor makes use of potentials and tunnelling; transistors are applied quantum devices.

The transistor is just one of a family of semiconductor devices that includes the field-effect transistor (FET), the metal-oxide-silicon field-effect transistor (MOSFET), the junction gate field-effect transistor (JFET), the insulated gate bipolar transistor (IGBT) and the unijunction transistor (UJT), but also the memristors, diode, the PIN diode, the Zener diode, the avalanche diode, the light-emitting diode (LED), the photodiode, the photovoltaic cell, the diac, the triac, the thyristor and finally, the integrated cicuit (IC). These are important in industrial applications: the semiconductor industry has at least 300 thousand million Euro sales every year (2010 value) and employ millions of people across the world.

MOTION WITHOUT FRICTION – SUPERCONDUCTIVITY AND SUPERFLUIDITY

We are used to thinking that friction is inevitable. We even learned that friction was an inevitable result of the particle structure of matter. It should come to the surprise of every physicist that motion *without* friction is indeed possible.

In 1911 Gilles Holst and Heike Kamerlingh Onnes discovered that at low temperatures, electric currents can flow with no resistance, i.e., with no friction, through lead. The observation is called *superconductivity*. In the century after that, many metals, alloys and ceramics have been found to show the same behaviour.

The condition for the observation of motion without friction is that quantum effects play an essential role. To ensure this, low temperature in usually needed. Despite a large amount of data, it took over 40 years to reach a full understanding of superconductivity. This happened in 1957, when Bardeen, Cooper and Schrieffer published their results. At low temperatures, electron behaviour in certain materials is dominated by an *attractive* interaction that makes them form pairs. These so-called *Cooper pairs* are effective bosons. And bosons can all be in the same state, and can thus effectively move without friction.

In superconductivity, the attractive interaction between electrons is due to the deformation of the lattice. At low temperature, two electrons attract each other in the same way as two masses attract each other due to deformation of the space-time mattress. However, in the case of solids, these deformations are quantized. With this approach, Bardeen, Cooper and Schrieffer explained the lack of electric resistance of superconducting materials, their complete diamagnetism ($\mu_r = 0$), the existence of an energy gap, the second-order transition to normal conductivity at a specific temperature, and the dependence of this temperature on the mass of the isotopes. As a result, they received the Nobel Prize in 1972.*

Another type of motion without friction is *superfluidity*. Already in 1937, Pyotr Kapitsa had predicted that normal helium (^{4}He), below a transition observed at the temperature of 2.17 K, would be a superfluid. In this temperature domain, the fluid moves without

Ref. 61

---

* For John Bardeen (1908–1991), this was his second, after he had got the first Nobel Prize in Physics in 1956, shared with William Shockley and Walter Brattain, for the discovery of the transistor. The first Nobel Prize was a problem for Bardeen, as he needed time to work on superconductivity. In an example to many, he reduced the tam-tam around himself to a minimum, so that he could work as much as possible on the problem of superconductivity. By the way, Bardeen is topped by Frederick Sanger and by Marie Curie. Sanger first won a Nobel Prize in Chemistry in 1958 by himself and then won a second one shared with Walter Gilbert in 1980; Marie Curie first won one with her husband and a second one by herself, though in two different fields.

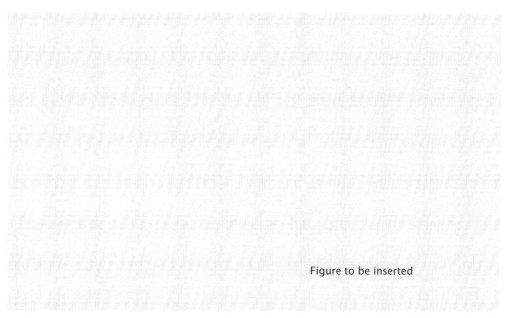

Figure to be inserted

FIGURE 66 Superfluidity.

friction through tubes. (In fact, the fluid remains a mixture of a superfluid component and a normal component.) Helium is even able, after an initial kick, to flow over obstacles, such as glass walls, or to flow out of bottles. ^{4}He is a boson, so no pairing is necessary for it to flow without friction. This research earned Kapitsa a Nobel Prize in 1978.

In 1972, Richardson, Lee and Osheroff found that even ^{3}He is superfluid, if temperatures are lowered below 2.7 mK. ^{3}He is a fermion, and requires *pairing* to become superfluid. In fact, below 2.2 mK, ^{3}He is even superfluid in two different ways; one speaks of phase A and phase B. They received the Nobel Prize in 1996 for this discovery.

In the case of ^{3}He, the theoreticians had been faster than the experimentalists. The theory for superconductivity through pairing had been adapted to superfluids already in 1958 – before any data were available – by Bohr, Mottelson and Pines. This theory was then adapted and expanded by Anthony Leggett.* The attractive interaction between ^{3}He atoms, the basic mechanism that leads to superfluidity, turns out to be the spin-spin interaction.

In superfluids, like in ordinary fluids, one can distinguish between laminar and turbulent flow. The transition between the two regimes is mediated by the behaviour of *vortices*. But in superfluids (as well as in cold gases and in superconductors), vortices have properties that do not appear in normal fluids. In the superfluid ^{3}He-B phase, vortices are *quantized*: vortices only exist in integer multiples of the elementary circulation $h/2m_{^3\text{He}}$. In superfluids, these quantized vortized flow forever. Present research is studying how these Ref. 62 vortices behave and how they induce the transition form laminar to turbulent flows.

In recent years, studying the behaviour of *gases* at lowest temperatures has become very popular. When the temperature is so low that the de Broglie wavelength is compa-

---

* Aage Bohr, son of Niels Bohr, and Ben Mottelson received the Nobel Prize in 1975, Anthony Leggett in 2003.

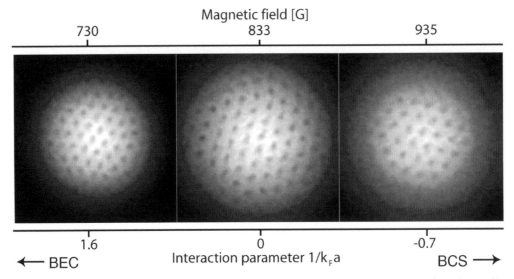

FIGURE 67 A vortex lattice in cold lithium gas, showing their quantized structure (© Andre Schirotzek).

rable to the atom-atom distance, bosonic gases form a Bose–Einstein condensate. The first one were realized in 1995 by several groups; the group around Eric Cornell and Carl Wieman used ^{87}Rb, Rand Hulet and his group used ^{7}Li and Wolfgang Ketterle and his group used ^{23}Na. For fermionic gases, the first degenerate gas, ^{40}K, was observed in 1999 by the group around Deborah Jin. In 2004, the same group observed the first gaseous Fermi condensate, after the potassium atoms paired up.

THE FRACTIONAL QUANTUM HALL EFFECT

The fractional quantum Hall effect is one of the most intriguing discoveries of materials science, and possibly, of physics as a whole. The effect concerns the flow of electrons in a two-dimensional surface. In 1982, Robert Laughlin predicted that in this system one should be able to observe objects with electrical charge $e/3$. This strange and fascinating prediction was indeed verified in 1997.

Ref. 63

Ref. 64

Page 84

Ref. 65

We encountered the (classical) Hall effect above. The story continues with the discovery by Klaus von Klitzing of the *quantum* Hall effect. In 1980, Klitzing and his collaborators found that in two-dimensional systems at low temperatures – about 1 K – the electrical conductance S, also called the Hall conductance, is quantized in multiples of the quantum of conductance

$$S = n \frac{e^2}{h} .$$ (10)

The explanation is straightforward: it is the quantum analogue of the classical Hall effect, which describes how conductance varies with applied magnetic field. The corresponding resistance values are

$$R = \frac{1}{n} \frac{\hbar}{e^2} = \frac{1}{n} 25\,812,807557(18)\,\Omega .$$ (11)

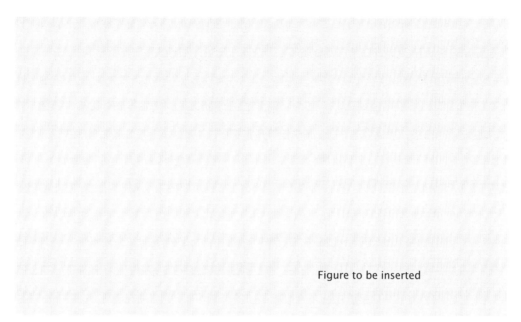

Figure to be inserted

FIGURE 68 The set-up of a quantum Hall effect measurement, and a typical result.

The values are independent of material, temperature, or magnetic field. They are constants of nature. Von Klitzing received the Nobel Prize for physics for the discovery, because the effect was unexpected, allows a highly precise measurement of the fine structure constant, and also allows one to build detectors for the smallest voltage variations measurable so far. His discovery started a large wave of subsequent research.

Only two years later, in 1982, it was found that in extremely strong magnetic fields and at extremely low temperatures, the conductance could vary in steps *one third* that size. Shortly afterwards, even stranger numerical fractions were also found. In fact, all fractions of the form $m/(2m+1)$ or of the form $(m+1)/(2m+1)$, m being an integer, are possible. This is the *fractional quantum Hall effect*. In a landmark paper, Robert Laughlin explained all these results by assuming that the electron gas could form collective states showing quasiparticle excitations with a charge $e/3$. This was confirmed experimentally 15 years later and earned him a Nobel Prize as well. We have seen in several occasions that quantization is best discovered through noise measurements; also in this case, the clearest confirmation came from electrical current noise measurements. Ref. 66 Ref. 63

Subsequent experiments confirmed Laughlin's deduction. He had predicted the appearance of a new form of a *composite* quasi-particle, built of electrons and of one or several magnetic flux quanta. If an electron bonds with an *even* number of quanta, the composite is a fermion, and leads to Klitzing's *integral* quantum Hall effect. If the electron bonds with an *odd* number of quanta, the composite is a boson, and the *fractional* quantum Hall effect appears. The experimental and theoretical details of these quasi-particles might well be the most complex and fascinating aspects of physics, but exploring them would lead us too far from the aim of our adventure.

In 2007, a new chapter in the story was opened by Andre Geim and his team, and a second team, when they discovered a new type of quantum Hall effect at room temperature. Ref. 67

TABLE 6 Matter at lowest temperatures.

PHASE	TYPE	LOW TEMPERATURE BEHAVIOUR	EXAMPLE
Solid	conductor	superconductivity	lead, MgB_2 (40 K)
		antiferromagnet	chromium, MnO
		ferromagnet	iron
	insulator	diamagnet	
Liquid	bosonic	Bose–Einstein condensation, i.e., superfluidity	^{4}He
	fermionic	pairing, then BEC, i.e., superfluidity	^{3}He
Gas	bosonic	Bose–Einstein condensation	^{87}Rb, ^{7}Li, ^{23}Na, H, ^{4}He, ^{41}K
	fermionic	pairing, then Bose–Einstein condensation	^{40}K, ^{6}Li

Page 86 They used graphene, i.e., single-atom layers of graphite, and found a relativistic analogue of the quantum Hall effect. This effect was even more unexpected than the previous ones, is equally interesting, and can be performed on a table top. The groups are good candidates for a trip to Stockholm.*

What do we learn from these results? Systems in two dimensions have states which follow different rules than systems in three dimensions. The fractional charges in superconductors have no relation to quarks. Quarks, the constituents of protons and neutrons, have charges $e/3$ and $2e/3$. Might the quarks have something to do with superconductivity? At this point we need to stand the suspense, as no answer is possible; we come back to this issue in the last part of this adventure.

HOW DOES MATTER BEHAVE AT THE LOWEST TEMPERATURES?

The low-temperature behaviour of matter has numerous experimental and theoretical aspects. The first issue is whether matter is always solid at low temperatures. The answer is no: all phases exist at low temperatures, as shown in Table 6.

Concerning the electric properties of matter at lowest temperatures, the present status is that matter is either insulating or superconducting. Finally, one can ask about the magnetic properties of matter at low temperatures. We know already that matter can not be paramagnetic at lowest temperatures. It seems that matter is either ferromagnetic, diamagnetic or antiferromagnetic at lowest temperatures.

LASERS AND OTHER SPIN-ONE VECTOR BOSON LAUNCHERS

Photons are vector bosons; a lamp is thus a vector boson launcher. All existing lamps fall
Ref. 68 into one of three classes. *Incandescent lamps* use emission from a hot solid, *gas discharge lamps* use excitation of atoms, ions or molecules through collision, and *recombination lamps* generate (cold) light through recombination of charges in semiconductors or liquids. The latter are the only lamp types found in living systems. The other main sources

* This prediction from the December 2008 edition became reality in December 2010.

FIGURE 69 The beauty of lasers: the fine mesh created by a green laser delay line (© Laser Zentrum Hannover).

of light are *lasers*. All light sources are based on quantum effects, but for lasers the connection is especially obvious. The following table gives an overview of the main types and their uses.

TABLE 7 A selection of lamps and lasers.

Lamp type, application	Wave-length	Bright-ness or power	Cost	Life-time
Incandescent lamps				
Oil lamps, candles, for illumination	white	up to 500 lm	1 cent/lm	5 h
Tungsten wire light bulbs, halogen lamps, for illumination	300 to 800 nm	5 to 25 lm/W	0.1 cent/lm	700 h
Stars, for production of heavy elements	full spectrum	up to 10^{44} W	free	up to thousands of millions of years
Gas discharge lamps				
Neon lamps, for advertising	red			up to 30 kh
Mercury lamps, for illumination	UV plus spectrum	45 to 110 lm/W	0.05 cent/lm	3000 to 24 000 h
Metal halogenide lamps (ScI_3 or 'xenon light', NaI, DyI_3, HoI_3, TmI_5) for car headlights and illumination	white	110 lm/W	1 cent/lm	up to 20 kh

TABLE 7 A selection of lamps and lasers (continued).

LAMP TYPE, APPLICATION	WAVE-LENGTH	BRIGHT-NESS OR POWER	COST	LIFE-TIME
Sodium low pressure lamps for street illumination	589 nm yellow	200 lm/W	0.2 cent/lm	up to 18 kh
Sodium high pressure lamps for street illumination	broad yellow	120 lm/W	0.2 cent/lm	up to 24 kh
Xenon arc lamps, for cinemas	white	30 to 150 lm/W, up to 15 kW		100 to 2500 h
Stars, for production of heavy elements	many lines	up to 10^{20} W	free	up to thousands of millions of years
Recombination lamps				
Foxfire in forests, e.g. due to *Armillaria mellea, Neonothopanus gardneri* or other bioluminescent fungi	green	just visible	free	years
Firefly, to attract mates	green-yellow		free	*c.* 10 h
Large deep sea squid, *Taningia danae*, producing light flashes, to confuse prey	red	*c.* 1 W	free	years
Deep-sea fish, such as *angler fish*, to attract prey or find mates	white	*c.* 1 μW	free	years
Deep-sea medusae, to produce attention so that predators of predators are attracted	blue and all other colours		free	years
Light emitting diodes, for measurement, illumination and communication	red, green, blue, UV	up to 150 lm/W, up to 5 W	10 cent/lm	15k to 100 kh
Synchrotron radiation sources				
Electron synchroton source	X-rays to radio waves	pulsed	many MEuro	years
Maybe some stars	broad spectrum		free	thousands of years
Ideal white lamp or laser	visible	*c.* 300 lm/W	0	∞
Ideal coloured lamp or laser	green	683 lm/W	0	∞

Gas lasers

TABLE 7 A selection of lamps and lasers (continued).

LAMP TYPE, APPLICATION	WAVE-LENGTH	BRIGHT-NESS OR POWER	COST	LIFE-TIME
He-Ne laser (obsolete), for school experiments	632.8 nm	550 lm/W	2000 cent/lm	300 h
Argon laser, for pumping and laser shows, now obsolete	several blue and green lines	up to 100 W	10 kEuro	
Krypton laser, for pumping and laser shows, now obsolete	several blue, green, red lines	50 W		
Xenon laser	many lines in the IR, visible and near UV	20 W		
Nitrogen (or 'air') laser, for pumping of other lasers, for hobbyists	337.1 nm	pulsed up to 1 MW	down to a few hundred Euro	limited by metal electrode lifetime
Water vapour laser, for research, now obsolete	many lines between 7 and 220 µm, often 118 µm	CW 0.5 W, pulsed much higher	a few kEuro	
CO_2 laser, for cutting, welding, glass welding and surgery	10.6 µm	CW up to 100 kW, pulsed up to 10 TW	c. 100 Euro/W	1500 h
Excimer laser, for lithography in silicon chip manufacturing, eye surgery, laser pumping, psoriasis treatment, laser deposition	193 nm (ArF), 248 nm (KrF), 308 nm (XeCl), 353 nm (XeF)	100 W	10 to 500 kEuro	years
Metal vapour lasers (Cu, Cd, Se, Ca, Ag, Au, Mn, Tl, In, Hg)				
Copper vapour laser, for pumping, photography, dermatology, laser cutting, hobby constructions and explorative research	248 nm, 511 nm and 578 nm	pulses up to 5 MW	10 kEuro	1 khour
Cadmium vapour laser, for printing, typesetting and recognition of forged US dollar notes	325 nm and 442 nm	up to 200 mW	12 kEuro	10 kh
Gold vapour laser, for explorative research, dermatology	627 nm	pulses up to 1 MW	from a few hundred Euro upwards	
Chemical gas lasers				

TABLE 7 A selection of lamps and lasers (continued).

LAMP TYPE, APPLICATION	WAVE-LENGTH	BRIGHT-NESS OR POWER	COST	LIFE-TIME
HF, DF and oxygen-iodine laser, used as weapons, pumped by chemical reactions, all obsolete	1.3 to 4.2 μm	up to MW in CW mode	over 10 MEuro	un-known
Liquid dye lasers				
Rhodamine, stilbene, coumarin etc. lasers, for spectroscopy and medical uses	tunable, range depends on dye in 300 to 1100 nm range	up to 10 W	10 kEuro	dye-dependent
Beer, vodka, whiskey and many other liquids work as laser material	IR, visible	usually mW	1 kEuro	a few minutes
Solid state lasers				
Ruby laser (obsolete), for holography and tattoo removal	694 nm		1 kEuro	
Nd:YAG (neodymium:yttrium aluminium granate) laser, for material processing, surgery, pumping, range finding, velocimetry, also used with doubled frequency (532 nm), with tripled frequency (355 nm) and with quadrupled frequency (266 nm), also used as slab laser	1064 nm	CW 10 kW, pulsed 300 MW	50 to 500 kEuro	1000 h
Er:YAG laser, for dermatology	2940 nm			
Ti:sapphire laser, for ultrashort pulses for spectroscopy, LIDAR, and research	650 to 1200 nm	CW 1 W, pulsed 300 TW	from 5 kEuro upwards	
Alexandrite laser, for laser machining, dermatology, LIDAR	700 to 840 nm			
Cr:LiSAF laser		pulsed 10 TW, down to 30 fs		
Cr:YAG laser	1.35 to 1.6 μm	pulsed, down to 100 fs		
Cr:Forsterite laser, optical tomography	1200 to 1300 nm	pulsed, below 100 fs		
Erbium doped glass fibre laser, used in optical communications (undersea cables) and optical amplifiers	1.53 to 1.56 μm			years

TABLE 7 A selection of lamps and lasers (continued).

LAMP TYPE, APPLICATION	WAVE-LENGTH	BRIGHT-NESS OR POWER	COST	LIFE-TIME
Perovskite laser, such as $Co:KZnF_3$, for research	NIR tunable, 1650 to 2070 nm	100 mW	2 kEuro	
F-centre laser, for spectroscopy (NaCl:OH-, KI:Li, LiF)	tuning ranges between 1.2 and 6 μm	100 mW	20 kEuro	
Semiconductor lasers				
GaN laser diode, for optical recording	355 to 500 nm, depending on doping	up to 150 mW	a few Euro to 5 kEuro	c. 10 000 h
AlGaAs laser diode, for optical recording, pointers, data communication, laser fences, bar code readers (normal or vertical cavity)	620 to 900 nm, depending on doping	up to 1 W	below 1 Euro to 100 Euro	c. 10 000 h
InGaAsP laser diode, for fiberoptic communication, laser pumping, material processing, medial uses (normal and vertical cavity or VCSEL)	1 to 2.5 μm	up to 100 W	below 1 Euro up to a few k Euro	up to 20 000 h
Lead salt (PbS/PbSe) laser diode, for spectroscopy and gas detection	3 to 25 μm	0.1 W	a few 100 Euro	
Quantum cascade laser, for research and spectroscopy	2.7 to 350 μm	up to 4 W	c. 10 kEuro	c. 1 000 h
Hybrid silicon lasers, for research	IR	nW	0.1 MEuro	
Free electron lasers				
Used for materials science	5 nm to 1 mm	CW 20 kW, pulsed in GW range	10 MEuro	years
Nuclear-reaction pumped lasers				
Have uses only in science fiction and for getting money from gullible military				

FROM LAMPS TO LASERS

Most solid state lamps are light emitting diodes. The large progress in brightness of light emitting diodes could lead to a drastic reduction in future energy consumption, if their cost is lowered sufficiently. Many engineers are working on this task. Since the cost is a good estimate for the energy needed for production, can you estimate which lamp is the most friendly to the environment?

Challenge 55 s

Nobody thought much about lamps, until Albert Einstein and a few other great physicists came along, such as Theodore Maiman and Hermann Haken. Many other researchers later received Nobel Prizes by building on their work. In 1916, Einstein showed that there are two types of sources of light – or of electromagnetic radiation in general – both of which actually 'create' light. He showed that every lamp whose brightness is turned up high enough will change behaviour when a certain intensity threshold is passed. The main mechanism of light emission then changes from spontaneous emission to *stimulated emission*. Nowadays such a special lamp is called a *laser*. (The letters 'se' in laser are an abbreviation of 'stimulated emission'.) After a passionate worldwide research race, in 1960 Maiman was the first to build a laser emitting visible light. (So-called *masers* emitting microwaves were already known for several decades.) In summary, Einstein and the other physicists showed that whenever a lamps is sufficiently turned up, it becomes a laser. Lasers consist of some light producing and amplifying material together with a mechanism to pump energy into it. The material can be a gas, a liquid or a solid; the pumping process can use electrical current or light. Usually, the material is put between two mirrors, in order to improve the efficiency of the light production. Common lasers are semiconductor lasers (essentially strongly pumped LEDs or light emitting diodes), He–Ne lasers (strongly pumped neon lamps), liquid lasers (essentially strongly pumped fire flies) and ruby lasers (strongly pumped luminescent crystals). Most materials can be used to make lasers for fun, including water, beer and vodka.

Lasers produce radiation in the range from microwaves and extreme ultraviolet. They have the special property of emitting *coherent* light, usually in a collimated beam. Therefore lasers achieve much higher light intensities than lamps, allowing their use as tools. In modern lasers, the coherence length, i.e., the length over which interference can be observed, can be thousands of kilometres. Such high quality light is used e.g. in gravitational wave detectors.

People have become pretty good at building lasers. Lasers are used to cut metal sheets up to 10 cm thickness, others are used instead of knives in surgery, others increase surface hardness of metals or clean stones from car exhaust pollution. Other lasers drill holes in teeth, measure distances, image biological tissue or grab living cells.

Some materials amplify light so much that end mirrors are not necessary. This is the case for nitrogen lasers, in which nitrogen, or simply air, is used to produce a UV beam. Even a laser made of a single atom (and two mirrors) has been built; in this example, only eleven photons on average were moving between the two mirrors. Quite a small lamp. Also lasers emitting light in two dimensions have been built. They produce a light plane instead of a light beam.

Ref. 69

THE THREE LIGHTBULB SCAMS

In the 1990s, all major light bulb producers in the world were fined large sums because they had agreed to keep the lifetimes of light bulbs constant. It is no technical problem to make light bulbs that last 2000 hours; however, the producers agreed not to increase the lifetime above 700 hours, thus effectively making every lightbulb three times as expensive as it should. This was the first world-wide light bulb scam.

Despite the fines, the crooks in the light bulb industry did not give up. In 2012, a large German light bulb maker explained in its advertising that its new light sources were much longer living than its conventional light bulbs, which, they explained on their ads, lasted only 500 hours. In other words, not only did the fines not help, the light bulb industry even *reduced* the lifetimes of the light bulbs from the 1990s to 2012. This was the second light bulb scam.

Parellel to the second scam, in the years around 2000, the light bulb industry started lobbying politics with the false statement that light bulbs were expensive and would waste energy. As a result of the false data provided from the other two scams, light bulbs were forbidden in Europe, with the result that consumers in Europe are now forced to buy other, much more expensive means of illumination. On top of this, many of these more expensive light sources are bad for the eyes. Indeed, flickering mercury or flickering LED lamps, together with their reduced colour spectrum, force the human visual system in overload mode, a situation that does not occur with the constantly glowing light bulbs. In other words, with this third scam, the light bulb industry increased their profits even more, while ruining the health of consumers at the same time. One day, maybe, parliaments will be less corrupt and more sensible. The situation will then again improve.

APPLICATIONS OF LASERS

As shown in Figure 70, lasers can be used to make beautiful parts – including good violins and personalized bicycle parts – via sintering of polymer or metal powders. Lasers are used in rapid prototyping machines and to build architectural models. Lasers can cut paper, metal, plastics and flesh.

Lasers are used to read out data from compact discs (CDs) and digital versatile discs (DVDs), are used in the production of silicon integrated circuits and for the transport telephone signals through optical fibres. In our adventure, we already encountered lasers that work as loudspeakers. Important advances in recent years came from the applica- Vol. I, page 348 tions of *femtosecond laser* pulses. Femtosecond pulses generate high-temperature plasmas in the materials they propagate; this happens even in air, if they are focused. The effect has been used to create luminous three-dimensional displays floating in mid-air, as shown in Figure 70, or in liquids. Such short pulses can also be used to cut material without heating it, for example to cut bones in skull operations. The lack of heating is so complete that femtosecond lasers can be used to engrave mateches and even dynamite without triggering a reaction. Femtosecond lasers have been used to make high resolution holograms of human heads within a single flash. Recently such lasers have been used to guide lightning along a predetermined path; they seem promising candidates for laser ligtning rods. A curious demonstration application of femtosecond lasers is the storage Ref. 70 of information in fingernails (up to 5 Mbit for a few months), in a way not unlike that used in recordable compact discs (CD-R). Ref. 71

Lasers are used in opthalmology, with a technique called *optical coherence tomopgraphy*, to diagnose eye and heart illnesses. Around 2025, there will finally be laser-based breast screening devices that use laser light to search for cancer without any danger to the patient. The race to produce the first working system is already ongoing since the 1990s. Additional medical laser applications will appear in the coming years.

Lasers have been used in recent demonstrators, together with imaging processing software, to kill mosquitos in flight; other lasers are burning weeds while the laser is moved over a field of crops. One day, such combined laser and vision systems will be used to

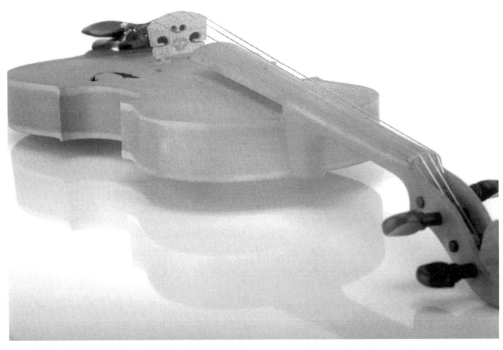

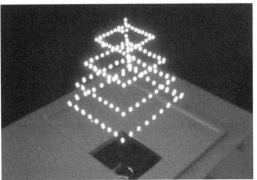

FIGURE 70 Some laser applications. Top: a violin, with excellent sound quality, made of a single piece of polymer (except for the chords and the black parts) through laser sintering of PEEK by EOS from Krailling, in Germany. Bottom: a display floating in mid-air produced with a galvanometer scanner and a fast focus shifter (© Franz Aichinger, Burton).

evaporate falling rain drops one by one; as soon as the first such *laser umbrella* will be available, it will be presented here. The feat should be possible before the year 2022.

Ref. 72

CHALLENGES, DREAMS AND CURIOSITIES ABOUT QUANTUM TECHNOLOGY

Nowadays, we carry many electronic devices in our jacket or trousers. Almost all use batteries. In the future, there is a high chance that some of these devices will extract energy from the human body. There are several options. One can extract thermal energy with thermoelements, or one can extract vibrational energy with piezoelectric, electrostatic or electromagnetic transducers. The challenge is to make these elements small and cheap.

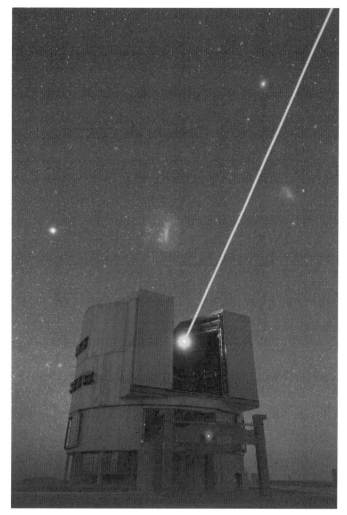

FIGURE 71 The most expensive laser pointer: a yellow 10 W laser that is frequency-stabilized at the wavelength of sodium lamps allows astronomers to improve the image quality of terrestrial telescopes. By exciting sodium atoms found at a height of 80 to 90 km, the laser provides an artificial guide star that is used to compensate for atmospheric turbulence, using the adaptive optics built into the telescope. (© ESO/Babak Tafreshi).

It will be interesting to find out which technology will arrive to the market first.

* *

In 2007, Humphrey Maris and his student Wei Guo performed an astonishing experiment: they filmed *single electrons* with a video camera. Actually the truth is a bit more complicated, but it is not a lie to summarize it in this way.

Ref. 73

Maris is an expert on superfluid helium. For many years he knew that free electrons in superfluid helium repel helium atoms, and can move, surrounded by a small vacuum bubble, about 2 nm across, through the fluid. He also discovered that under negative pressure, these bubbles can grow and finally explode. When they explode, they are able

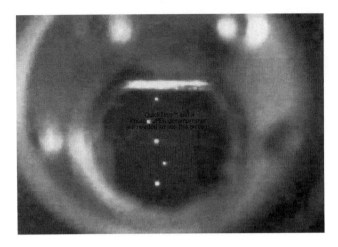

FIGURE 72 How to image single electrons with a video camera: isolated electrons surrounded by bubbles that explode in liquid helium under negative pressure produce white spots (mpg film © Humphrey Maris).

to scatter light. With his student Wei Guo, he then injected electrons into superfluid helium through a tungsten needle under negative voltage, produced negative pressure by focussing waves from two piezoelectric transducers in the bulk of the helium, and shone light through the helium. When the pressure became negative enough they saw the explosions of the bubbles. Figure 72 shows the video. The experiment is one of the highlights of experimental physics in the last decade.

* *

Challenge 56 d   Is it possible to make A4-size, thin and flexible colour displays for an affordable price?

* *

Challenge 57 r   Will there ever be desktop laser engravers for 1000 euro?

* *

Challenge 58 r   Will there ever be room-temperature superconductivity?

* *

Challenge 59 s   Will there ever be teleportation of everyday objects?

* *

One process that quantum physics does not allow is telephathy. An unnamed space agency found this out during the Apollo 14 mission, when, during the flight to the moon, cosmonaut Edgar Mitchell tested telepathy as communication means. Unsurprisingly, he Ref. 74   found that it was useless. It is unclear why the unnamed space agency spent so much money for a useless experiment – an experiment that could have been performed, at a cost of a phone call, also down here on earth.

* *

Challenge 60 d   Will there ever be applied quantum cryptology?

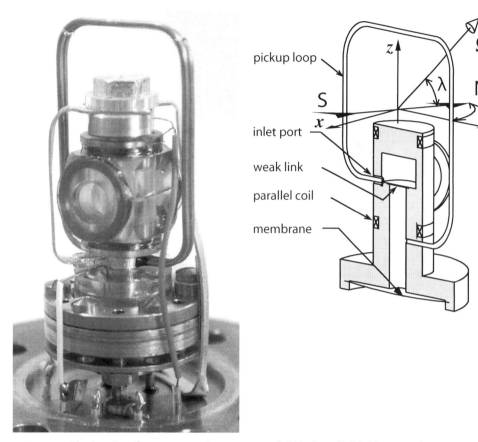

FIGURE 73 The interior of a gyroscope that uses superfluid helium (© Eric Varoquaux).

* *

Will there ever be printable polymer electronic circuits, instead of lithographically patterned silicon electronics as is common now?

Challenge 61 d

* *

Will there ever be radio-controlled flying toys in the size of insects?

Challenge 62 r

* *

In 1997, Eric Varoquaux and his group built a quantum version of the Foucault pendulum, using the superfluidity of helium. In this beautiful piece of research, they cooled a small ring of fluid helium below the temperature of 0.28 K, below which the helium moves without friction. In such situations it thus can behave like a Foucault pendulum. With a clever arrangement, it was possible to measure the rotation of the helium in the ring using phonon signals, and to show the rotation of the Earth.

Ref. 99

SUMMARY ON CHANGING THE WORLD WITH QUANTUM EFFECTS

Atoms form bonds. Quantum effects thus produce molecules, gases, liquids, solids and all effects properties of all materials. In the past, quantum effects have been used to develop numerous materials with desired properties, such as new steel types, new carbon fibre composites, new colourants, new magnetic materials and new polymers.

Quantum effects have been used to develop modern electronics, lasers, light detectors, data storage devices, superconducting magnets, new measurement systems and new production machines. Magnetic resonance imaging, computers, polymers, telecommunication and the internet resulted from applying quantum effects to technology.

Quantum effects will continue to be used to design new materials and systems: new nanoparticles to deliver drugs inside the body, new polymers, new crystals, new environmentally friendly production processes and new medical devices, among others.

Chapter 3

QUANTUM ELECTRODYNAMICS
– THE ORIGIN OF VIRTUAL REALITY

T HE central concept that quantum filed theory adds to the description of nature is
he idea of *virtual particles*. Virtual particles are short-lived particles; they owe
heir existence exclusively to the quantum of action. Because of the quantum of
action, they do not need to follow the energy-mass relation that special relativity requires
of usual, *real* particles. Virtual particles can move faster than light and can move backward in time. Despite these strange properties, they have many observable effects. We
explore the most spectacular ones.

Ships, mirrors and the Casimir effect

When two parallel ships roll in a big swell, *without* even the slightest wind blowing, they
will attract each other. The situation is illustrated in Figure 74. It might be that this effect
was known before the nineteenth century, when many places still lacked harbours.*

Waves induce oscillations of ships because a ship absorbs energy from the waves.
When oscillating, the ship also emits waves. This happens mainly towards the two sides
of the ship. As a result, for a single ship, the wave emission has no net effect on its position. Now imagine that two parallel ships oscillate in a long swell, with a wavelength
much larger than the distance between the ships. Due to the long wavelength, the two
ships will oscillate in phase. The ships will thus not be able to absorb energy from each
other. As a result, the energy they radiate towards the outside will push them towards
each other.

The effect is not difficult to calculate. The energy of a rolling ship is

$$E = mgh\, \alpha^2/2 \qquad (12)$$

where α is the roll angle amplitude, m the mass of the ship and $g = 9,8\,\text{m/s}^2$ the acceleration due to gravity. The *metacentric height h* is the main parameter characterizing a ship,
especially a sailing ship; it tells with what torque the ship returns to the vertical when
inclined by an angle α. Typically, one has $h = 1.5\,\text{m}$.

When a ship is inclined, it will return to the vertical by a damped oscillation. A
damped oscillation is characterized by a period T and a quality factor Q. The *quality
factor* is the number of oscillations the system takes to reduce its amplitude by a factor

---

* Sipko Boersma published a paper in which he gave his reading of shipping manuals, advising captains to
let the ships be pulled apart using a well-manned rowing boat. This reading has been put into question by
subsequent research, however.

Ref. 75
Ref. 76

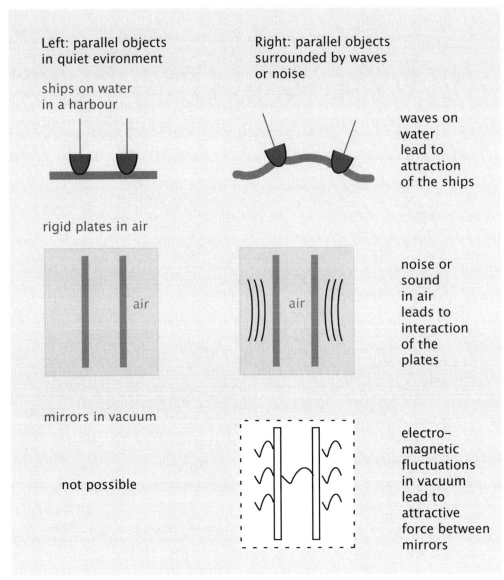

FIGURE 74 The analogy between ships in a harbour, metal plates in air and metal mirrors in vacuum.

$e = 2.718$. If the quality factor Q of an oscillating ship and its oscillation period T are given, the radiated power W is

$$W = 2\pi \frac{E}{QT} .$$

(13)

Vol. III, page 109 We saw above that radiation force (radiation pressure times area) is W/c, where c is the wave propagation velocity. For gravity water waves in deep water, we have the well-
Vol. III, page 256 known relation

$$c = \frac{gT}{2\pi} .$$

(14)

Assuming that for two nearby ships each one completely absorbs the power emitted from the other, we find that the two ships are attracted towards each other following

$$ma = m2\pi^2 \frac{h\alpha^2}{QT^2} \, .$$ (15)

Inserting typical values such as $Q = 2.5$, $T = 10\,\text{s}$, $\alpha = 0.14\,\text{rad}$ and a ship mass of 700 tons, we get about 1.9 kN. Long swells thus make ships attract each other. The strength of the attraction is comparatively small and could be overcome with a rowing boat. On the other hand, even the slightest wind will damp the oscillation amplitude and have other effects that will hide or overshadow this attraction.

Sound waves or noise in air show the same effect. It is sufficient to suspend two metal plates in air and surround them by loudspeakers. The sound will induce attraction (or [Ref. 77] repulsion) of the plates, depending on whether the sound wavelength cannot (or can) be taken up by the other plate.

In 1948, the Dutch physicist Hendrik Casimir made one of the most spectacular predictions of quantum theory: he predicted a similar effect for metal plates in vacuum. Casimir, who worked at the Dutch Electronics company Philips, wanted to understand why it was so difficult to build television tubes. The light-emitting surface in a cathode ray tube – or today, in a plasma display – of a television, the phosphor, is made by depositing small neutral, but conductive particles on glass. Casimir observed that the particles somehow attracted each other. Casimir got interested in understanding how neutral particles interact. During these theoretical studies he discovered that two neutral metal plates (or metal mirrors) would attract each other even in complete vacuum. This is the famous *Casimir effect*. Casimir also determined the attraction strength between a sphere and a plate, and between two spheres. In fact, all *conducting* neutral bodies attract each other in vacuum, with a force depending on their geometry. [Ref. 78, Ref. 79]

In all these situations, the role of the sea is taken by the zero-point fluctuations of the electromagnetic field, the role of the ships by the conducting bodies. Casimir understood that the space between two parallel conducting mirrors, due to the geometrical constraints, had different zero-point fluctuations than the free vacuum. Like in the case of two ships, the result would be the attraction of the two mirrors.

Casimir predicted that the attraction for two mirrors of mass m and surface A is given by

$$\frac{ma}{A} = \frac{\pi^3}{120} \frac{\hbar c}{d^4} \, .$$ (16)

The effect is a pure quantum effect; in classical electrodynamics, two neutral bodies do not attract. The effect is small; it takes some dexterity to detect it. The first experimental confirmation was by Derjaguin, Abrikosova and Lifshitz in 1956; the second experimen- [Ref. 80] tal confirmation was by Marcus Sparnaay, Casimir's colleague at Philips, in 1958. Two [Ref. 81] beautiful high-precision measurements of the Casimir effect were performed in 1997 by Lamoreaux and in 1998 by Mohideen and Roy; they confirmed Casimir's prediction with [Ref. 82] a precision of 5 % and 1 % respectively. (Note that at very small distances, the dependence is not $1/d^4$, but $1/d^3$.) In summary, uncharged bodies attract through electromagnetic [Ref. 83] field fluctuations.

The Casimir effect thus confirms the existence of the zero-point fluctuations of the electromagnetic field. It confirms that quantum theory is valid also for electromagnetism.

The Casimir effect between two spheres is proportional to $1/r^7$ and thus is much weaker than between two parallel plates. Despite this strange dependence, the fascination of the Casimir effect led many amateur scientists to speculate that a mechanism similar to the Casimir effect might explain gravitational attraction. Can you give at least three arguments why this is impossible, even if the effect had the correct distance depen-

Challenge 63 s dence?

Like the case of sound, the Casimir effect can also produce repulsion instead of attraction. It is sufficient that one of the two materials be perfectly permeable, the other a perfect conductor. Such combinations repel each other, as Timothy Boyer discovered in

Ref. 84 1974.

In a cavity, spontaneous emission is suppressed, if it is smaller than the wavelength of the emitted light! This effect has also been observed. It confirms that spontaneous emission is emission stimulated by the zero point fluctuations.

Ref. 85 The Casimir effect bears another surprise: between two metal plates, the speed of light changes and can be larger than c. Can you imagine what exactly is meant by 'speed of

Challenge 64 s light' in this context?

In 2006, the Casimir effect provided another surprise. The ship story just presented is beautiful, interesting and helps understanding the effect; but it seems that the story is based on a misunderstanding. Alas, the interpretation of the old naval text given by Sipko Boersma seems to be wishful thinking. There might be such an effect for ships, but

Ref. 76 it has never been observed nor put into writing by seamen, as Fabrizio Pinto has pointed out after carefully researching naval sources. As an analogy however, it remains valid.

The Lamb shift

In the old days, it was common that a person receives the Nobel Prize in Physics for observing the colour of a lamp – if the observation was sufficiently careful. In 1947, Willis Lamb performed such a careful measurement of the spectrum of hydrogen. He found that the $2S_{1/2}$ energy level in atomic hydrogen lies slightly above the $2P_{1/2}$ level. This

Vol. IV, page 174 observation is in contrast to the calculation performed earlier on, where the two levels are predicted to have the same energy. In contrast, the measured energy difference is 1057.864 MHz, or 4.3 μeV. This discovery had important consequences for the description of quantum theory and yielded Lamb a share of the 1955 Nobel Prize in Physics. Why?

The reason for contrast between calulcation and observation is an approximation performed in the relativistic calculation of the hydrogen levels that took over twenty years to clarify. There are two equivalent explanations. One explanation is to say that the relativistic calculation neglects the coupling terms between the Dirac equation and the Maxwell equations. This explanation lead to the first calculations of the Lamb shift, around the year 1950. The other, equivalent explanation is to say that the calculation *neglects virtual particles*. In particular, the calculation neglects the virtual photons emitted and absorbed during the motion of the electron around the nucleus. This second explanation is in line with the modern vocabulary of quantum electrodynamics. Quantum electrodynamics, or QED, is the perturbative approach to solve the coupled Dirac and Maxwell equations.

In short, Lamb discovered the first effect due to virtual particles. In fact, Lamb used microwaves for his experiments; only in the 1970 it became possible to see the Lamb shift with optical means. This and similar feats Arthur Schawlow received the Nobel Prize for Physics in 198....

THE QED LAGRANGIAN AND ITS SYMMETRIES

In simplified terms, *quantum electrodynamics* is the description of electron motion. This implies that the description is fixed by the effects of mass and charge, and by the quantum of action. The QED Lagrangian density is given by:

$$
\mathcal{L}_{\text{EW}} = \quad
\begin{aligned}
& \overline{\psi}(i\slashed{\partial} - m_k)\psi \quad &\rbrace \quad & \text{the matter term} \\
& -\frac{1}{4\mu_0}F_{\mu\nu}F^{\mu\nu} \quad &\rbrace \quad & \text{the electromagnetic field term} \\
& -q\overline{\psi}\gamma^{\mu}\psi A_{\mu} \quad &\rbrace \quad & \text{the eelectromagnetic interaction term}
\end{aligned}
\qquad (17)
$$

We know the matter term from the Dirac equation for free particles; it describes the kinetic energy of free electrons. We know the term of the electromagnetic field from the Maxwell's equations; it describes the kinetic energy of photons. The interaction term is the term that encodes the gauge symmetry of electromagnetism, also called 'minimal coupling'; it encodes the potential energy. In other words, the Lagrangian describes the motion of electrons and photons.

All experiments ever performed agree with the prediction by this Lagrangian. In other words, this Lagrangian is the final and correct description of the motion of electrons and photons. In particular, the Lagrangian describes the size, shape and colour of atoms, the size, shape and colour of molecules, as well as all interactions of molecules. In short, the Lagrangian describes all of materials science, all of chemistry and all of biology. Exaggerating a bit, this is the Lagrangian that describes life. (In fact, the description of atomic nuclei must be added; we will explore it below.)

Page 146

All electromagnetic effects, including the growth of the coloured spots on butterfly wings, the functioning of the transistor or the cutting of paper with scissors, are completely described by the QED Lagrangian. In fact, the Lagrangian also describes the motion of muons, tau leptons and all other charged particles. Since the Lagrangian is part of the final description of motion, it is worth thinking about it in more detail.

Which requirements are necessary to deduce the QED Lagrangian? This issue has been explored in great detail. The answer is given by the following list:

— compliance with the observer-invariant quantum of action for the motion of electrons and photons,
— symmetry under the permutation group among many electrons, i.e., fermion behaviour of electrons,
— compliance with the invariance of the speed of light, i.e., symmetry under transformations of special relativity,
— symmetry under U(1) gauge transformations for the motion of photons and of charged electrons,
— symmetry under renormalization group,

— low-energy interaction strength described by the *fine structure constant*, the electromagnetic coupling constant, $\alpha \approx 1/137.036$.

The last two points require some comments. As in all cases of motion, the action is the time-volume integral of the Lagrangian density. All fields, be they matter and radiation, move in such a way that this action remains minimal. In fact there are no known differences between the prediction of the least action principle based on the QED Lagrangian density and observations. Even though the Lagrangian density is known since 1926, it took another twenty years to learn how to calculate with it. Only in the years around 1947 it became clear, through the method of *renormalization*, that the Lagrangian density of QED is the *final* description of all motion of matter due to electromagnetic interaction in flat space-time. The details were developed independently by Julian Schwinger, Freeman Dyson, Richard Feynman and Tomonaga Shin'ichiro, four among the smartest physicists ever. *

The QED Lagrangian density contains the strength of the electromagnetic interaction in the form of the fine structure constant α. This number is part of the Lagrangian; no explanation for its value is given, and the explanation was still unknown in the year 2012. It is one of the hardest puzzles of physics. Also the U(1) gauge group is specific to electromagnetism. All others requirements are valid for every type of interaction. Indeed, the search for the Lagrangians of the two nuclear interactions became really focused and finally successful only when the necessary requirements were clearly spelled out, as we will discover in the rest of this volume.

The Lagrangian density retains all symmetries that we know from classical physics. Motion is continuous, it conserves energy–momentum and angular momentum, it is relative, it is right–left symmetric, it is reversible, i.e., symmetric under change of velocity sign, and it is lazy, i.e., it minimizes action. In short, within the limits given by the quantum of action, also motion due to QED remains predictable.

INTERACTIONS AND VIRTUAL PARTICLES

The electromagnetic interaction is exchange of virtual photons. So how can the interaction be attractive? At first sight, any exchange of virtual photons should drive the electrons from each other. However, this is not correct. The momentum of virtual photons does not have to be in the direction of its energy flow; it can also be in opposite direction.** Obviously, this is only possible within the limits provided by the indeterminacy principle.

But virtual particles have also other surprising properties: virtual photons cannot be counted.

---

* Tomonaga Shin'ichiro (1906–1979) Japanese developer of quantum electrodynamics, winner of the 1965 Nobel Prize for physics together with Feynman and Schwinger. Later he became an important figure of science politics; together with his class mate from secondary school and fellow physics Nobel Prize winner, Yukawa Hidei, he was an example to many scientists in Japan.

** One of the most beautiful booklets on quantum electrodynamics which makes this point remains the text by RICHARD FEYNMAN, *QED: the Strange Theory of Light and Matter*, Penguin Books, 1990.

Vacuum energy: infinite or zero?

The strangest result of quantum field theory is the energy density of the vacuum. On one side, the vacuum has, to an excellent approximation, no mass and no energy content. The vacuum energy of vacuum is thus measured and expected to be *zero* (or at least extremely small).*

On the other side, the energy density of the zero-point fluctuations of the electromagnetic field is given by

$$\frac{E}{V} = \frac{4\pi h}{c^3} \int_0^\infty v^3 \mathrm{d}v \; . \tag{18}$$

The result of this integration is infinite. Quantum field theory thus predicts an *infinite* energy density of the vacuum.

We can try to moderate the problem in the following way. As we will discover in the last part of our adventure, there are good arguments that a smallest measurable distance exists in nature; this smallest length appears when gravity is taken into account. The minimal distance is of the order of the *Planck length*

Vol. VI, page 36

$$l_{\mathrm{Pl}} = \sqrt{\hbar G / c^3} \approx 1.6 \cdot 10^{-35}\,\mathrm{m} \; . \tag{19}$$

A minimal distance leads to a maximum cut-off frequency. But even in this case the vacuum density that follows is still a huge number, and is much larger than observed by over 100 orders of magnitude. In other words, QED seems to predict an infinite, or, when gravity is taken into account, a huge vacuum energy. But measurements show a tiny value. What exactly is wrong in this simple calculation? The answer cannot be given at this point; it will become clear in the last volume of our adventure.

Vol. VI, page 36

Moving mirrors

Mirrors also work when they or the light source is in motion. In contrast, walls, i.e., sound mirrors, do *not* produce echoes for every sound source or for every wall speed. For example, experiments show that walls do not produce echoes if the wall or the sound source moves faster than sound. Walls do not produce echoes even if the sound source moves with them, if both objects move faster than sound. On the other hand, light mirrors *always* produce an image, whatever the involved speed of the light source or the mirror may be. These observations confirm that the speed of light is the same for all observers: it is *invariant* and a limit speed. (Can you detail the argument?) In contrast, the speed of sound in air depends on the observer; it is *not* invariant.

Challenge 66 s

Light mirrors also differ from tennis rackets. (Rackets are tennis ball mirrors, to continue the previous analogy.) We have seen that light mirrors cannot be used to change the speed of the light they hit, in contrast to what tennis rackets can do with balls. This observation shows that the speed of light is a *limit* speed. In short, the simple existence of mirrors and of their properties are sufficient to derive special relativity.

Vol. II, page 21

---

* In 1998, this side of the issue was confused even further. Astrophysical measurements, confirmed in the subsequent years, have found that the vacuum energy has a small, but non-zero value, of the order of $0.5\,\mathrm{nJ/m^3}$. The reason for this value is not yet understood, and is one of the open issues of modern physics.

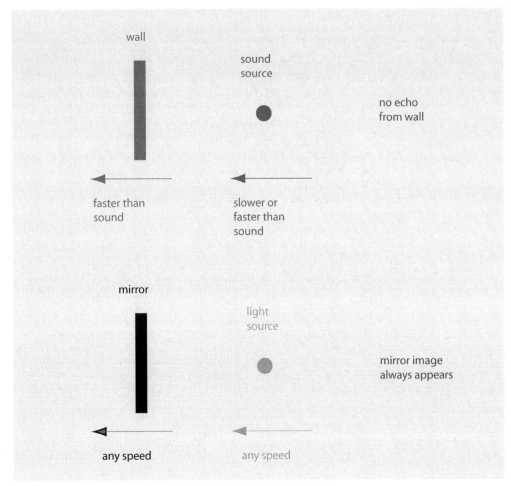

FIGURE 75 A fast wall does not produce an echo; a fast mirror does.

But there are more interesting things to be learned from mirrors. We only have to ask whether mirrors work when they undergo *accelerated* motion. This issue yields a surprising result.

In the 1970s, quite a number of researchers independently found that there is no vacuum for accelerated observers. This effect is called *Fulling–Davies–Unruh effect*. (The incorrect and rarely used term *dynamical Casimir effect* has been abandoned.) For an accelerated observer, the vacuum is full of heat radiation. We will discuss this below. This fact has an interesting consequence for accelerated mirrors: a mirror in accelerated motion reflects the heat radiation it encounters. In short, *an accelerated mirror emits light*! Unfortunately, the intensity of this so-called *Unruh radiation* is so weak that it has not been measured directly, up to now. We will explore the issue in more detail below. (Can you explain why accelerated mirrors emit light, but not matter?)

Page 130

Challenge 67 s

PHOTONS HITTING PHOTONS

When virtual particles are taken into account, light beams can 'bang' onto each other. This result is in full contrast to classical electrodynamics. Indeed, QED shows that the appearance of virtual electron-positron pairs allow photons to hit each other. And such pairs are found in any light beam.

However, the cross-section for photons banging onto each other is small. In fact, the bang is extremely weak. When two light beams cross, most photons will pass undisturbed. The cross-section A is approximately

$$A \approx \frac{973}{10\,125\pi}\, \alpha^4 \left(\frac{\hbar}{m_e c}\right)^2 \left(\frac{\hbar\omega}{m_e c^2}\right)^6 \tag{20}$$

for the everyday case that the energy $\hbar\omega$ of the photon is much smaller than the rest energy $m_e c^2$ of the electron. This low-energy value is about 18 orders of magnitude smaller than what was measurable in 1999; the future will show whether the effect can be observed for visible light. However, for high energy photons these effects are observed daily in particle accelerators. In these settings one observes not only interaction through virtual electron–antielectron pairs, but also through virtual muon–antimuon pairs, virtual quark–antiquark pairs, and much more.

Everybody who consumes science fiction knows that matter and antimatter annihilate and transform into pure light. More precisely, a matter particle and an antimatter particle annihilate into two or more photons. Interestingly, quantum theory predicts that the opposite process is also possible: photons hitting photons can produce matter! In 1997, this prediction was also confirmed experimentally.   Ref. 86

At the Stanford particle accelerator, photons from a high energy laser pulse were bounced off very fast electrons. In this way, the reflected photons acquired a large energy, when seen in the inertial frame of the experimenter. The green laser pulse, of 527 nm wavelength or 2.4 eV photon energy, had a peak power density of 10^{22} W/m^2, about the highest achievable so far. That is a photon density of 10^{34} /m^3 and an electric field of 10^{12} V/m, both of which were record values at the time. When this green laser pulse was reflected off a 46.6 GeV electron beam, the returning photons had an energy of 29.2 GeV and thus had become high-energy gamma rays. These gamma rays then collided with   Challenge 68 e
other, still incoming green photons and produced electron–positron pairs through the reaction

$$\gamma_{29.2\,\text{GeV}} + n\, \gamma_{\text{green}} \rightarrow e^+ + e^- \tag{21}$$

for which both final particles were detected by special apparatuses. The experiment thus showed that light can hit light in nature, and above all, that doing so can produce matter. This is the nearest we can get to the science fiction fantasy of light swords or of laser swords banging onto each other.

IS THE VACUUM A BATH?

If the vacuum is a sea of virtual photons and particle–antiparticle pairs, vacuum could be suspected to act as a bath. In general, the answer is negative. Quantum field theory works because the vacuum is *not* a bath for single particles. However, there is always an exception. For dissipative systems made of many particles, such as electrical conductors, the vacuum *can* act as a viscous fluid. Irregularly shaped, neutral, but conducting bodies can emit photons when accelerated, thus damping such type of motion. This is due to the Fulling–Davies–Unruh effect, also called the *dynamical Casimir effect*, as described above. The damping depends on the shape and thus also on the direction of the body's motion.

Vol. I, page 332 In 1998, Gour and Sriramkumar even predicted that Brownian motion should also appear for an imperfect, i.e., partly absorbing *mirror* placed in vacuum. The fluctuations of the vacuum should produce a mean square displacement

$$\langle d^2 \rangle = \hbar/mt \tag{22}$$

that increases linearly with time; however, the extremely small displacement produced in this way is out of experimental reach so far. But the result is not a surprise. Are you able to give another, less complicated explanation for it?

RENORMALIZATION – WHY IS AN ELECTRON SO LIGHT?

In classical physics, the field energy of a point-like charged particle, and hence its mass, was predicted to be infinite. QED effectively *smears out* the charge of the electron over its Compton wavelength; as a result, the field energy contributes only a small correction to its total mass. Can you confirm this?

However, in QED, many intermediate results in the perturbation expansion are divergent integrals, i.e., integrals with infinite value. The divergence is due to the assumption that infinitely small distances are possible in nature. The divergences thus are artefacts that can be eliminated; the elimination procedure is called renormalization.

Sometimes it is claimed that the infinities appearing in quantum electrodynamics in the intermediate steps of the calculation show that the theory is incomplete or wrong. However, this type of statement would imply that classical physics is also incomplete or wrong, on the ground that in the definition of the velocity v with space x and time t, namely

$$v = \frac{dx}{dt} = \lim_{\Delta t \to 0} \frac{\Delta x}{\Delta t} = \lim_{\Delta t \to 0} \Delta x \frac{1}{\Delta t} \, , \tag{23}$$

one gets an infinity as intermediate step. Indeed, dt being vanishingly small, one could argue that one is dividing by zero. Both arguments show the difficulty to accept that the result of a limit process can be a finite quantity even if infinite quantities appear in the calculation. The parallel between electron mass and velocity is closer than it seems; both intermediate 'infinities' stem from the assumption that space-time is continuous, i.e., infinitely divisible. The infinities necessary in limit processes for the definition of differentiation, integration or for renormalization appear only when space-time is approximated, as physicists say, as a 'continuous' set, or as mathematicians say, as a 'complete' set.

On the other hand, the conviction that the appearance of an infinity might be a sign of incompleteness of a theory was an interesting development in physics. It shows how uncomfortable many physicists had become with the use of infinity in our description

Ref. 87

Ref. 88

Challenge 69 ny

Vol. III, page 199

Challenge 70 s

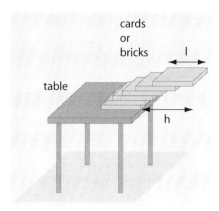

FIGURE 76 What is the maximum possible value of h/l?

of nature. Notably, this was the case for Paul Dirac himself, who, after having laid in his Ref. 89 youth the basis of quantum electrodynamics, has tried for the rest of his life to find a way, without success, to change the theory so that intermediate infinities are avoided.

Renormalization is a procedure that follows from the requirement that continuous space-time and gauge theories must work together. In particular, renormalization follows form the requirement that the particle concept is consistent, i.e., that perturbation expansions are possible. Intermediate infinities are not an issue. In a bizarre twist, a few decades after Dirac's death, his wish has been fulfilled after all, although in a different manner than he envisaged. The final part of this mountain ascent will show the way out of the issue. Vol. VI, page 33

CURIOSITIES AND FUN CHALLENGES OF QUANTUM ELECTRODYNAMICS

Motion is an interesting topic, and when a curious person asks a question about it, most of the time quantum electrodynamics is needed for the answer. Together with gravity, quantum electrodynamics explains almost all of our everyday experience, including numerous surprises. Let us have a look at some of them.

∗ ∗

A famous riddle, illustrated in Figure 76, asks how far the last card (or the last brick) of a stack can hang over the edge of a table. Of course, only gravity, no glue or any other means is allowed to keep the cards on the table. After you solved the riddle, can you give the solution in case that the quantum of action is taken into account? Challenge 71 s

∗ ∗

Quantum electrodynamics explains why there are only a *finite* number of different atom types. In fact, it takes only two lines to prove that pair production of electron– Ref. 90 antielectron pairs make it impossible that a nucleus has more than about 137 protons. Can you show this? The effect at the basis of this limit, the polarization of the vacuum, Challenge 72 s also plays a role in much larger systems, such as charged black holes, as we will see shortly. Page 137

∗ ∗

Stripping 91 of the 92 electrons off an uranium atom allows researchers to check with high precision whether the innermost electron still is described by QED. The electric field near the uranium nucleus, 1 EV/m, is the highest achievable in the laboratory; the field value is near the threshold for spontaneous pair production. The field is the highest constant field producible in the laboratory, and an ideal testing ground for precision QED experiments. The effect of virtual photons is to produce a Lamb shift; but even for these extremely high fields, the value matches the calculation.

Ref. 91

* *

Is there a *critical magnetic field* in nature, like there is a critical electric field, limited by spontaneous pair production?

Challenge 73 ny

* *

Microscopic evolution can be pretty slow. Light, especially when emitted by single atoms, is always emitted by some metastable state. Usually, the decay times, being induced by the vacuum fluctuations, are much shorter than a microsecond. However, there are metastable atomic states with a lifetime of ten years: for example, an ytterbium ion in the $^2F_{7/2}$ state achieves this value, because the emission of light requires an octupole transition, in which the angular momentum changes by $3\hbar$; this is an extremely unlikely process.

Ref. 92

In radioactive decay, the slowness record is held by ^{209}Bi, with over 10^{19} years of half-life.

Page 317

* *

Microscopic evolution can be pretty fast. Can you imagine how to deduce or to measure the speed of electrons inside atoms? And inside metals?

Challenge 74 s

* *

If an electrical wire is sufficiently narrow, its electrical conductance is quantized in steps of $2e^2/\hbar$. The wider the wire, the more such steps are added to its conductance. Can you explain the effect? By the way, quantized conductance has also been observed for light and for phonons.

Challenge 75 s
Ref. 100

* *

The Casimir effect, as well as other experiments, imply that there is a specific and *finite* energy density that can be ascribed to the vacuum. Does this mean that we can apply the Banach–Tarski effect to pieces of vacuum?

Challenge 76 d

* *

Challenge 77 s   Can you explain why mud is not clear?

* *

The instability of the vacuum also yields a (trivial) limit on the fine structure constant. The fine structure constant value of around 1/137.036 cannot be explained by quantum electrodynamics. However, it can be deduced that it must be lower than 1 to lead to a

Ref. 190

consistent theory. Indeed, if its value were larger than 1, the vacuum would become unstable and would spontaneously generate electron-positron pairs.

* *

Can the universe ever have been smaller than its own Compton wavelength?   Challenge 78 s

* *

In the past, the description of motion with formulae was taken rather seriously. Before
computers appeared, only those examples of motion were studied that could be described
with simple formulae. But this narrow-minded approach turns out to be too restrictive.
Indeed, mathematicians showed that Galilean mechanics cannot solve the three-body
problem, special relativity cannot solve the two-body problem, general relativity the one-
body problem and quantum field theory the zero-body problem. It took some time to
the community of physicists to appreciate that understanding motion does not depend
on the description by formulae, but on the description by clear equations based on space
and time.

* *

In fact, quantum electrodynamics, or QED, provides a vast number of curiosities and
every year there is at least one interesting new discovery. We now conclude the theme
with a more general approach.

How can one move on perfect ice? – The ultimate physics test

In our quest, we have encountered motion of many sorts. Therefore, the following test –
not to be taken too seriously – is the ultimate physics test, allowing to check your understanding and to compare it with that of others.

Imagine that you are on a perfectly frictionless surface and that you want to move
to its border. How many methods can you find to achieve this? Any method, so tiny its
effect may be, is allowed.

Classical physics provided quite a number of methods. We saw that for rotating ourselves, we just need to turn our arm above the head. For translation motion, throwing a
shoe or inhaling vertically and exhaling horizontally are the simplest possibilities. Can
you list at least six additional methods, maybe some making use of the location of the   Challenge 79 s
surface on Earth? What would you do in space?

Electrodynamics and thermodynamics taught us that in vacuum, heating one side of
the body more than the other will work as motor; the imbalance of heat radiation will
push you, albeit rather slowly. Are you able to find at least four other methods from these
two domains?   Challenge 80 s

General relativity showed that turning one arm will emit gravitational radiation un-
symmetrically, leading to motion as well. Can you find at least two better methods?   Challenge 81 s

Quantum theory offers a wealth of methods. Of course, quantum mechanics shows
that we actually are always moving, since the indeterminacy relation makes rest an impossibility. However, the average motion can be zero even if the spread increases with
time. Are you able to find at least four methods of moving on perfect ice due to quantum
effects?   Challenge 82 s

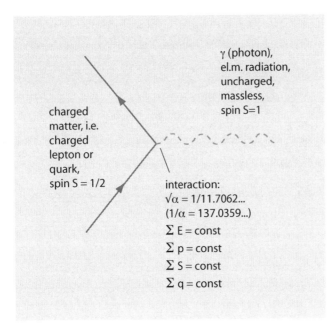

FIGURE 77 QED as perturbation theory in space-time.

Materials science, geophysics, atmospheric physics and astrophysics also provide ways to move, such as cosmic rays or solar neutrinos. Can you find four additional methods?

Self-organization, chaos theory and biophysics also provide ways to move, when the inner workings of the human body are taken into account. Can you find at least two methods?

Assuming that you read already the section following the present one, on the effects of semiclassical *quantum gravity*, here is an additional puzzle: is it possible to move by accelerating a pocket mirror, using the emitted Unruh radiation? Can you find at least two other methods to move yourself using quantum gravity effects? Can you find one from string theory?

If you want points for the test, the marking is simple. For students, every working method gives one point. Eight points is ok, twelve points is good, sixteen points is very good, and twenty points or more is excellent. For graduated physicists, the point is given only when a back-of-the-envelope estimate for the ensuing momentum or acceleration is provided.

A SUMMARY OF QUANTUM ELECTRODYNAMICS

The shortest possible summary of quantum electrodynamics is the following:

> ▷ *Everyday matter is made of charged elementary particles that interact through photon exchange in the way described by Figure 77.*

No additional information is necessary. In a bit more detail, quantum electrodynamics starts with *elementary particles* – characterized by their mass, spin, charge, and parities – and with the *vacuum*, essentially a sea of virtual particle–antiparticle pairs. *Interactions*

Challenge 83 s

Challenge 84 s

Challenge 85 s

between charged particles are described as the exchange of virtual photons, and electromagnetic *decay* is described as the interaction with the virtual photons of the vacuum.

> ▷ The Feynman diagram of Figure 77 provides an *exact* description of *all* electromagnetic phenomena and processes.

No contradiction between observation and calculation are known. In particular, the Feynman diagram is equivalent to the QED Lagrangian of equation (17). Because QED Page 111 is a perturbative theory, the Feynman diagram directly describes the first order effects; its composite diagrams describe effects of higher order. QED is a perturbative theory.

QED describes all everyday properties of matter and radiation. It describes the *divisibility* down to the smallest constituents, the *isolability* from the environment and the *impenetrability* of matter. It also describes the *penetrability* of radiation. All these properties are due to electromagnetic interactions of constituents and follow from Figure 77. Matter is divisible because the interactions are of finite strength, matter is also divisible because the interactions are of finite range, and matter is impenetrable because interactions among the constituents increase in intensity when they approach each other, in particular because matter constituents are fermions. Radiation is divisible into photons, and is penetrable because photons are bosons and first order photon-photon interactions do not exist.

Both matter and radiation are made of *elementary* constituents. These elementary constituents, whether bosons or fermions, are indivisible, isolable, indistinguishable, and point-like.

It is necessary to use quantum electrodynamics in all those situations for which the characteristic dimensions d are of the order of the Compton wavelength

$$d \approx \lambda_{\mathrm{C}} = \frac{h}{m\,c}\ .\tag{24}$$

In situations where the dimensions are of the order of the de Broglie wavelength, or equivalently, where the action is of the order of the Planck value, simple quantum mechanics is sufficient:

$$d \approx \lambda_{\mathrm{dB}} = \frac{h}{m\,v}\ .\tag{25}$$

For even larger dimensions, classical physics will do.

Together with gravity, quantum electrodynamics explains almost all observations of motion on Earth; QED unifies the description of matter and electromagnetic radiation in daily life. All everyday objects and all images are described, including their properties, their shape, their transformations and their other changes. This includes self-organization and chemical or biological processes. In other words, QED gives us full grasp of the effects and the variety of motion due to electromagnetism.

OPEN QUESTIONS IN QED

Even though QED describes motion due to electromagnetism without any discrepancy from experiment, that does not mean that we understand every detail of every example

FIGURE 78 The rainbow can only be explained fully if the fine structure constant α can be calculated (© ed g2s, Christophe Afonso).

of such motion. For example, nobody has described the motion of an animal with QED yet.* In fact, there is beautiful and fascinating research going on in many branches of electromagnetism.

---

* On the other hand, outside QED, there is beautiful work going on how humans move their limbs; it seems that any general human movement is constructed in the brain by combining a small set of fundamental movements.

Ref. 101

Atmospheric physics still provides many puzzles and regularly delivers new, previously unknown phenomena. For example, the detailed mechanisms at the origin of aurorae are still controversial; and the recent unexplained discoveries of discharges *above* clouds should not make one forget that even the precise mechanism of charge separation *inside* clouds, which leads to lightning, is not completely clarified. In fact, all examples of electrification, such as the charging of amber through rubbing, the experiment which gave electricity its name, are still poorly understood.

Ref. 102
Ref. 103

Materials science in all its breadth, including the study of solids, fluids, and plasmas, as well as biology and medicine, still provides many topics of research. In particular, the twenty-first century will undoubtedly be the century of the life sciences.

The study of the interaction of atoms with intense light is an example of present research in atomic physics. Strong lasers can strip atoms of many of their electrons; for such phenomena, there are not yet precise descriptions, since they do not comply to the weak field approximations usually assumed in physical experiments. In strong fields, new effects take place, such as the so-called Coulomb explosion.

Ref. 104

But also the skies have their mysteries. In the topic of cosmic rays, it is still not clear how rays with energies of 10^{22} eV are produced outside the galaxy. Researchers are intensely trying to locate the electromagnetic fields necessary for their acceleration and to understand their origin and mechanisms.

Ref. 105

In the theory of quantum electrodynamics, discoveries are expected by all those who study it in sufficient detail. For example, Dirk Kreimer has discovered that higher order interaction diagrams built using the fundamental diagram of Figure 77 contain relations to the theory of knots. This research topic will provide even more interesting results in the near future. Relations to knot theory appear because QED is a *perturbative* description, with the vast richness of its non-perturbative effects still hidden. Studies of QED at high energies, where perturbation is *not* a good approximation and where particle numbers are not conserved, promise a wealth of new insights.

Ref. 106

If we want to be very strict, we need to add that we do *not* fully understand any colour, because we still do not know the origin of the fine structure constant. In particular, the fine structure constant determines the refractive index of water, and thus the formation of a rainbow, as pictured in Figure 78.

Ref. 107

Many other open issues of more practical nature have not been mentioned. Indeed, by far the largest numbers of physicists get paid for some form of applied QED. In our adventure however, our quest is the description of the *fundamentals* of motion. And so far, we have not achieved it. In particular, we still need to understand motion in the realm of atomic nuclei and the effect of the quantum of action in the domain of gravitation. We start with the latter topic.

CHAPTER 4

QUANTUM MECHANICS WITH GRAVITATION – FIRST STEPS

Gravitation is a weak effect. Indeed, every seaman knows that storms, not gravity, cause the worst accidents. Despite its weakness, the inclusion of gravity into quantum theory raises a number of issues. We must solve them all in order to complete our mountain ascent.

Gravity acts on quantum systems: in the chapter on general relativity we already mentioned that light frequency changes with height. Thus gravity has a simple and measurable effect on photons. But gravity also acts on all other quantum systems, such as atoms and neutrons, as we will see. And the quantum of action plays an important role in the behaviour of black holes. We explore these topics now.

FALLING ATOMS

In 2004 it finally became possible to repeat Galileo's leaning tower experiment with single atoms instead of steel balls. This is not an easy experiment because even the smallest effects disturb the motion. The result is as expected: single atoms do fall like stones. In particular, atoms of different mass fall with the same acceleration, within the experimental precision of one part in 6 million.

The experiment was difficult to perform, but the result is not surprising, because all falling everyday objects are made of atoms. Indeed, Galileo himself had predicted that atoms fall like stones, because parts of a body have to fall with the same acceleration as the complete body. But what is the precise effect of gravity on wave functions? This question is best explored with the help of neutrons.

PLAYING TABLE TENNIS WITH NEUTRONS

The gravitational potential also has directly measurable effects on quantum particles. Classically, a table tennis ball follows a parabolic path when bouncing over a table tennis table, as long as friction can be neglected. The general layout of the experiment is shown in Figure 79. How does a quantum particle behave in the same setting?

In the gravitational field, a bouncing quantum particle is still described by a wave function. In contrast to the classical case however, the possible energy values of a falling quantum particle are discrete. Indeed, the quantization of the action implies that for a

Ref. 108

Vol. I, page 179

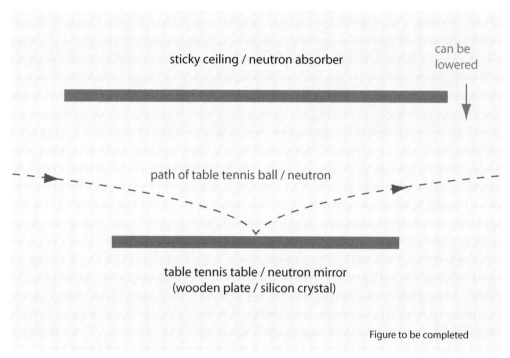

FIGURE 79 Table tennis and neutrons.

bounce of energy E_n and duration t_n,

Challenge 86 e

$$n\hbar \sim E_n \, t_n \sim \frac{E_n^{3/2}}{g m^{1/2}} \; . \tag{26}$$

In other words, only discrete bounce heights, distinguished by the number n, are possible in the quantum case. The discreteness leads to an expected probability density that changes with height in discrete steps, as shown in Figure 79.

The best way to realize the experiment with quantum particles is to produce an intense beam of neutral particles, because neutral particles are not affected by the stray electromagnetic fields that are present in every laboratory. Neutrons are ideal, as they are produced in large quantities by nuclear reactors. The experiment was first performed in 2002, by Hartmut Abele and his group, after years of preparations. Using several clever tricks, they managed to slow down neutrons from a nuclear reactor to the incredibly small value of 8 m/s, comparable to the speed of a table tennis ball. (The equivalent temperature of these ultracold neutrons is 1 mK, or 100 neV.) They then directed the neutrons onto a neutron mirror made of polished glass – the analogue of the table tennis table – and observed the neutrons bouncing back up. To detect the bouncing, they lowered an absorber – the equivalent of a sticky ceiling – towards the table tennis table, i.e., towards the neutron mirror, and measured how many neutrons still reached the other end of the table. (Both the absorber and the mirror were about 20 cm in length.)

Why did the experiment take so many years of work? The lowest energy levels for neutrons due to gravity are $2.3 \cdot 10^{-31}$ J, or 1.4 peV, followed by 2.5 peV, 3.3 peV, 4.1 peV,

Ref. 109

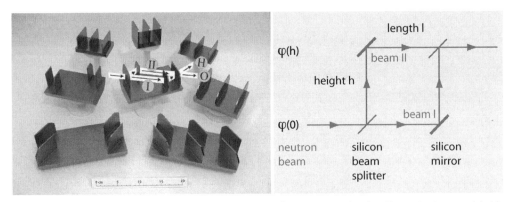

FIGURE 80 The weakness of gravitation. A neutron interferometer made of a silicon single crystal (with the two neutron beams I and II) can be used to detect the effects of gravitation on the phase of wave functions (photo © Helmut Rauch and Erwin Seidl).

and so forth. To get an impression of the smallness of these values, we can compare it to the value of $2.2 \cdot 10^{-18}$ J or 13.6 eV for the lowest state in the hydrogen atom. Despite these small energy values, the team managed to measure the first few discrete energy levels. The results confirmed the prediction of the Schrödinger equation, with the gravitational potential included, to the achievable measurement precision.

In short, gravity influences wave functions. In particular, gravity changes the phase of wave functions, and does so as expected.

THE GRAVITATIONAL PHASE OF WAVE FUNCTIONS

Not only does gravity change the shape of wave functions; it also changes their *phase*. Can you imagine why? The prediction was first confirmed in 1975, using a device invented by Helmut Rauch and his team. Rauch had developed neutron interferometers based on single silicon crystals, shown in Figure 80, in which a neutron beam – again from a nuclear reactor – is split into two beams and the two beams are then recombined and brought to interference.

By rotating the interferometer mainly around the horizontal axis, Samuel Werner and his group let the two neutron beams interfere after having climbed a small height h at two different locations. The experiment is shown schematically on the right of Figure 80. The neutron beam is split; the two beams are deflected upwards, one directly, one a few centimetres further on, and then recombined.

For such a experiment in gravity, quantum theory predicts a phase difference $\Delta\varphi$ between the two beams given by

$$\Delta\varphi = \frac{mghl}{\hbar v} \, ,\tag{27}$$

where l is the horizontal distance between the two climbs and v and m are the speed and mass of the neutrons. All experiments – together with several others of similar simple elegance – have confirmed the prediction by quantum theory within experimental errors.

In the 1990s, similar experiments have even been performed with complete atoms. These atom interferometers are so sensitive that local gravity g can be measured with a

Challenge 87 s

Ref. 110

Ref. 111

Challenge 88 ny

Ref. 112

precision of more than eight significant digits.

In short, neutrons, atoms and photons show no surprises in gravitational fields. Gravity can be included into all quantum systems of everyday life. By including gravity in the potential, the Schrödinger and Dirac equations can thus be used, for example, to describe the growth and the processes inside trees. Trees can mostly be described with quantum electrodynamics in weak gravity.

The gravitational Bohr atom

Can gravity lead to *bound* quantum systems? A short calculation shows that an electron circling a proton due to gravity alone, without electrostatic attraction, would do so at a gravitational Bohr radius of

Challenge 89 ny

$$r_{\text{gr.B.}} = \frac{\hbar^2}{G\, m_{\text{e}}^2\, m_{\text{p}}} = 1.1 \cdot 10^{29}\ \text{m} \tag{28}$$

which is about a thousand times the distance to the cosmic horizon. A gravitational Bohr atom would be larger than the universe. This enormous size is the reason that in a normal hydrogen atom there is not a *single way* to measure gravitational effects between its components. (Are you able to confirm this?)

Challenge 90 e

But why is gravity so weak? Or equivalently, why are the universe and normal atoms so much smaller than a gravitational Bohr atom? At the present point of our quest these questions cannot be answered. Worse, the weakness of gravity even means that with high probability, future experiments will provide little additional data helping to decide among competing answers. The only help is careful thought.

We might conclude from all this that gravity does not require a quantum description. Indeed, we stumbled onto quantum effects because classical electrodynamics implies, in stark contrast with reality, that atoms decay in about 0.1 ns. Classically, an orbiting electron would emit radiation until it falls into the nucleus. Quantum theory is thus necessary to explain the existence of matter.

When the same stability calculation is performed for the emission of gravitational radiation by orbiting electrons, one finds a decay time of around 10^{37} s. (True?) This extremely large value, trillions of times longer than the age of the universe, is a result of the low emission of gravitational radiation by rotating masses. Therefore, the existence of normal atoms does not require a quantum theory of gravity.

Challenge 91 ny

Vol. II, page 160

Curiosities about quantum theory and gravity

Due to the influence of gravity on phases of wave functions, some people who do not believe in bath induced decoherence have even studied the influence of gravity on the decoherence process of usual quantum systems in flat space-time. Predictably, the calculated results do not reproduce experiments.

Ref. 136

∗ ∗

Despite its weakness, gravitation provides many puzzles. Most famous are a number of curious coincidences that can be found when quantum mechanics and gravitation are combined. They are usually called 'large number hypotheses' because they usually involve large dimensionless numbers. A pretty, but less well known version connects the Planck length, the cosmic horizon R_0, and the number of baryons N_b:

$$(N_b)^3 \approx \left(\frac{R_0}{l_{Pl}}\right)^4 = \left(\frac{t_0}{t_{Pl}}\right)^4 \approx 10^{244} \tag{29}$$

in which $N_b = 10^{81}$ and $t_0 = 1.2 \cdot 10^{10}$ a were used. There is no known reason why the number of baryons and the horizon size R_0 should be related in this way. This coincidence is equivalent to the one originally stated by Dirac,* namely

$$m_p^3 \approx \frac{\hbar^2}{Gct_0} . \tag{31}$$

where m_p is the proton mass. This approximate equality seems to suggest that certain microscopic properties, namely the mass of the proton, is connected to some general properties of the universe as a whole. This has lead to numerous speculations, especially since the time dependence of the two sides differs. Some people even speculate whether relations (29) or (31) express some long-sought relation between local and global topological properties of nature. Up to this day, the only correct statement seems to be that they are *coincidences* connected to the time at which we happen to live, and that they should not be taken too seriously.

* *

Photons not travelling parallel to each other attract each other through gravitation and thus deflect each other. Could two such photons form a bound state, a sort of atom of light, in which they would circle each other, provided there were enough empty space for this to happen?

In summary, quantum gravity is unnecessary in every single domain of everyday life. However, we will see now that quantum gravity is necessary in domains which are more remote, but also more fascinating.

---

* The equivalence can be deduced using $Gn_b m_p = 1/t_0^2$, which, as Weinberg explains, is required by several cosmological models. Indeed, this can be rewritten simply as

$$m_0^2/R_0^2 \approx m_{Pl}^2/R_{Pl}^2 = c^4/G^2 . \tag{30}$$

Together with the definition of the baryon density $n_b = N_b/R_0^3$ one gets Dirac's large number hypothesis, substituting protons for pions. Note that the Planck time and length are defined as $\sqrt{\hbar G/c^5}$ and $\sqrt{\hbar G/c^3}$ and are the natural units of length and time. We will study them in detail in the last part of the mountain ascent.

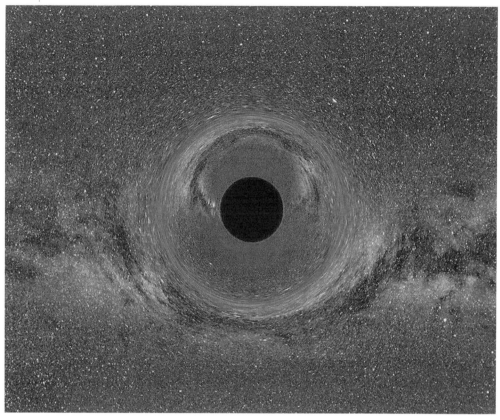

FIGURE 81 A simplified simulated image – not a photograph – of how a black hole of ten solar masses, with Schwarzschild radius of 30 km, seen from a constant distance of 600 km, will distort an image of the Milky Way in the background (image © Ute Kraus at www.tempolimit-lichtgeschwindigkeit.de).

GRAVITATION AND LIMITS TO DISORDER

> Die Energie der Welt ist constant.
> Die Entropie der Welt strebt einem Maximum zu.*
> Rudolph Clausius

We have already encountered the famous statement by Clausius, the father of the term 'entropy'. We have also found that the Boltzmann constant k is the smallest entropy value found in nature. Vol. I, page 237

What is the influence of gravitation on entropy, and on thermodynamics in general? For a long time, nobody was interested in this question. In parallel, for many decades nobody asked whether there also exists a theoretical *maximum* for entropy. The situations changed dramatically in 1973, when Jacob Bekenstein discovered that the two issues are related.

Bekenstein was investigating the consequences gravity has for quantum physics. He Ref. 113

---

* 'The energy of the universe is constant. Its entropy tends towards a maximum.'

found that the entropy S of an object of energy E and size L is bound by

$$S \leqslant EL\frac{k\pi}{\hbar c} \qquad (32)$$

Vol. II, page 243

for all physical systems, where k is the Boltzmann constant. In particular, he deduced that (nonrotating) *black holes* saturate the bound. We recall that black holes are the densest systems for a given mass. They occur when matter collapses completely. Figure 81 shows an artist's impression.

Challenge 93 s

Bekenstein found that black holes have an entropy given by

$$S = A\,\frac{kc^3}{4G\hbar} = M^2\,\frac{4\pi kG}{\hbar c} \qquad (33)$$

where A is now the area of the *horizon* of the black hole. It is given by $A = 4\pi R^2 = 4\pi(2GM/c^2)^2$. In particular, the result implies that every black hole has an entropy. Black holes are thus disordered systems described by thermodynamics. In fact, *black holes are the most disordered systems known.**

As an interesting note, the maximum entropy also implies an upper memory limit for

Challenge 94 s

memory chips. Can you find out how?

Black hole entropy is somewhat mysterious. What are the different microstates leading to this macroscopic entropy? It took many years to convince physicists that the microstates are due to the various possible states of the black hole horizon itself, and that

Ref. 115

they are somehow due to the diffeomorphism invariance at this boundary. As Gerard 't Hooft explains, the entropy expression implies that the number of degrees of freedom

Challenge 95 s

of a black hole is about (but not exactly) one per Planck area of the horizon.

If black holes have entropy, they must have a temperature. What does this temperature mean? In fact, nobody believed this conclusion until two unrelated developments confirmed it within a short time.

MEASURING ACCELERATION WITH A THERMOMETER: FULLING–DAVIES–UNRUH RADIATION

Independently, Stephen Fulling in 1973, Paul Davies in 1975 and William Unruh in 1976

Ref. 116

made the same theoretical discovery while studying quantum theory: if an inertial observer observes that he is surrounded by vacuum, a second observer *accelerated* with respect to the first does not: he observes black body radiation. The appearance of radiation for an accelerated observer in vacuum is called the *Fulling–Davies–Unruh effect*. All

---

* The precise discussion that black holes are the most disordered systems in nature is quite subtle. The

Ref. 114

issue is summarized by Bousso. Bousso claims that the area appearing in the maximum entropy formula cannot be taken naively as the area at a given time, and gives four arguments why this should be not allowed. However, all four arguments are wrong in some way, in particular because they assume that lengths smaller than the Planck length or larger than the universe's size can be measured. Ironically, he brushes aside some of the arguments himself later in the paper, and then deduces an improved formula, which is exactly the same as the one he criticizes first, just with a different interpretation of the area A. Later in his career, Bousso revised his conclusions; he now supports the maximum entropy bound. In short, the expression of black hole entropy is indeed the maximum entropy for a physical system with surface A.

these results about black holes were waiting to be discovered since the 1930s; incredibly, nobody had thought about them for the subsequent 40 years.

The radiation has a spectrum corresponding to the temperature

$$T = a \, \frac{\hbar}{2\pi kc} \,, \tag{34}$$

where a is the value of the acceleration. The result means that there is no vacuum on Earth, because any observer on its surface can maintain that he is accelerated with $9.8\,\mathrm{m/s^2}$, thus leading to $T = 40\,\mathrm{zK}$! We can thus measure gravity, at least in principle, using a thermometer. However, even for the largest practical accelerations the temperature values are so small that it is questionable whether the effect will ever be confirmed experimentally in this precise way. But if it will, it will be a beautiful experimental result. Ref. 117

When this effect was predicted, people explored all possible aspects of the argument. For example, also an observer in rotational motion detects radiation following expression (34). But that was not all. It was found that the simple acceleration of a *mirror* leads to radiation *emission*! Mirrors are thus harder to accelerate than other bodies of the same mass.

When the acceleration is high enough, also *matter* particles can be emitted and detected. If a particle counter is accelerated sufficiently strongly across the vacuum, it will start counting particles! We see that the difference between vacuum and matter becomes fuzzy at large accelerations. This result will play an important role in the search for unification, as we will discover later on. Vol. VI, page 59

Surprisingly, at the end of the twentieth century it became clear that the Fulling–Davies–Unruh effect had already been observed *before* it was predicted! The Fulling–Davies–Unruh effect turned out be equivalent to a well-established observation: the so-called *Sokolov–Ternov effect*. In 1963, the Russian physicist Igor Ternov, together with Ref. 118 Arsenji Sokolov, had used the Dirac equation to predict that electrons in circular accelerators and in storage rings that circulate at high energy would automatically polarize. The prediction was first confirmed by experiments at the Russian Budker Institute of Nuclear Physics in 1971, and then confirmed by experiments in Orsay, in Stanford and in Hamburg. Nowadays, the effect is used routinely in many accelerator experiments. In the 1980s, Bell and Leinaas realized that the Sokolov–Ternov effect is the *same* effect as the Fulling–Davies–Unruh effect, but seen from a different reference frame! The equivalence is somewhat surprising, but is now well-established. In charges moving in a storage ring, Ref. 118 the emitted radiation is not thermal, so that the analogy is not obvious or simple. But the effect that polarizes the beam – namely the difference in photon emission for spins that are parallel and antiparallel to the magnetic field – is the same as the Fulling–Davies–Unruh effect. We thus have another case of a theoretical discovery that was made much later than necessary.

Black holes aren't black

In 1973 and 1974, Jacob Bekenstein, and independently, Stephen Hawking, famous for the intensity with which he fights a disease which forces him into the wheelchair, surprised the world of general relativity with a fundamental theoretical discovery. They found that

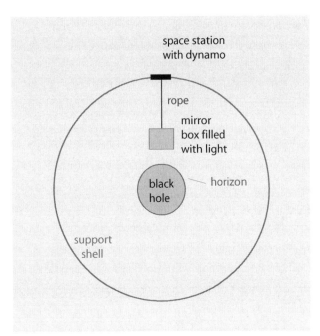

FIGURE 82 A thought experiment allowing to deduce the existence of black hole radiation.

if a virtual particle–antiparticle pair appeared in the vacuum near the horizon, there is a finite chance that one particle escapes as a real particle, while the virtual antiparticle is captured by the black hole. The virtual antiparticle is thus of negative energy, and reduces the mass of the black hole. The mechanism applies both to fermions and bosons. From far away this effect looks like the emission of a particle. A detailed investigation showed that the effect is most pronounced for photon emission. In short, Bekenstein and Hawking showed that *black holes radiate as black bodies.*

Black hole radiation confirms both the result on black hole entropy by Bekenstein and the effect for observers accelerated in vacuum found by Fulling, Davies and Unruh. When all this became clear, a beautiful thought experiment was published by William Unruh and Robert Wald, showing that the whole result could have been deduced already 50 years earlier!

Ref. 119

Shameful as this delay of the discovery is for the community of theoretical physicists, the story itself remains beautiful. It starts in the early 1970s, when Robert Geroch studied the issue shown in Figure 82. Imagine a mirror box full of heat radiation, thus full of light. The mass of the box is assumed to be negligible, such as a box made of thin aluminium paper. We lower the box, with all its contained radiation, from a space station towards a black hole. On the space station, lowering the weight of the heat radiation allows generating energy. Obviously, when the box reaches the black hole horizon, the heat radiation is red-shifted to infinite wavelength. At that point, the full amount of energy originally contained in the heat radiation has been provided to the space station. We can now do the following: we can open the box on the horizon, let drop out whatever is still inside, and wind the empty and massless box back up again. As a result, we have completely converted heat radiation into mechanical energy. Nothing else has changed: the black hole has the same mass as beforehand.

But the lack of change contradicts the second principle of thermodynamics! Geroch concluded that something must be wrong. We must have forgotten an effect which makes this process impossible.

In the 1980s, William Unruh and Robert Wald showed that black hole radiation is precisely the forgotten effect that puts everything right. Because of black hole radiation, the box feels buoyancy, so that it cannot be lowered down to the horizon completely. The box floats somewhat above the horizon, so that the heat radiation inside the box has not yet *zero* energy when it falls out of the opened box. As a result, the black hole does increase in mass and thus in entropy when the box is opened. In summary, when the empty box is pulled up again, the final situation is thus the following: only part of the energy of the heat radiation has been converted into mechanical energy, part of the energy went into the increase of mass and thus of entropy of the black hole. The second principle of thermodynamics is saved.

Well, the second principle of thermodynamics is only saved if the heat radiation has precisely the right energy density at the horizon and above. Let us have a look. The centre of the box can only be lowered up to a hovering distance d above the horizon. At the horizon, the acceleration due to gravity is $g_{\text{surf}} = c^4/4GM$. The energy E gained by lowering the box is

$$E = c^2 m - m g_{\text{surf}} \frac{d}{2} = c^2 m \left(1 - \frac{dc^2}{8GM} \right) \tag{35}$$

The efficiency of the process is $\eta = E/c^2 m$. To be consistent with the second law of thermodynamics, this efficiency must obey

$$\eta = \frac{E}{c^2 m} = 1 - \frac{T_{\text{BH}}}{T} , \tag{36}$$

where T is the temperature of the radiation inside the box. We thus find a black hole temperature T_{BH} that is determined by the hovering distance d. The hovering distance is roughly given by the size of the box. The box size in turn must be at least the wavelength of the thermal radiation; in first approximation, Wien's relation gives $d \approx \hbar c/kT$. A precise calculation introduces a factor π, giving the result

$$T_{\text{BH}} = \frac{\hbar c^3}{8\pi kGM} = \frac{\hbar c}{4\pi k} \frac{1}{R} = \frac{\hbar}{2\pi kc} \, g_{\text{surf}} \quad \text{with} \quad g_{\text{surf}} = \frac{c^4}{4GM} \tag{37}$$

where R and M are the radius and the mass of the black hole. The quantity T_{BH} is either called the *black-hole temperature* or the *Bekenstein–Hawking temperature*. As an example, a black hole with the mass of the Sun would have the rather small temperature of 62 nK, whereas a smaller black hole with the mass of a mountain, say 10^{12} kg, would have a temperature of 123 GK. That would make quite a good oven. All known black hole candidates have masses in the range from a few to a few million solar masses. The radiation is thus extremely weak.

The reason for the weakness of black hole radiation is that the emitted wavelength is of the order of the black hole radius, as you might want to check. The radiation emitted Challenge 96 ny

TABLE 8 The principles of thermodynamics and those of horizon mechanics.

PRINCIPLE	THERMODYNAMICS	HORIZONS
Zeroth principle	the temperature T is constant in a body at equilibrium	the surface gravity a is constant on the horizon
First principle	energy is conserved: $dE = TdS - pdV + \mu dN$	energy is conserved: $d(c^2 m) = \frac{ac^2}{8\pi G}dA + \Omega dJ + \Phi dq$
Second principle	entropy never decreases: $dS \geqslant 0$	surface area never decreases: $dA \geqslant 0$
Third principle	$T = 0$ cannot be achieved	$a = 0$ cannot be achieved

Ref. 120
by black holes is often also called *Bekenstein–Hawking radiation*.

Challenge 97 ny
Black hole radiation is thus so weak that we must speak of an academic effect! It leads to a luminosity that increases with decreasing mass or size as

$$L \sim \frac{1}{M^2} \sim \frac{1}{R^2} \quad \text{or} \quad L = nA\sigma T^4 = n\frac{c^6\hbar}{G^2 M^2}\frac{\pi^2}{15 \cdot 2^7} \tag{38}$$

Vol. III, page 196
where σ is the Stefan–Boltzmann or *black body radiation constant*, n is the number of particle degrees of freedom that can be radiated; as long as only photons are radiated –
Ref. 121
the only case of practical importance – we have $n = 2$.

Black holes thus shine, and the more the smaller they are. This is a genuine *quantum* effect, since classically, black holes, as the name says, cannot emit any light. Even though the effect is academically weak, it will be of importance later on. In actual systems, many other effects around black holes increase the luminosity far above the Bekenstein–Hawking value; indeed, black holes are usually brighter than normal stars, due to the radiation emitted by the matter falling into them. But that is another story. Here we are only treating isolated black holes, surrounded only by pure vacuum.

THE LIFETIME OF BLACK HOLES

Due to the emitted radiation, black holes gradually lose mass. Therefore their theoretical
Challenge 98 ny
lifetime is *finite*. A calculation shows that the lifetime is given by

$$t = M^3 \frac{20\,480\,\pi\,G^2}{\hbar c^4} \approx M^3\,3.4 \cdot 10^{-16}\,\text{s/kg}^3 \tag{39}$$

as function of their initial mass M. For example, a black hole with mass of 1 g would have a lifetime of $3.4 \cdot 10^{-25}$ s, whereas a black hole of the mass of the Sun, $2.0 \cdot 10^{30}$ kg, would have a lifetime of about 10^{68} years. Again, these numbers are purely academic. The important point is that *black holes evaporate*. However, this extremely slow process for usual black holes determines their lifetime only if no other, faster process comes into play. We will present a few such processes shortly. Bekenstein–Hawking radiation is the weakest of all known effects. It is not masked by stronger effects only if the black hole is non-rotating, electrically neutral and with no matter falling into it from the surroundings.

So far, none of these quantum gravity effects has been confirmed experimentally, as the values are much too small to be detected. However, the deduction of a Hawking temperature has been beautifully confirmed by a theoretical discovery of William Unruh, who found that there are configurations of fluids in which sound waves cannot escape, so-called 'silent holes'. Consequently, these silent holes radiate sound waves with a temperature satisfying the same formula as real black holes. A second type of analogue system, namely *optical black holes*, are also being investigated.

Ref. 122

Ref. 123

BLACK HOLES ARE ALL OVER THE PLACE

Around the year 2000, astronomers amassed a large body of evidence that showed something surprising: there seems to be a *supermassive* black hole at the centre of almost all galaxies. The most famous of all is of course the black hole at the centre of our own galaxy. Also quasars, active galactic nuclei and gamma ray bursters seem to be due to supermassive black holes at the centre of galaxies. The masses of these black holes are typically higher than a million solar masses.

Astronomers also think that many other, smaller astrophysical objects contain black holes: ultraluminous X-ray sources and x-ray binary stars are candidates for black holes of *intermediate* mass.

Finally, one candidate explanation for dark matter on the outskirts of galaxies is a hypothetical cloud of *small* black holes.

In short, black holes seem to be quite common across the universe. Whenever astronomers observe a new class of objects, two questions arise directly: how do the objects form? And how do they disappear? We have seen that quantum mechanics puts an upper limit to the life time of a black hole. The upper limit is academic, but that is not important. The main point is that it exists. Indeed, astronomers think that most black holes disappear in other ways, and much before the Bekenstein–Hawking limit, for example through mergers. All this is still a topic of research. The detectors of gravitational waves might clarify these processes in the future.

Vol. II, page 155

How are black holes born? It turns out that the birth of black holes can actually be observed.

FASCINATING GAMMA RAY BURSTS

Nuclear explosions produce flashes of γ rays, or gamma rays. In the 1960s, several countries thought that detecting γ ray flashes, or better, their absence, using satellites, would be the best way to ensure that nobody was detonating nuclear bombs above ground. But when the military sent satellites into the sky to check for such flashes, they discovered something surprising. They observed about two γ flashes *every* day. For fear of being laughed at, the military kept this result secret for many years.

It took the military six years to understand what an astronomer could have told them in five minutes: the flashes, today called *gamma ray bursts*, were coming from outer space. Finally, the results were published; this is probably the only discovery about nature that was made by the military. Another satellite, this time built by scientists, the Comp-

Ref. 125

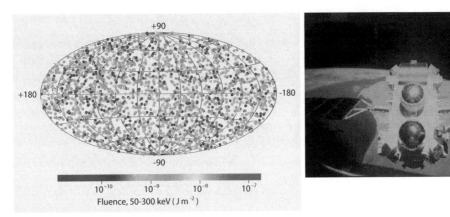

FIGURE 83 The location and energy of the 2704 γ ray bursts observed in the sky between 1991 and 2000 by the BATSE experiment on board of the Compton Gamma Ray Observatory, a large satellite deployed by the space shuttle after over 20 years of planning and construction. The Milky Way is located around the horizontal line running from +180 to −180 (NASA).

ton Gamma Ray Observatory, confirmed that the bursts were *extragalactic* in origin, as proven by the map of Figure 83.

Measurements of gamma ray burst measurements are done by satellites because most gamma rays do not penetrate the atmosphere. In 1996, the Italian-Dutch BeppoSAX satellite started mapping and measuring gamma ray bursts systematically. It discovered that they were followed by an *afterglow* in the X-ray domain, lasting many hours, sometimes even days. In 1997, afterglow was discovered also in the optical domain. The satellite also allowed to find the corresponding X-ray, optical and radio sources for each burst. These measurements in turn allowed to determine the distance of the burst sources; red-shifts between 0.0085 and 4.5 were measured. In 1999 it finally became possible to detect the *optical* bursts corresponding to gamma ray bursts.*

All this data together showed that gamma ray bursts have durations between milliseconds and about an hour. Gamma ray bursts seem to fall into (at least) two classes: the *short bursts*, usually below 3 s in duration and emitted from closer sources, and the *long bursts*, emitted from distant galaxies, typically with a duration of 30 s and more, and with a softer energy spectrum. The long bursts produce luminosities estimated to be up to 10^{45} W. This is about one hundredth of the brightness all stars of the whole visible universe taken together! Put differently, it is the same amount of energy that is released when converting several solar masses into radiation within a few seconds.

In fact, the measured luminosity of long bursts is near the theoretical maximum luminosity a body can have. This limit is given by

$$L < L_{\mathrm{Pl}} = \frac{c^5}{4G} = 0.9 \cdot 10^{52}\,\mathrm{W} \;, \qquad (40)$$

as you might want to check yourself. In short, the sources of gamma ray bursts are the biggest bombs found in the universe. The are explosions of almost unimaginable propor-

Ref. 126

Ref. 128

Challenge 99 s

Vol. II, page 99

Challenge 100 e

---

* For more about this fascinating topic, see the www.aip.de/~jcg/grb.html website by Jochen Greiner.

tions. Recent research seems to suggest that long gamma ray bursts are not isotropic, but that they are *beamed*, so that the huge luminosity values just mentioned might need to be divided by a factor of 1000.

However, the mechanism that leads to the emission of gamma rays is still unclear. It is often speculated that short bursts are due to merging neutron stars, whereas long bursts are emitted when a black hole is formed in a supernova or hypernova explosion. In this case, long gamma ray bursts would be 'primal screams' of black holes *in formation*. However, a competing explanation states that long gamma ray bursts are due to the *death* of black holes.

Ref. 126

Indeed, already 1975, a powerful radiation emission mechanism was predicted for dying *charged* black holes by Damour and Ruffini. Charged black holes have a much shorter lifetime than neutral black holes, because during their formation a second process takes place. In a region surrounding them, the electric field is larger than the so-called vacuum polarization value, so that large numbers of electron-positron pairs are produced, which then almost all annihilate. This process effectively reduces the charge of the black hole to a value for which the field is below critical everywhere, while emitting large amounts of high energy light. It turns out that the mass is reduced by up to 30 % in a time of the order of seconds. That is quite shorter than the 10^{68} years predicted by Bekenstein–Hawking radiation! This process thus produces an extremely intense gamma ray burst.

Ref. 124

Ruffini took up his 1975 model again in 1997 and with his collaborators showed that the gamma ray bursts generated by the annihilation of electron-positrons pairs created by vacuum polarization, in the region they called the *dyadosphere*, have a luminosity and a duration exactly as measured, if a black hole of about a few up to 30 solar masses is assumed. Charged black holes therefore reduce their charge and mass through the vacuum polarization and electron positron pair creation process. (The process reduces the mass because it is one of the few processes which is *reversible*; in contrast, most other attempts to reduce charge on a black hole, e.g. by throwing in a particle with the opposite charge, increase the mass of the black hole and are thus irreversible.) The left over remnant then can lose energy in various ways and also turns out to be responsible for the afterglow discovered by the BeppoSAX satellite. Among others, Ruffini's team speculates that the remnants are the sources for the high energy cosmic rays, whose origin had not been localized so far. All these exciting studies are still ongoing.

Ref. 127

Understanding long gamma ray bursts is one of the most fascinating open questions in astrophysics. The relation to black holes is generally accepted. But many processes leading to emission of radiation from black holes are possible. Examples are matter falling into the black hole and heating up, or matter being ejected from rotating black holes through the Penrose process, or charged particles falling into a black hole. These mechanisms are known; they are at the origin of *quasars*, the extremely bright quasi-stellar sources found all over the sky. They are assumed to be black holes surrounded by matter, in the development stage following gamma ray bursters. But even the details of what happens in quasars, the enormous voltages (up to 10^{20} V) and magnetic fields generated, as well as their effects on the surrounding matter are still object of intense research in astrophysics.

Vol. II, page 252

MATERIAL PROPERTIES OF BLACK HOLES

Once the concept of entropy of a black hole was established, people started to think about black holes like about any other material object. For example, black holes have a matter density, which can be defined by relating their mass to a fictitious volume defined by $4\pi R^3/3$, where R is their radius. This density is then given by

$$\rho = \frac{1}{M^2} \frac{3c^6}{32\pi G^3} \tag{41}$$

and can be quite low for large black holes. For the largest black holes known, with 1000 million solar masses or more, the density is of the order of the density of air. Nevertheless, even in this case, the density is the highest possible in nature for that mass.

Challenge 101 e

By the way, the gravitational acceleration at the horizon is still appreciable, as it is given by

$$g_{\text{surf}} = \frac{1}{M} \frac{c^4}{4G} = \frac{c^2}{2R} \tag{42}$$

Challenge 102 ny

which is still $15 \, \text{km/s}^2$ for an air density black hole.

Obviously, the black hole temperature is related to the entropy S by its usual definition

$$\frac{1}{T} = \left.\frac{\partial S}{\partial E}\right|_\rho = \left.\frac{\partial S}{\partial (c^2 M)}\right|_\rho \tag{43}$$

All other thermal properties can be deduced by the standard relations from thermostatics.

Challenge 103 e

In particular, black holes are the matter states with the largest possible entropy. Can you confirm this statement?

It also turns out that black holes have a *negative* heat capacity: when heat is added, they cool down. In other words, black holes cannot achieve equilibrium with a bath. This is not a real surprise, since *any* gravitationally bound material system has negative specific heat. Indeed, it takes only a bit of thinking to see that any gas or matter system collapsing under gravity follows $dE/dR > 0$ and $dS/dR > 0$. That means that while collapsing, the energy and the entropy of the system shrink. (Can you find out where they go?) Since temperature is defined as $1/T = dS/dE$, temperature is always positive; from the temperature increase $dT/dR < 0$ during collapse one deduces that the specific heat dE/dT is negative.

Challenge 104 ny
Challenge 105 s

Ref. 129

Black holes, like any object, oscillate when slightly perturbed. These vibrations have also been studied; their frequency is proportional to the mass of the black hole.

Ref. 130

Nonrotating black holes have no magnetic field, as was established already in the 1960s by Russian physicists. On the other hand, black holes have something akin to a finite electrical conductivity and a finite viscosity. Some of these properties can be understood if the horizon is described as a membrane, even though this model is not always applicable. In any case, we can study and describe isolated macroscopic black holes like any other macroscopic material body. The topic is not closed.

Ref. 121

Ref. 131

HOW DO BLACK HOLES EVAPORATE?

When a nonrotating and uncharged black hole loses mass by radiating Hawking radiation, eventually its mass reaches values approaching the Planck mass, namely a few micrograms. Expression (39) for the lifetime, applied to a black hole of Planck mass, yields a value of over sixty thousand Planck times. A surprising large value. What happens in those last instants of evaporation?

A black hole approaching the Planck mass at some time will get smaller than its own Compton wavelength; that means that it behaves like an elementary particle, and in particular, that quantum effects have to be taken into account. It is still unknown how these final evaporation steps take place, whether the mass continues to diminish smoothly or in steps (e.g. with mass values decreasing as $\sqrt{n}$ when n approaches zero), how its internal structure changes, whether a stationary black hole starts to rotate (as the author predicts), or how the emitted radiation deviates from black body radiation. There is still enough to study. However, one important issue *has* been settled.

THE INFORMATION PARADOX OF BLACK HOLES

When the thermal radiation of black holes was discovered, one question was hotly debated for many years. The matter forming a black hole can contain lots of information; e.g., we can imagine the black hole being formed by a large number of books collapsing onto each other. On the other hand, a black hole radiates thermally until it evaporates. Since thermal radiation carries no information, it seems that information somehow disappears, or equivalently, that entropy increases.

An incredible number of papers have been written about this problem, some even claiming that this example shows that physics as we know it is incorrect and needs to be changed. As usual, to settle the issue, we need to look at it with precision, laying all prejudice aside. Three intermediate questions can help us finding the answer.
— What happens when a book is thrown into the Sun? When and how is the information radiated away?
— How precise is the sentence that black hole radiate *thermal* radiation? Could there be a slight deviation?
— Could the deviation be measured? In what way would black holes radiate information?

You might want to make up your own mind before reading on.

Challenge 106 e

Let us walk through a short summary. When a book or any other highly complex – or low entropy – object is thrown into the Sun, the information contained is radiated away. The information is contained in some slight deviations from black hole radiation, namely in slight correlations between the emitted radiation emitted over the burning time of the Sun. A short calculation, comparing the entropy of a room temperature book and the information contained in it, shows that these effects are extremely small and difficult to measure.

A clear exposition of the topic was given by Don Page. He calculated what information would be measured in the radiation if the system of black hole and radiation *together* would be in a *pure* state, i.e., a state containing specific information. The result is simple. Even if a system is large – consisting of many degrees of freedom – and in pure state, any smaller *sub*system nevertheless looks almost perfectly thermal. More specifically, if

Ref. 132

a total system has a Hilbert space dimension $N = nm$, where n and $m \leqslant n$ are the dimensions of two subsystems, and if the total system is in a pure state, the subsystem m would have an entropy S_m given by

$$S_m = \frac{1-m}{2n} + \sum_{k=n+1}^{mn} \frac{1}{k} \tag{44}$$

which is approximately given by

$$S_m = \ln m - \frac{m}{2n} \quad \text{for} \quad m \gg 1 . \tag{45}$$

To discuss the result, let us think of n and m as counting degrees of freedom, instead of Hilbert space dimensions. The first term in equation (45) is the usual entropy of a mixed state. The second term is a small deviation and describes the amount of specific information contained in the original pure state; inserting numbers, one finds that it is extremely small compared to the first. In other words, the subsystem m is almost indistinguishable from a mixed state; it looks like a thermal system even though it is not.

A calculation shows that the second, small term on the right of equation (45) is indeed sufficient to radiate away, during the lifetime of the black hole, any information contained in it. Page then goes on to show that the second term is so small that not only it is lost in measurements; it is also lost in the usual, perturbative calculations for physical systems.

The question whether any radiated information could be measured can now be answered directly. As Don Page showed, even measuring half of the system only gives about one half of a bit of the radiated information. It is thus necessary to measure almost the *complete* radiation to obtain a sizeable chunk of the radiated information. In other words, it is extremely hard to determine the information contained in black hole radiation.

In summary, at any given instant, the amount of information radiated by a black hole is negligible when compared with the total black hole radiation; it is practically impossible to obtain valuable information through measurements or even through calculations that use usual approximations.

More paradoxes

A black hole is a macroscopic object, similar to a star. Like all objects, it can interact with its environment. It has the special property to swallow everything that falls into them. This immediately leads us to ask if we can use this property to cheat around the usual everyday 'laws' of nature. Some attempts have been studied in the section on general relativity and above; here we explore a few additional ones.

* *

Apart from the questions of entropy, we can look for methods to cheat around conservation of energy, angular momentum, or charge. But every thought experiment comes to the same conclusions. No cheats are possible. Every reasoning confirms that the maximum number of degrees of freedom in a region is proportional to the surface area of the region, and *not* to its volume. This intriguing result will keep us busy for quite some

Challenge 107 ny

Ref. 133

Vol. II, page 256

Challenge 108 ny

time.

* *

A black hole transforms matter into antimatter with a certain efficiency. Indeed, a black hole formed by collapsing matter also radiates antimatter. Thus one might look for departures from particle number conservation. Are you able to find an example?   Challenge 109 ny

* *

Black holes deflect light. Is the effect polarization dependent? Gravity itself makes no difference of polarization; however, if virtual particle effects of QED are included, the story might change. First calculations seem to show that such an effect exists, so that   Ref. 134
gravitation might produce rainbows. Stay tuned.

* *

If lightweight boxes made of mirrors can float in radiation, one might deduce a strange consequence: such a box could self-accelerate in free space. In a sense, an accelerated box could float on the Fulling–Davies–Unruh radiation it creates by its own acceleration. Are you able to show the that this situation is impossible because of a small but significant difference between gravity and acceleration, namely the absence of tidal effects? (Other   Challenge 110 ny
reasons, such as the lack of perfect mirrors, also make the effect impossible.)

* *

In 2003, Michael Kuchiev has made the spectacular prediction that matter and radiation with a wavelength *larger* than the diameter of a black hole is partly reflected when it hits a black hole. The longer the wavelength, the more efficient the reflection would be. For   Ref. 135
stellar or even larger black holes, he predicts that only photons or gravitons are reflected. Black holes would thus not be complete trash cans. Is the effect real? The discussion is still ongoing.

QUANTUM MECHANICS OF GRAVITATION

Let us take a conceptual step at this stage. So far, we looked at quantum theory *with* gravitation; now we have a glimpse at quantum theory *of* gravitation.

If we bring to our mind the similarity between the electromagnetic field and the gravitational 'field,' our next step should be to find the quantum description of the gravitational field. However, despite attempts by many brilliant minds for almost a century, this search was not successful. Indeed, modern searches take another direction, as will be explained in the last part of our adventure. But let us see what was achieved and why the results are not sufficient.

DO GRAVITONS EXIST?

Quantum theory says that everything that moves is made of particles. What kind of particles are gravitational waves made of? If the gravitational field is to be treated quantum mechanically like the electromagnetic field, its waves should be quantized. Most properties of these quanta of gravitation can be derived in a straightforward way.

The $1/r^2$ dependence of universal gravity, like that of electricity, implies that the quanta of the gravitational field have *vanishing* mass and move at light speed. The independence of gravity from electromagnetic effects implies a *vanishing* electric charge.

Ref. 140 We observe that gravity is always attractive and never repulsive. This means that the field quanta have *integer* and *even* spin. Vanishing spin is ruled out, since it implies no coupling to energy. To comply with the property that 'all energy has gravity', spin $S = 2$ is needed. In fact, it can be shown that *only* the exchange of a massless spin 2 particle leads, in the classical limit, to general relativity.

The coupling strength of gravity, corresponding to the fine structure constant of electromagnetism, is given either by

$$\alpha_{G1} = \frac{G}{\hbar c} = 2.2 \cdot 10^{-15}\,\text{kg}^{-2} \quad \text{or by} \quad \alpha_{G2} = \frac{Gmm}{\hbar c} = \left(\frac{m}{m_{Pl}}\right)^2 = \left(\frac{E}{E_{Pl}}\right)^2 \qquad (46)$$

However, the first expression is not a pure number; the second expression is, but depends on the mass we insert. These difficulties reflect the fact that gravity is not properly speaking an interaction, as became clear in the section on general relativity. It is often argued that m should be taken as the value corresponding to the energy of the system in question. For everyday life, typical energies are 1 eV, leading to a value $\alpha_{G2} \approx 1/10^{56}$. Gravity is indeed weak compared to electromagnetism, for which $\alpha_{em} = 1/137.036$.

If all this is correct, *virtual* field quanta would also have to exist, to explain static gravitational fields. However, up to this day, the so-called *graviton* has not yet been detected, and there is in fact little hope that it ever will. On the experimental side, nobody knows Challenge 111 s yet how to build a graviton detector. Just try! On the theoretical side, the problems with the coupling constant probably make it impossible to construct a *renormalizable* theory of gravity; the lack of renormalization means the impossibility to define a perturbation expansion, and thus to define particles, including the graviton. It might thus be that relations such as $E = \hbar\omega$ or $p = \hbar/2\pi\lambda$ are not applicable to gravitational waves. In short, it may be that the particle concept has to be changed before applying quantum theory to gravity. The issue is still open at this point.

SPACE-TIME FOAM

The indeterminacy relation for momentum and position also applies to the gravitational field. As a result, it leads to an expression for the indeterminacy of the metric tensor g in a region of size L, which is given by

$$\Delta g \approx 2\frac{l_{Pl}{}^2}{L^2}\,, \qquad (47)$$

Challenge 112 ny where $l_{Pl} = \sqrt{\hbar G/c^3}$ is the Planck length. Can you deduce the result? Quantum theory thus shows that like the momentum or the position of a particle, also the metric tensor g is a *fuzzy* observable.

But that is not all. Quantum theory is based on the principle that actions below $\hbar$ cannot be observed. This implies that the observable values for the metric g in a region of size L are bound by

$$g \geqslant \frac{2\hbar G}{c^3} \frac{1}{L^2} \,.$$

(48)

Can you confirm this? The result has far-reaching consequences. A minimum value for the metric depending inversely on the region size implies that it is impossible to say what happens to the shape of space-time at extremely small dimensions. In other words, at extremely high energies, the concept of space-time itself becomes fuzzy. John Wheeler introduced the term *space-time foam* to describe this situation. The term makes clear that space-time is not continuous nor a manifold in those domains. But this was the basis on which we built our description of nature so far! We are forced to deduce that our description of nature is built on sand. This issue will be essential in the last volume of our mountain ascent.

Vol. VI, page 54

Decoherence of space-time

General relativity taught us that the gravitational field and space-time are the same. If the gravitational field evolves like a quantum system, we may ask why no superpositions of different macroscopic space-times are observed.

The discussion is simplified for the simplest case of all, namely the superposition, in a vacuum region of size l, of a homogeneous gravitational field with value g and one with value g'. As in the case of a superposition of macroscopic distinct wave functions, such a superposition *decays*. In particular, it decays when particles cross the volume. A short calculation yields a decay time given by

Ref. 141

Vol. IV, page 136

Challenge 113 ny

$$t_{\mathrm{d}} = \left(\frac{2kT}{\pi m} \right)^{3/2} \frac{nl^4}{(g - g')^2} \,,$$

(49)

where n is the particle number density, kT their kinetic energy and m their mass. Inserting typical numbers, we find that the variations in gravitational field strength are *extremely* small. In fact, the numbers are so small that we can deduce that the gravitational field is the *first* variable which behaves classically in the history of the universe. Quantum gravity effects for space-time will thus be extremely hard to detect.

Challenge 114 e

In short, matter not only tells space-time how to curve, it also tells it to behave with class.

Quantum theory as the enemy of science fiction

How does quantum theory change our ideas of space-time? The end of the twentieth century has brought several unexpected but strong results in semiclassical quantum gravity.

In 1995 Ford and Roman found that *worm holes*, which are imaginable in general relativity, cannot exist if quantum effects are taken into account. They showed that macroscopic worm holes require unrealistically large negative energies. (For microscopic worm holes the issue is still unclear.)

Ref. 142

In 1996 Kay, Radzikowski and Wald showed that *closed time-like curves* do not exist in semiclassical quantum gravity; there are thus no time machines in nature.

Ref. 143

In 1997 Pfenning and Ford showed that *warp drive* situations, which are also imag-

Ref. 144

FIGURE 84 Every tree, such as this beautiful Madagascar baobab (*Adansonia grandidieri*), shows that nature, in contrast to physicists, is able to combine quantum theory and gravity (© Bernard Gagnon).

inable in general relativity, cannot exist if quantum effects are taken into account. Such situations require unrealistically large negative energies.

In short, the inclusion of quantum effects destroys all those fantasies which were started by general relativity.

No vacuum means no particles

Vol. IV, page 106 Gravity has an important consequence for quantum theory. To count and define particles, quantum theory needs a defined vacuum state. However, the vacuum state cannot be defined when the curvature radius of space-time, instead of being larger than the Compton wavelength, becomes comparable to it. In such highly curved space-times, particles cannot be defined. The reason is the impossibility to distinguish the environment from the particle in these situations: in the presence of strong curvatures, the vacuum is full of spontaneously generated matter, as black holes show. Now we just saw that at small dimensions, space-time fluctuates wildly; in other words, space-time is highly curved at small dimensions or high energies. In other words, strictly speaking particles cannot be defined; the particle concept is only a low energy approximation! We will explore this strange conclusion in more detail in the final part of our mountain ascent.

SUMMARY ON QUANTUM THEORY AND GRAVITY

Every tree tells us: everyday, *weak* gravitational fields can be included in quantum theory. *Weak* gravitational fields have measurable and predictable effects on wave functions: quantum particles fall and their phases change in gravitational fields. Conversely, the inclusion of quantum effects into general relativity leads to space-time foam, space-time superpositions, and probably of gravitons. The inclusion of quantum effects into gravity prevents the existence of wormholes, time-like curves and negative energy regions.

The inclusion of *strong* gravitational fields into quantum theory works for practical situations but leads to problems with the particle concept. Conversely, the inclusion of quantum effects into situations with high space-time curvature leads to problems with the concept of space-time.

In summary, the combination of quantum theory and gravitation leads to problems with both the particle concept and the space-time concept. The combination of quantum theory and general relativity puts into question the foundations of the description of nature that we used so far. As shown in Figure 84, nature is smarter than we are.

In fact, up to now we hid a simple fact: conceptually, quantum theory and general relativity *contradict* each other. This contradiction was one of the reasons that we stepped back to special relativity before we started exploring quantum theory. By stepping back we avoided many problems, because quantum theory does not contradict *special* relativity, but only *general* relativity. The issues are dramatic, changing everything from the basis of classical physics to the results of quantum theory. There will be surprising consequences for the nature of space-time, for the nature of particles, and for motion itself. Before we study these issues, however, we complete the theme of the present, quantum part of the mountain ascent, namely exploring motion inside matter, and in particular the motion of and in nuclei.

Vol. VI, page 16

CHAPTER 5

THE STRUCTURE OF THE NUCLEUS – THE DENSEST CLOUDS

Ref. 145

N UCLEAR physics was born in 1896 in France, but is now a small activity. ot many researchers are working on the topic now. The field produced ot more than one daughter, experimental high energy physics, which was born around 1930. All these activities have been in strong decline across the world since 1985, with the exception of the latest CERN experiment, the *Large Hadron Collider*. Given the short time nuclear physics has been in existence, the history of the field is impressive: it discovered why stars shine, how powerful bombs work, how cosmic evolution produced the atoms we are made of and how medical doctors can dramatically improve their healing rate.

((Nuclear physics is just low-density astrophysics.))
Anonymous

A PHYSICAL WONDER – MAGNETIC RESONANCE IMAGING

Arguably, the most spectacular tool that physical research produced in the twentieth century was *magnetic resonance imaging*, or MRI for short. This imaging technique allows one to image the interior of human bodies with a high resolution and with no damage or danger to the patient, in strong contrast to X-ray imaging. The technique is based on moving atomic nuclei. Though the machines are still expensive – costing up to several million euro – there is hope that they will become cheaper in the future. Such an MRI machine, shown in Figure 85, consists essentially of a large magnetic coil, a radio transmitter and a computer. Some results of putting part of a person into the coil are shown in Figure 86. The images allow detecting problems in bones, in the spine, in the beating heart and in general tissue. Many people and owe their life and health to these machines; in many cases the machines allow making precise diagnoses and thus choosing the appropriate treatment for patients.

In MRI machines, a radio transmitter emits radio waves that are absorbed because hydrogen nuclei – protons – are small spinning magnets. The magnets can be parallel or antiparallel to the magnetic field produced by the coil. The transition energy E is absorbed from a radio wave whose frequency ω is tuned to the magnetic field B. The energy absorbed by a single hydrogen nucleus is given by

$$E = \hbar\omega = \hbar\gamma B \qquad (50)$$

The material constant $\gamma/2\pi$ has a value of 42.6 MHz/T for hydrogen nuclei; it results from

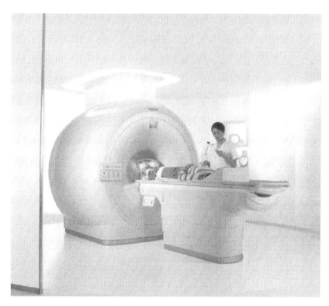

FIGURE 85 A commercial MRI machine (© Royal Philips Electronics).

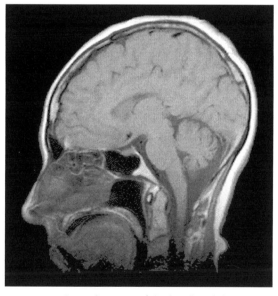 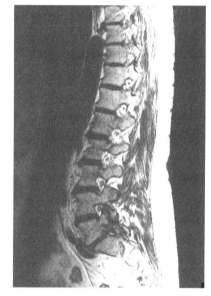

FIGURE 86 Sagittal images of the head and the spine (used with permission from Joseph P. Hornak, *The Basics of MRI*, www.cis.rit.edu/htbooks/mri, Copyright 2003).

the non-vanishing spin of the proton. The absorption is a pure quantum effect, as shown by the appearance of the quantum of action h. Using some cleverly applied magnetic fields, typically with a strength between 0.3 and 7 T for commercial and up to 21 T for experimental machines, the absorption for each volume element can be measured separately. Interestingly, the precise absorption *level* depends on the chemical compound the nucleus is built into. Thus the absorption value will depend on the chemical substance. When the intensity of the absorption is plotted as grey scale, an image is formed that

Ref. 146

retraces the different chemical compositions. Figure 86 shows two examples. Using additional tricks, modern machines can picture blood flow in the heart or air flow in lungs; they now routinely make films of the heart beat. Other techniques show how the location of sugar metabolism in the brain depends on what you are thinking about.*

Ref. 147
Ref. 148

Magnetic resonance imaging can even image the great wonders of nature. The first video of a human birth, taken in 2010 and published in 2012, is shown in Figure 87. MRI scans are loud, but otherwise harmless for unborn children. The first scan of a couple making love has been taken by Willibrord Weijmar Schultz and his group in 1999. It is shown on page 375.

Every magnetic resonance image thus proves that

> ▷ Many – but not all – atoms have nuclei that spin.

Like any other object, nuclei have size, shape, colour, composition and interactions. Let us explore them.

THE SIZE OF NUCLEI AND THE DISCOVERY OF RADIOACTIVITY

The magnetic resonance signal shows that hydrogen nuclei are sensitive to magnetic fields. The g-factor of protons, defined using the magnetic moment μ, their mass m and charge e as $g = \mu 4m/eh$, is about 5.6. Using the expression that relates the g-factor and the radius of a composite object, we deduce that the radius of the proton is about 0.9 fm; this value is confirmed by many experiments and other measurement methods. Protons are thus much smaller, about 30 000 times, than hydrogen atoms, whose radius is about 30 pm. The proton is the smallest of all nuclei; the largest known nuclei have radii about 7 times the proton value.

Vol. IV, page 99

The small size of nuclei is no news. It is known since the beginning of the twentieth century. The story starts on the first of March in 1896, when Henri Becquerel** discovered a puzzling phenomenon: minerals of uranium potassium sulphate blacken photographic plates. Becquerel had heard that the material is strongly fluorescent; he conjectured that fluorescence might have some connection to the X-rays discovered by Conrad Röntgen the year before. His conjecture was wrong; nevertheless it led him to an important new discovery. Investigating the reason for the effect of uranium on photographic plates, Becquerel found that these minerals emit an undiscovered type of radiation, different from anything known at that time; in addition, the radiation is emitted by any substance containing uranium. In 1898, Gustave Bémont named the property of these minerals *radioactivity*.

Radioactive rays are also emitted from many elements other than uranium. This radiation can be 'seen': it can be detected by the tiny flashes of light that appear when the

---

* The website www.cis.rit.edu/htbooks/mri by Joseph P. Hornak gives an excellent introduction to magnetic resonance imaging, both in English and Russian, including the physical basis, the working of the machines, and numerous beautiful pictures. The method of studying nuclei by putting them at the same time into magnetic and radio fields is also called *nuclear magnetic resonance*.

** Henri Becquerel (b. 1852 Paris, d. 1908 Le Croisic), important French physicist; his primary topic was the study of radioactivity. He was the thesis adviser of Marie Curie, the wife of Pierre Curie, and was central to bringing her to fame. The SI unit for radioactivity is named after him. For his discovery of radioactivity he received the 1903 Nobel Prize for physics; he shared it with the Curies.

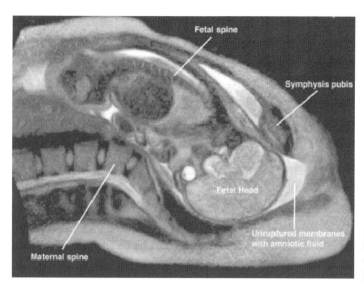

FIGURE 87 An image of the first magnetic resonance video of a human birth (© C. Bamberg).

FIGURE 88 Henri Becquerel (1852–1908)

rays hit a scintillation screen. The light flashes are tiny even at a distance of several metres from the source; thus the rays must be emitted from point-like sources. In short, radioactivity has to be emitted from single atoms. Thus radioactivity confirmed unambiguously that atoms do exist. In fact, radioactivity even allows *counting* atoms: in a diluted radioactive substance, the flashes can be counted, either with the help of a photographic film or with a photon counting system.

Vol. IV, page 41

The intensity of radioactivity cannot be influenced by magnetic or electric fields; it does not depend on temperature or light irradiation. In short, radioactivity does not depend on electromagnetism and is not related to it. Also the high energy of the emitted radiation cannot be explained by electromagnetic effects. Radioactivity must thus be due to another, new type of force. In fact, it took 30 years and a dozen of Nobel Prizes to fully understand the details. It turns out that several types of radioactivity exist; the types of emitted radiation behave differently when they fly through a magnetic field or when they encounter matter. The types of radiation are listed in Table 9. In the meantime, all these rays have been studied in great detail, with the aim to understand the nature of the emitted entity and its interaction with matter.

FIGURE 89 Marie Curie (1867–1934)

In 1909, radioactivity inspired the 37 year old physicist Ernest Rutherford,* who had won the Nobel Prize just the year before, to another of his smart experiments. He asked his collaborator Hans Geiger to take an emitter of α radiation – a type of radioactivity which Rutherford had identified and named 10 years earlier – and to point the radiation at a thin metal foil. The quest was to find out where the α rays would end up. The experiment is shown in Figure 90. The research group followed the path of the particles by using scintillation screens; later on they used an invention by Charles Wilson: the cloud chamber. A *cloud chamber*, like its successor, the bubble chamber, produces white traces along the path of charged particles; the mechanism is the same as the one than leads to the white lines in the sky when an aeroplane flies by. Both cloud chambers and bubble chambers thus allow *seeing* radioactivity, as shown in the example of Figure 91.

The radiation detectors gave a consistent, but strange result: most α particles pass through the thin metal foil undisturbed, whereas a few are scattered and a few are reflected. In addition, those few which are reflected are not reflected by the surface, but in the inside of the foil. (Can you imagine how they showed this?) Rutherford and Geiger deduced from their scattering experiment that first of all, the atoms in the metal foil are mainly transparent. Only transparency of atoms explains why most α particles pass the foil without disturbance, even though it was over 2000 atoms thick. But some particles were scattered by large angles or even reflected. Rutherford showed that the reflections must be due to a single scattering point. By counting the particles that were reflected (about 1 in 20000 for his 0.4 μm gold foil), Rutherford was also able to deduce the *size* of the reflecting entity and to estimate its *mass*. (This calculation is now a standard exercise in the study of physics at universities.) He found that the reflecting entity contains practically all of the mass of the atom in a diameter of around 1 fm. Rutherford thus named this concentrated mass the atomic *nucleus*.

Challenge 115 s

* Ernest Rutherford (b. 1871 Brightwater, d. 1937 Cambridge), important New Zealand physicist. He emigrated to Britain and became professor at the University of Manchester. He coined the terms α particle, β particle, proton and neutron. A gifted experimentalist, he discovered that radioactivity transmutes the elements, explained the nature of α rays, discovered the nucleus, measured its size and performed the first nuclear reactions. Ironically, in 1908 he received the Nobel Prize for chemistry, much to the amusement of himself and of the world-wide physics community; this was necessary as it was impossible to give enough physics prizes to the numerous discoverers of the time. He founded a successful research school of nuclear physics and many famous physicists spent some time at his institute. Ever an experimentalist, Rutherford deeply disliked quantum theory, even though it was and is the only possible explanation for his discoveries.

TABLE 9 The main types of radioactivity and rays emitted by matter.

TYPE	PARTICLE	EXAMPLE	RANGE	DANGER	SHIELD	USE
α rays 3 to 10 MeV	helium nuclei	^{235}U, ^{238}U, ^{238}Pu, ^{238}Pu, ^{241}Am	a few cm in air	when eaten, inhaled, touched	any material, e.g. paper	thickness measurement
β rays 0 to 5 MeV	electrons and antineutrinos	^{14}C, ^{40}K, ^{3}H, ^{101}Tc	< 1 mm in metal / light years	serious / none	metals / none	cancer treatment research
β⁺ rays	positrons and neutrinos	^{40}K, ^{11}C, ^{11}C, ^{13}N, ^{15}O	less than β / light years	medium / none	any material / none	tomography research
γ rays	high energy photons	^{110}Ag	several m in air	high	thick lead	preservation of herbs, disinfection
n reactions c. 1 MeV	neutrons	^{252}Cf, Po-Li (α,n), ^{38}Cl-Be (γ,n)	many m in air	high	0.3 m of paraffin	nuclear power, quantum gravity experiments
n emission typ. 40 MeV	neutrons	^{9}He, ^{24}N, ^{254}Cf	many m in air	high	0.3 m of paraffin	research experiments
p emission typ. 20 MeV	protons	^{5}Be, ^{161}Re	like α rays	small	solids	
spontaneous fission typ. 100 MeV	nuclei	^{232}Cm, ^{263}Rf	like α rays	small	solids	detection of new elements

Using the knowledge that atoms contain electrons, Rutherford then deduced from this experiment that atoms consist of an electron cloud that determines the size of atoms – of the order of 0.1 nm – and of a tiny but heavy nucleus at the centre. If an atom had the size of a basketball, its nucleus would have the size of a dust particle, yet contain 99.9 % of the basketball's mass. Atoms resemble thus candy floss around a heavy dust particle. Even though the candy floss – the electron cloud – around the nucleus is extremely thin and light, it is strong enough to avoid that two atoms interpenetrate. In solids, liquids and gases, the candy floss, i.e., the electron cloud, keeps the neighbouring nuclei at constant distance. For the tiny and massive α particles however, the candy floss is essentially empty space, so that they simply fly through the electron clouds until they either exit on the other side of the foil or hit a nucleus. Figure 92 shows the correct way to picture an atom.

The density of the nucleus is impressive: about $5.5 \cdot 10^{17}$ kg/m^3. At that density, the

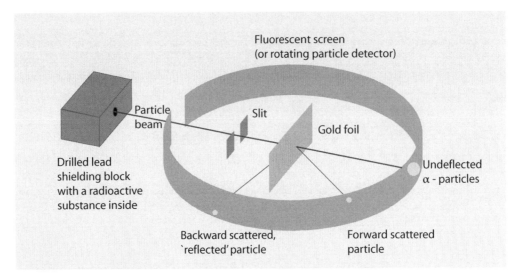

FIGURE 90 The schematics of the Rutherford–Geiger scattering experiment.

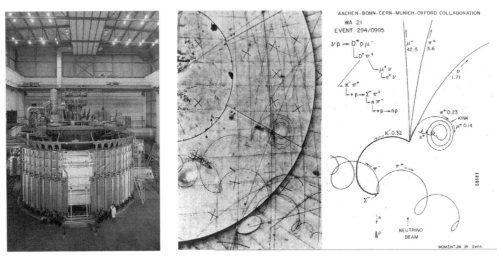

FIGURE 91 The 'Big European Bubble Chamber' – the biggest bubble chamber ever built – and an example of tracks of relativistic particles it produced, with the momentum values deduced from the photograph (© CERN).

Challenge 116 e mass of the Earth would fit in a sphere of 137 m radius and a grain of sand would have a mass larger than the largest existing oil tanker. (Can you confirm this?)

> I now know how an atom looks like!
>
> Ernest Rutherford

Nuclei are composed

Magnetic resonance images also show that nuclei are *composed*. Indeed, images can also be taken using heavier nuclei instead of hydrogen, such as certain fluorine or oxygen nuclei. The g-factors of these nuclei depart from the value 2 characteristic of point particles;

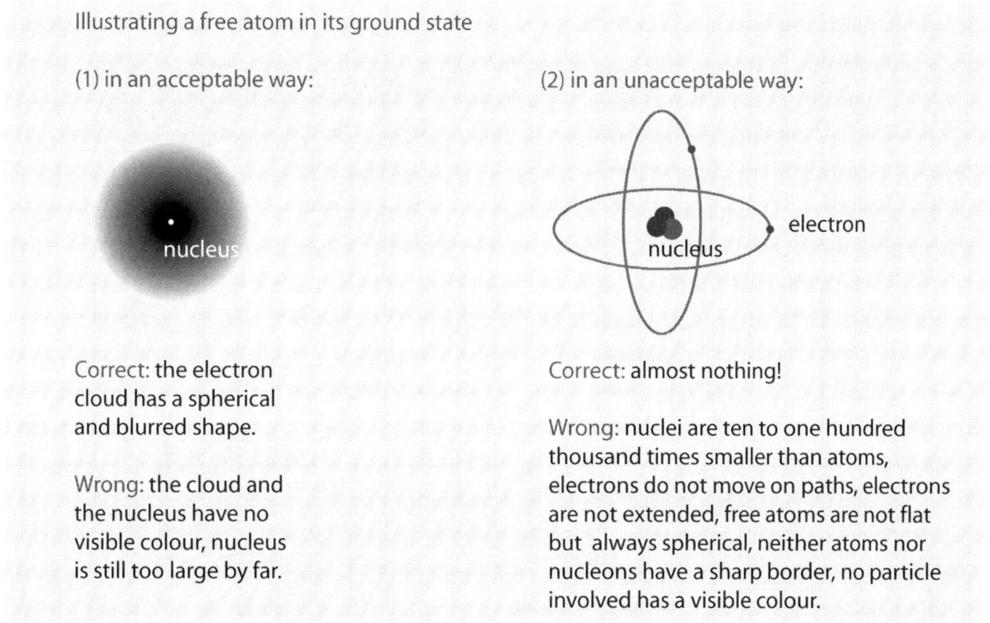

Illustrating a free atom in its ground state

(1) in an acceptable way:

(2) in an unacceptable way:

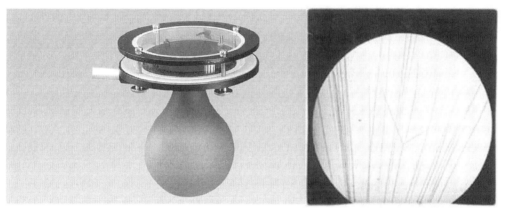

nucleus

nucleus   electron

Correct: the electron
cloud has a spherical
and blurred shape.

Wrong: the cloud and
the nucleus have no
visible colour, nucleus
is still too large by far.

Correct: almost nothing!

Wrong: nuclei are ten to one hundred
thousand times smaller than atoms,
electrons do not move on paths, electrons
are not extended, free atoms are not flat
but always spherical, neither atoms nor
nucleons have a sharp border, no particle
involved has a visible colour.

FIGURE 92 A reasonably realistic (left) and a misleading illustration of an atom (right) as is regularly found in school books. *Atoms in the ground state are spherical electron clouds with a tiny nucleus, itself a cloud, at its centre.* Interacting atoms, chemically bound atoms and some, but not all excited atoms have electron clouds of different shapes.

FIGURE 93 A Wilson cloud chamber, diameter c. 99 mm, and one of the first pictures of α rays taken with it, by Patrick Blackett, showing also a collision with an atom in the chamber (© , Royal Society)

the more massive the nuclei are, the bigger the departure. All nuclei have a finite size; in particular, the size of nuclei can actually be measured, as in the Rutherford–Geiger experiment. The measured values confirm the values predicted by the g-factor. In short, both the values of the g-factor and the non-vanishing sizes show that nuclei are composed.

Interestingly, the idea that nuclei are composed is older than the concept of nucleus itself. Already in 1815, after the first mass measurements of atoms by John Dalton and others, researchers noted that the mass of the various chemical elements seem to be almost

Vol. IV, page 99

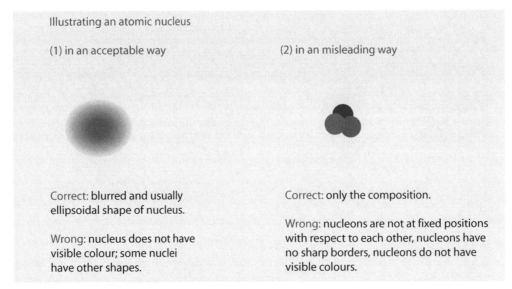

Illustrating an atomic nucleus

(1) in an acceptable way   (2) in an misleading way

Correct: blurred and usually   Correct: only the composition.
ellipsoidal shape of nucleus.
                                        Wrong: nucleons are not at fixed positions
Wrong: nucleus does not have   with respect to each other, nucleons have
visible colour; some nuclei   no sharp borders, nucleons do not have
have other shapes.   visible colours.

FIGURE 94 A reasonably realistic (left) and a misleading illustration of a nucleus (right) as is regularly found in school books. *Nuclei are spherical nucleon clouds.*

perfect multiples of the weight of the hydrogen atom. William Prout then formulated the hypothesis that all elements are composed of hydrogen. When the nucleus was discovered, knowing that it contains almost all mass of the atom, it was therefore first thought that all nuclei are made of hydrogen nuclei. Being at the origin of the list of constituents, the hydrogen nucleus was named *proton*, from the greek term for 'first' and reminding the name of Prout at the same time. Protons carry a positive unit of electric charge, just the opposite of that of electrons, but are about 1836 times as heavy. More details on the proton are listed in Table 10.

However, the charge multiples and the mass multiples of the other nuclei do not match. On average, a nucleus that has n times the charge of a proton, has a mass that is about 2.6 n times than of the proton. Additional experiments confirmed an idea formulated by Werner Heisenberg: all nuclei heavier than hydrogen nuclei are made of positively charged *protons* and of neutral *neutrons*. Neutrons are particles a tiny bit more massive than protons (the difference is less than a part in 700, as shown in Table 10), but without any electrical charge. Since the mass is almost the same, the mass of nuclei – and thus that of atoms – is still an (almost perfect) integer multiple of the proton mass. But since neutrons are neutral, the mass and the charge number of nuclei differ. Being neutral, neutrons do not leave tracks in clouds chambers and are more difficult to detect than protons, charged hadrons or charged leptons. For this reason, neutron were discovered later than several other, more exotic subatomic particles.

Today it is possible to keep single neutrons suspended between suitably shaped coils, with the aid of teflon 'windows'. Such traps were proposed in 1951 by Wolfgang Paul. They work because neutrons, though they have no charge, do have a small magnetic moment. (By the way, this implies that neutrons are themselves composed of charged particles.) With a suitable arrangement of magnetic fields, neutrons can be kept in place, in other words, they can be levitated. Obviously, a trap only makes sense if the trapped particle can

TABLE 10 The properties of the nucleons: proton and neutron (source: pdg.web.cern.ch).

PROPERTY	PROTON	NEUTRON
Mass		
	$1.672\,621\,777(74) \cdot 10^{-27}$ kg	$1.674\,927\,351(74) \cdot 10^{-27}$ kg
	$0.150\,327\,7484(66)$ nJ	$0.150\,534\,9631(66)$ nJ
	$938,272\,046(21)$ MeV	$939,565\,379(21)$ MeV
	$1.007\,276\,466\,812(90)$ u	$1.008\,664\,916\,00(43)$ u
	$1836.152\,6675(39) \cdot m_e$	$1838.683\,6605(11) \cdot m_e$
Spin	1/2	1/2
P parity	+1	+1
Antiparticle	antiproton $\bar{p}$	antineutron $\bar{n}$
Quark content	uud	udd
Electric charge	1 e	0
Charge radius	$0.88(1)$ fm	$0.12(1)$ fm^2
Electric dipole moment	$< 5.4 \cdot 10^{-26}$ e · m	$< 2.9 \cdot 10^{-28}$ e · m
Electric polarizability	$1.20(6) \cdot 10^{-3}$ fm^3	$1.16(15) \cdot 10^{-3}$ fm^3
Magnetic moment	$1.410\,606\,743(33) \cdot 10^{-26}$ J/T	$-0.966\,236\,47(23) \cdot 10^{-26}$ J/T
g-factor	$5.585\,694\,713(46)$	$-3.826\,085\,45(90)$
	$2.792\,847\,356\,(23) \cdot \mu_N$	$-1.913\,042\,72(45) \cdot \mu_N$
Gyromagnetic ratio	$0.267\,522\,2005(63)$ 1/nsT	
Magnetic polarizability	$1.9(5) \cdot 10^{-4}$ fm^3	$3.7(2.0) \cdot 10^{-4}$ fm^3
Mean life (free particle)	$> 2.1 \cdot 10^{29}$ a	$880.1(1.1)$ s
Shape (quadrupole moment)	oblate	oblate
Excited states	more than ten	more than ten

be observed. In case of neutrons, this is achieved by the radio waves absorbed when the magnetic moment switches direction with respect to an applied magnetic field. The result of these experiments is simple: the lifetime of free neutrons is 885.7(8) s. Nevertheless, we all know that inside most nuclei we are made of, neutrons do not decay for millions of years, because the decay products do not lead to a state of lower energy. (Why not?) Challenge 117 s

Magnetic resonance images also show that most elements have different types of atoms. These elements have atoms with the same number of protons, but with different numbers of neutrons. One says that these elements have several *isotopes*.* This result

---

* The name is derived from the Greek words for 'same' and 'spot', as the atoms are on the same spot in the periodic table of the elements.

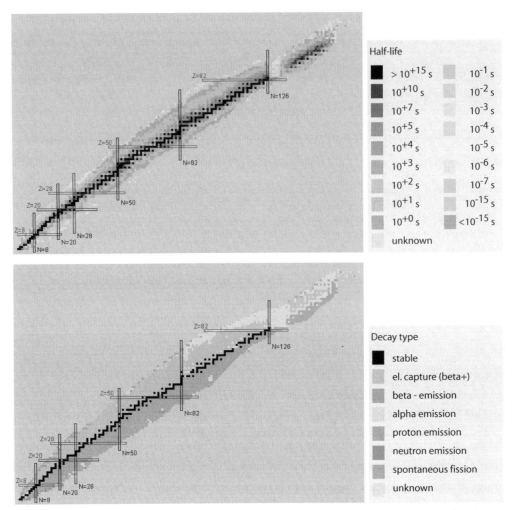

FIGURE 95 All known nuclides with their lifetimes (above) and main decay modes (below). The data are from www.nndc.bnl.gov/nudat2.

also explains why some elements radiate with a mixture of different decay times. Though chemically isotopes are (almost) indistinguishable, they can differ strongly in their nuclear properties. Some elements, such as tin, caesium, or polonium, have over thirty isotopes each. Together, the 118 known elements have over 2000 isotopes. They are shown in Figure 95. (Isotopes without electrons, i.e., specific nuclei with a given number of neutrons and protons, are called *nuclides*.)

Since nuclei are so extremely dense despite containing numerous positively charged protons, there must be a force that keeps everything together against the electrostatic repulsion. We saw that the force is not influenced by electromagnetic or gravitational fields; it must be something different. The force must be short range; otherwise nuclei would not decay by emitting high energy α rays. The additional force is called the *strong nuclear interaction*. We shall study it in detail shortly.

Page 198

The strong nuclear interaction binds protons and neutrons in the nucleus. It is essen-

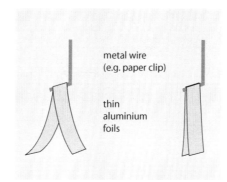

metal wire
(e.g. paper clip)

thin
aluminium
foils

FIGURE 96 An electroscope (or electrometer) (© Harald Chmela) and its charged (middle) and uncharged state (right).

FIGURE 97 Viktor Heß (1883–1964)

tial to recall that inside a nucleus, the protons and neutrons – they are often collectively called *nucleons* – move in a similar way to the electrons moving in atoms. Figure 94 illustrates this. The motion of protons and neutrons inside nuclei allows us to understand the shape, the spin and the magnetic moment of nuclei.

NUCLEI CAN MOVE ALONE – COSMIC RAYS

In everyday life, nuclei are mostly found inside atoms. But in some situations, they move all by themselves, without surrounding electron clouds. The first to discover an example was Rutherford; with a clever experiment he showed that the α particles emitted by many radioactive substance are helium nuclei. Like all nuclei, α particles are small, so that they are quite useful as projectiles.

Then, in 1912, Viktor Heß* made a completely unexpected discovery. Heß was intrigued by electroscopes (also called electrometers). These are the simplest possible detectors of electric charge. They mainly consist of two hanging, thin metal foils, such as two

Vol. III, page 22

---

* Viktor Franz Heß, (b. 1883 Waldstein, d. 1964 Mount Vernon), Austrian nuclear physicist, received the Nobel Prize for physics in 1936 for his discovery of cosmic radiation. Heß was one of the pioneers of research into radioactivity. Heß' discovery also explained why the atmosphere is always somewhat charged, a result important for the formation and behaviour of clouds. Twenty years after the discovery of cosmic radiation, in 1932 Carl Anderson discovered the first antiparticle, the positron, in cosmic radiation; in 1937 Seth Neddermeyer and Carl Anderson discovered the muon; in 1947 a team led by Cecil Powell discovered the pion; in 1951, the Λ^0 and the kaon K^0 are discovered. All discoveries used cosmic rays and most of these discoveries led to Nobel Prizes.

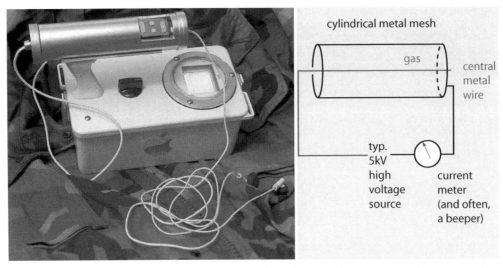

FIGURE 98 A Geiger–Müller counter with the detachable detection tube, the connection cable to the counter electronics, and, for this model, the built-in music player (© Joseph Reinhardt).

strips of aluminium foil taken from a chocolate bar. When the electroscope is charged, the strips repel each other and move apart, as shown in Figure 96. (You can build one easily yourself by covering an empty glass with some transparent cellophane foil and suspending a paper clip and the aluminium strips from the foil. You can charge the electroscope with the help of a rubber balloon and a woollen pullover.) An electroscope thus measures electrical charge. Like many before him, Heß noted that even for a completely isolated electroscope, the charge disappears after a while. He asked: why? By careful study he eliminated one explanation after the other. Heß (and others) were left with only one possibility: that the discharge could be due to charged rays, such as those of the recently discovered radioactivity, emitted from the environment. To increase the distance to the environment, Heß prepared a sensitive electrometer and took it with him on a balloon flight.

Challenge 118 e

As expected, the balloon flight showed that the discharge effect diminished with height, due to the larger distance from the radioactive substances on the Earth's surface. But above about 1000 m of height, the discharge effect increased again, and the higher he flew, the stronger it became. Risking his health and life, he continued upwards to more than 5000 m; there the discharge was several times faster than on the surface of the Earth. This result is exactly what is expected from a radiation coming from outer space and absorbed by the atmosphere. In one of his most important flights, performed during an (almost total) solar eclipse, Heß showed that most of the 'height radiation' did not come from the Sun, but from further away. He thus called the radiation *cosmic rays*. One also speaks of *cosmic radiation*. During the last few centuries, many people have drunk from a glass and eaten chocolate covered by aluminium foil; but only Heß combined these activities with such careful observation and deduction that he earned a Nobel Prize.*

Today, the most common detectors for cosmic rays are *Geiger–Müller counters* and *spark chambers*. Both share the same idea; a high voltage is applied between two metal

---

* In fact, Hess used gold foils in his electrometer, not aluminium foils.

FIGURE 99 A modern spark chamber showing the cosmic rays that constantly arrive on Earth (QuickTime film © Wolfgang Rueckner).

TABLE 11 The main types of cosmic radiation.

PARTICLE	ENERGY	ORIGIN	DETECTOR	SHIELD
At high altitude, the primary particles:				
Protons (90 %)	10^9 to 10^{22} eV	stars, supernovae, extragalactic, unknown	scintillator	in mines
α rays (9 %)	typ. $5 \cdot 10^6$ eV	stars, galaxy	ZnS, counters	1 mm of any material
Other nuclei, such as Le, Be, B, Fe (1 %)	10^9 to 10^{19} eV	stars, novae	1 mm of any material	counters, films
Neutrinos	MeV, GeV	Sun, stars	chlorine, gallium, water	none
Electrons (0.1 %)	10^6 to $> 10^{12}$ eV	supernova remnants		
Gammas (10^{-6})	1 eV to 50 TeV	stars, pulsars, galactic, extragalactic	semiconductor detectors	in mines
At sea level, secondary particles are produced in the atmosphere:				
Muons	3 GeV, 150/ m^2s	protons hit atmosphere, produce pions which decay into muons	drift chamber, bubble chamber, scintillation detector	15 m of water or 2.5 m of soil
Oxygen, radiocarbon and other nuclei	varies	e.g., $n + {}^{16}O \rightarrow p + {}^{14}C$	soil	
Positrons	varies		counters	soil
Neutrons	varies	reaction product when proton hits ${}^{16}O$ nucleus	counters	soil
Pions	varies	reaction product when proton hits ${}^{16}O$ nucleus	counters	soil
In addition, there are slowed down primary beam particles.				

parts kept in a thin and suitably chosen gas (a wire and a cylindrical mesh for the Geiger-Müller counter, two plates or wire meshes in the spark chambers). When a high energy ionizing particle crosses the counter, a spark is generated, which can either be observed through the generated spark (as you can do yourself by watching the spark chamber in the entrance hall of the CERN main building), or detected by the sudden current flow. Historically, the current was first amplified and sent to a loudspeaker, so that the particles can be heard by a 'click' noise. In short, with a Geiger counter one cannot see ions or particles, but one can hear them. Later on, with the advances in electronics, ionized atoms or particles could be counted.

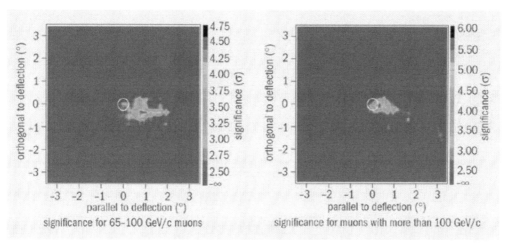

FIGURE 100 The cosmic ray moon shadow, observed with the L3 detector at CERN. The shadow is shifted with respect of the position of the moon, indicated by a white circle, because the Earth's magnetic field deflects the charged particles making up cosmic rays (© CERN Courier).

Finding the right gas mixture for a Geiger–Müller counter is tricky; it is the reason that the counter has a double name. One needs a gas that extinguishes the spark after a while, to make the detector ready for the next particle. Müller was Geiger's assistant; he made the best counters by adding the right percentage of alcohol to the gas in the chamber. Nasty rumours maintained that this was discovered when another assistant tried, without success, to build counters while Müller was absent. When Müller, supposedly a heavy drinker, came back, everything worked again. However, the story is apocryphal. Today, Geiger–Müller counters are used around the world to detect radioactivity; the smallest versions fit in mobile phones and inside wrist watches. An example is shown in Figure 98.

If you can ever watch a working spark chamber, do so. The one in the CERN entrance hall is about $0.5\,\mathrm{m}^3$ in size. A few times per minute, you can see the pink sparks showing the traces of cosmic rays. The rays appear in groups, called *showers*. And they hit us all the time.

Various particle detectors also allow measuring the energy of particles. The particle energy in cosmic rays spans a range between 10^3 eV and at least 10^{20} eV; the latter is the same energy as a tennis ball after serve, but for a single ion. This is a huge range in energy. Understanding the origin of cosmic rays is a research field on its own. Some cosmic rays are galactic in origin, some are extragalactic. For most energies, supernova remnants – pulsars and the like – seem the best candidates. However, the source of the highest energy particles is still unknown; black holes might be involved in their formation.

Cosmic rays are probably the only type of radiation discovered without the help of shadows. But in the meantime, such shadows have been found. In a beautiful experiment performed in 1994, the shadow thrown by the Moon on high energy cosmic rays (about 10 TeV) was measured, as shown in Figure 100. When the position of the shadow is compared with the actual position of the Moon, a shift is found. And indeed, due to the magnetic field of the Earth, the cosmic ray Moon shadow is expected to be shifted westwards for protons and eastwards for antiprotons. The data are consistent with a ratio of antiprotons in cosmic rays between 0 % and 30 %. By studying the shadow's position, Ref. 149

FIGURE 101 An aurora borealis, produced by charged particles in the night sky (© Jan Curtis).

the experiment thus showed that high energy cosmic rays are mainly positively charged and thus consist mainly of matter, and only in small part, if at all, of antimatter.

Page 160

Detailed observations showed that cosmic rays arrive on the surface of the Earth as a mixture of many types of particles, as shown in Table 11. They arrive from outside the atmosphere as a mixture of which the largest fraction are protons, followed by α particles, iron and other nuclei. And, as mentioned above, most rays do not originate from the Sun. In other words, nuclei can thus travel alone over large distances. In fact, the distribution of the incoming direction of cosmic rays shows that many rays must be extragalactic in origin. Indeed, the typical nuclei of cosmic radiation are ejected from stars and accelerated by supernova explosions. When they arrive on Earth, they interact with the atmosphere before they reach the surface of the Earth. The detailed acceleration mechanisms at the origin of cosmic rays are still a topic of research.

The flux of *charged* cosmic rays arriving at the surface of the Earth depends on their energy. At the lowest energies, charged cosmic rays hit the human body many times a second. Measurements also show that the rays arrive in irregular groups, called *showers*. In fact, the *neutrino* flux is many orders of magnitude higher than the flux of charged rays, but does not have any effect on human bodies.

Page 234
Vol. III, page 178

Cosmic rays have several effects on everyday life. Through the charges they produce in the atmosphere, they are probably responsible for the start and for the jagged, non-straight propagation of lightning. (Lightning advances in pulses, alternating fast propagation for about 30 m with slow propagation, until they hit connect. The direction they take at the slow spots depends on the wind and the charge distribution in the atmosphere.) Cosmic rays are also important in the creation of rain drops and ice particles inside clouds, and thus indirectly in the charging of the clouds. Cosmic rays, together with ambient radioactivity, also start the Kelvin generator.

Vol. III, page 17

If the magnetic field of the Earth would not exist, we might get sick from cosmic rays. The magnetic field diverts most rays towards the magnetic poles. Also the upper atmo-

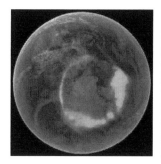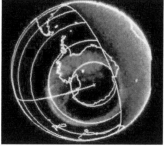

FIGURE 102 Two aurorae australes on Earth, seen from space (a composed image with superimposed UV intensity, and a view in the X-ray domain) and a double aurora on Saturn (all NASA).

sphere helps animal life to survive, by shielding life from the harmful effects of cosmic rays. Indeed, aeroplane pilots and airline employees have a strong radiation exposure that is not favourable to their health. Cosmic rays are also one of several reasons that long space travel, such as a trip to Mars, is not an option for humans. When cosmonauts get too much radiation exposure, the body weakens and eventually they die. Space heroes, including those of science fiction, would not survive much longer than two or three years.

Cosmic rays also produce beautifully coloured flashes inside the eyes of cosmonauts; they regularly enjoy these events in their trips. But cosmic rays are not only dangerous and beautiful. They are also useful. If cosmic rays would not exist, we would not exist either. Cosmic rays are responsible for mutations of life forms and thus are one of the causes of biological evolution. Today, this effect is even used artificially; putting cells into a radioactive environment yields new strains. Breeders regularly derive new mutants in this way.

Cosmic rays cannot be seen directly, but their cousins, the 'solar' rays, can. This is most spectacular when they arrive in high numbers. In such cases, the particles are inevitably deviated to the poles by the magnetic field of the Earth and form a so-called *aurora borealis* (at the North Pole) or an *aurora australis* (at the South pole). These slowly moving and variously coloured curtains of light belong to the most spectacular effects in the night sky. (See Figure 101 or www.nasa.gov/mov/105423main_FUV_2005-01_v01. mov.) Visible light and X-rays are emitted at altitudes between 60 and 1000 km. Seen from space, the aurora curtains typically form a circle with a few thousand kilometres diameter around the magnetic poles. Aurorae are also seen in the rest of the solar system. Aurorae due to core magnetic fields have been observed on Jupiter, Saturn, Uranus, Neptune, Earth, Io and Ganymede. For an example, see Figure 102. Aurorae due to other mechanisms have been seen on Venus and Mars.

Cosmic rays are mainly free nuclei. With time, researchers found that nuclei appear without electron clouds also in other situations. In fact, the vast majority of nuclei in the universe have no electron clouds at all: in the inside of stars, no nucleus is surrounded by bound electrons; similarly, a large part of intergalactic matter is made of protons. It is known today that most of the matter in the universe is found as protons or α particles inside stars and as thin gas between the galaxies. In other words, in contrast to what the Greeks said, matter is not usually made of atoms; it is mostly made of bare nuclei. Our everyday environment is an exception when seen on cosmic scales. In nature, atoms are rare, bare nuclei are common.

Incidentally, nuclei are in no way forced to move; nuclei can also be stored with almost no motion. There are methods – now commonly used in research groups – to superpose electric and magnetic fields in such a way that a single nucleus can be kept floating in mid-air; we discussed this possibility in the section on levitation earlier on.

Vol. III, page 185

NUCLEI DECAY – MORE ON RADIOACTIVITY

Not all nuclei are stable over time. The first measurement that provided a hint was the decrease of radioactivity with time. It is observed that the number N of emitted rays *decreases*. More precisely, radioactivity follows an exponential decay with time t:

$$N(t) = N(0)\, e^{-t/\tau} \tag{51}$$

The parameter τ, the so-called *life time* or *decay time*, depends on the type of nucleus emitting the rays. Life times can vary from much less than a microsecond to millions of millions of years. The expression has been checked for as long as 34 multiples of the duration τ; its validity and precision is well-established by experiments. Obviously, formula (51) is an approximation for large numbers of atoms, as it assumes that $N(t)$ is a continuous variable. Despite this approximation, deriving this expression from quantum theory is not a simple exercise, as we saw above. Though in principle, the quantum Zeno effect could appear for small times t, for the case of radioactivity it has not yet been observed.

Page 45

Instead of the life-time, often the half-life is used. The *half-time* is the time during which radioactivity decreases to *half* the starting value. Can you deduce how the two times are related?

Challenge 119 s

Radioactivity is the decay of unstable nuclei. Most of all, radioactivity allows us to count the number of atoms in a given mass of material. Imagine to have measured the mass of radioactive material at the beginning of your experiment; you have chosen an element that has a lifetime of about a day. Then you put the material inside a scintillation box. After a few weeks the number of flashes has become so low that you can count them; using expression (51) you can then determine how many atoms have been in the mass to begin with. Radioactivity thus allows us to determine the number of atoms, and thus their size, in addition to the size of nuclei.

The exponential decay (51) and the release of energy is typical of *metastable* systems. In 1903, Rutherford and Soddy discovered what the state of lower energy is for α and β emitters. In these cases, radioactivity changes the emitting atom; it is a spontaneous transmutation of the atom. An atom emitting α or β rays changes its chemical nature. Radioactivity thus implies, for the case of nuclei, the same result that statistical mechanics of gases implies for the case of atoms: they are quantum particles with a structure that can change over time.

Vol. I, page 346

— In *α decay* – or *alpha decay* – the radiating nucleus emits a (doubly charged) helium nucleus, also called an *α particle*. The kinetic energy is typically a handful of MeV. After the emission, the nucleus has changed to a nucleus situated two places earlier in the periodic system of the elements. α decay occurs mainly for nuclei that are rich in protons. An example of α decay is the decay of the ^{238}U isotope of uranium.

— In *β decay* – or *beta decay* – a neutron transforms itself into a proton, emitting an electron – also called a *β particle* – and an antineutrino. Also β decay changes the chemical nature of the atom, but to the place following the original atom in the periodic table of the elements. Example of β emitters are radiocarbon, ^{14}C, ^{38}Cl, and ^{137}Cs, the isotope expelled by damaged nuclear reactors. We will explore β decay below. A variant is the $β^+$ decay, in which a proton changes into a neutron and emits a neutrino and a positron. It occurs in proton-rich nuclei. An example is ^{22}Na. Another variant is electron capture; a nucleus sometimes captures an orbital electron, a proton is transformed into a neutron and a neutrino is emitted. This happens in ^{7}Be. Also bound β decay, as seen in ^{187}Re, is a variant of β decay. Page 218

— In *γ decay* – or *gamma decay* – the nucleus changes from an excited to a lower energy state by emitting a high energy photon, or *γ particle*. In this case, the chemical nature is not changed. Typical energies are in the MeV range. Due to the high energy, γ rays ionize the material they encounter; since they are not charged, they are not well absorbed by matter and penetrate deep into materials. γ radiation is thus by far the most dangerous type of (environmental) radioactivity. An example of γ decay is ^{99m}Tc. A variant of γ decay is *isomeric transition*. Still another variant is *internal conversion*, observed, for example, in ^{137m}Ba.

— In *neutron emission* the nucleus emits a neutron. The decay is rare on Earth, but occurs in the stellar explosions. Most neutron emitters have half-lives below a few seconds. Fxamples of neutron emitters are ^{5}He and ^{17}N.

— The process of *spontaneous fission* was discovered in 1940. The decay products vary, even for the same starting nucleus. ^{256}Fe but also ^{235}He can decay through spontaneous fission, though with a small probability.

— In *proton emission* the nucleus emits a proton. This decay is comparatively rare, and occurs only for about a hundred nuclides, for example for ^{53m}Co and ^{4}Li. The first example was discovered only in 1970. Around 2000, the simultaneous emission of two protons was also observed for the first time.

— In 1984, *cluster emission* or *heavy ion emission* was discovered. A small fraction of ^{223}Ra nuclei decay by emitting a ^{14}C nucleus. This decay occurs for half a dozen nuclides. Emission of ^{18}O has also been observed.

Many combined and mixed decays also exist. These decays are studied by nucleal physicists. Radioactivity is a common process. As an example, in every human body about nine thousand radioactive decays take place every second, mainly 4.5 kBq (0.2 mSv/a) from ^{40}K and 4 kBq from ^{14}C (0.01 mSv/a). Why is this not dangerous? Ref. 150 Challenge 120 s

All radioactivity is accompanied by emission of energy. The energy emitted by an atom through radioactive decay or reactions is regularly a million time larger than that emitted by a chemical process. More than a decay, a radioactive process is thus a microscopic explosion. A highly radioactive material thus emits a large amount of energy. That is the reason for the danger of nuclear weapons.

What distinguishes those atoms that decay from those which do not? An exponential decay law implies that the probability of decay is independent of the age of the atom. Age or time plays no role. We also know from thermodynamics, that all atoms have exactly Challenge 121 e Vol. IV, page 104

FIGURE 103 A modern accelerator mass spectrometer for radiocarbon dating, at the Hungarian
Academy of Sciences (© HAS).

identical properties. So how is the decaying atom singled out? It took around 30 years
to discover that radioactive decays, like all decays are a quantum effects. All decays are
triggered by the statistical fluctuations of the vacuum, more precisely, by the quantum
fluctuations of the vacuum. Indeed, radioactivity is one of the clearest observations that
classical physics is not sufficient to describe nature.

Radioactivity, like all decays, is a pure quantum effect. Only a finite quantum of action makes it possible that a system remains unchanged until it suddenly decays. Indeed,
in 1928 George Gamow explained α decay with the tunnelling effect. He found that the
tunnelling effect explains the relation between the lifetime and the range of the rays, as
well as the measured variation of lifetimes – between 10 ns and 10^{17} years – as the consequence of the varying potentials to be overcome in different nuclei.

RADIOMETRIC DATING

Page 167 As a result of the chemical effects of radioactivity, the composition ratio of certain elements in minerals allows us to determine the *age* of the mineral. Using radioactive decay
to deduce the age of a sample is called *radiometric dating*. With this technique, geologists
determined the age of mountains, the age of sediments and the age of the continents.
They determined the time that continents moved apart, the time that mountains formed
when the continents collided and the time when igneous rocks were formed. Where there

surprises? No. The times found with radiometric dating are consistent with the relative time scale that geologists had defined independently for centuries before the technique appeared. Radiometric dating confirmed what had been deduced before.

Radiometric dating is a science of its own. An overview of the isotopes used, together with their specific applications in dating of specimen, is given in Table 12. The table shows how the technique of radiometric dating has deeply impacted astronomy, geology, evolutionary biology, archaeology and history. (And it has reduced the number of violent believers.) Radioactive life times can usually be measured to within one or two per cent of accuracy, and they are known both experimentally and theoretically not to change over geological time scales. As a result, radiometric dating methods can be surprisingly precise. Can you imagine how one measure half-lives of thousands of millions of years to high precision? Ref. 152
Page 168
Ref. 151
Challenge 122 s

Radiometric dating was even more successful in the field of ancient history. With the *radiocarbon dating method* historians determined the age of civilizations and the age of human artefacts.* Many false beliefs were shattered. In some belief communities the shock is still not over, even though over hundred years have passed since these results became known. Ref. 151

Radiocarbon dating uses the β decay of the radioactive carbon isotope ^{14}C, which has a decay time of 5730 a. This isotope is continually created in the atmosphere through the influence of cosmic rays. This happens through the reaction $^{14}N + n \rightarrow p + {}^{14}C$. As a result, the concentration of radiocarbon in air is relatively constant over time. Inside living plants, the metabolism thus (unknowingly) maintains the same concentration. In dead plants, the decay sets in. The life time value of a few thousand years is particularly useful to date historic material. Therefore, radiocarbon dating has been used to determine the age of mummies, the age of prehistoric tools and the age of religious relics. The original version of the technique measured the radiocarbon content through its radioactive decay and the scintillations it produced. A quality jump was achieved when accelerator mass spectroscopy became commonplace. It was not necessary any more to wait for decays: it is now possible to determine the ^{14}C content directly. As a result, only a tiny amount of carbon, as low as 0.2 mg, is necessary for a precise dating. Such small amounts can be detached from most specimen without big damage. Accelerator mass spectroscopy showed that numerous religious relics are forgeries, such as a cloth in Turin, and that, in addition, several of their wardens are crooks.

Researchers have even developed an additional method to date stones that uses radioactivity. Whenever an α ray is emitted, the emitting atom gets a recoil. If the atom is part of a crystal, the crystal is damaged by the recoil. In many materials, the damage can be seen under the microscope. By counting the damaged regions it is possible to date the time at which rocks have been crystallized. In this way it has been possible to determine when the liquid material from volcanic eruptions has become rock.

With the advent of radiometric dating, for the first time it became possible to reliably date the age of rocks, to compare it with the age of meteorites and, when space travel became fashionable, with the age of the Moon. The result was beyond all previous estimates and expectations: the oldest rocks and the oldest meteorites, studied independently us-

---

* In 1960, the developer of the radiocarbon dating technique, Willard Libby, received the Nobel Prize for chemistry.

TABLE 12 The main natural isotopes used in radiometric dating.

ISOTOPE	DECAY PRODUCT	HALF-LIFE	METHOD USING IT	EXAMPLES
^{147}Sm	^{143}Nd	106 Ga	samarium–neodymium method	rocks, lunar soil, meteorites
^{87}Rb	^{87}Sr	48.8 Ga	rubidium–strontium method	rocks, lunar soil, meteorites
^{187}Re	^{187}Os	42 Ga	rhenium–osmium method	rocks, lunar soil, meteorites
^{176}Lu	^{176}Hf	37 Ga	lutetium–hafnium method	rocks, lunar soil, meteorites
^{40}K	^{40}Ar	1.25 Ga	potassium–argon method, argon–argon method	rocks, lunar soil, meteorites
^{40}K	^{40}Ca	1.25 Ga	potassium–calcium method	granite dating, not precise
^{232}Th	^{208}Pb	14 Ga	thorium–lead method, lead–lead method	rocks, lunar soil, meteorites
^{238}U	^{206}Pb	4.5 Ga	uranium–lead method, lead–lead method	rocks, lunar soil, meteorites
^{235}U	^{207}Pb	0.7 Ga	uranium–lead method, lead–lead method	rocks, lunar soil, meteorites
^{234}U	^{230}Th	248 ka	uranium–thorium method	corals, stalactites, bones, teeth
^{230}Th	^{226}Ra	75.4 ka	thorium-radon method	plant dating
^{26}Al	^{26}Mg	0.72 Ma	supernova debris dating, cosmogenic	checking that nucleosynthesis still takes place in the galaxy
^{10}Be	^{10}B	1.52 Ma	cosmogenic radiometric dating	ice cores
^{60}Fe	^{60}Ni	2.6 Ma	supernova debris dating	deep sea crust; lifetime updated in 2009 from the previously accepted 1.5 Ma
^{36}Cl	^{36}Ar	0.3 Ma	cosmogenic radiometric dating	ice cores
^{53}Mn	^{53}Cr	3.7 Ma	cosmogenic radiometric dating	meteorites, K/T boundary
^{182}Hf	^{182}W	9 Ma	cosmogenic radiometric dating	meteorites, sediments
^{14}C	^{14}N	5730 a	radiocarbon method, cosmogenic	wood, clothing, bones, organic material, wine
^{137}Cs	^{137}Ba	30 a	γ-ray counting	dating food and wine after nuclear accidents
^{210}Pb		22 a	γ-ray counting	dating wine
^{3}H	^{3}He	12.3 a	γ-ray counting	dating wine

FIGURE 104 The lava sea in the volcano Erta Ale in Ethiopia (© Marco Fulle).

ing different dating methods, are 4570(10) million years old. From this data, the age of Ref. 153 the Earth is estimated to be 4540(50) million years. The Earth is indeed *old*.

But if the Earth is so old, why did it not cool down in its core in the meantime?

WHY IS HELL HOT?

The lava seas and streams found in and around volcanoes are the origin of the imagery that many cultures ascribe to hell: fire and suffering. Because of the high temperature of lava, hell is inevitably depicted as a hot place located at the centre of the Earth. A striking example is the volcano Erta Ale, shown in Figure 104. But why is lava still hot, after so many million years?

A straightforward calculation shows that if the Earth had been a hot sphere in the beginning, it should have cooled down and solidified already long time ago. The Earth Challenge 123 ny should be a solid object, like the moon: the Earth should not contain any lava, and hell would not be hot.

The solution to the riddle is provided by radioactivity: the centre of the Earth contains an oven that is fuelled by radioactive potassium ^{40}K, radioactive uranium ^{235}U and ^{238}U and radioactive thorium ^{232}Th. The radioactivity of these elements, and a few oth- Ref. 154 ers to a minor degree, keeps the centre of the Earth glowing. More precise investigations, taking into account the decay times and material concentrations, show that this mecha- Page 168 nism indeed explains the internal heat of the Earth. (In addition, the decay of radioactive potassium is the origin for the 1 % of argon found in the Earth's atmosphere.)

In short, radioactivity keeps lava hot. Radioactivity is the reason that we depict hell

Challenge 124 s
as hot. This brings up a challenge: why is the radioactivity of lava and of the Earth in general not dangerous to humans?

NUCLEI CAN FORM COMPOSITES

Nuclei are highly unstable when they contain more than about 280 nucleons. Nuclei with higher number of nucleons inevitably decay into smaller fragments. In short, heavy nuclei are unstable. But when the mass is above 10^{57} nucleons, they are stable again: such systems are called *neutron stars*. This is the most extreme example of pure nuclear matter found in nature. Neutron stars are left overs of (type II) supernova explosions. They do not run any fusion reactions any more, as other stars do; in first approximation neutron stars are simply large nuclei.

Neutron stars are made of degenerate matter. Their density of 10^{18} kg/m^3 is a few times that of a nucleus, as gravity compresses the star. This density value means that a tea spoon of such a star has a mass of several hundred million tons. Neutron stars are about 10 km in diameter. They are never much smaller, as such smaller stars are unstable. They are never much larger, because much larger neutron stars turn into black holes.

NUCLEI HAVE COLOURS AND SHAPES

In everyday life, the colour of objects is determined by the wavelength of light that is least absorbed, or, if they shine, by the wavelength that is emitted. Also nuclei can absorb photons of suitably tuned energies and get into an excited state. In this case, the photon energy is converted into a higher energy of one or several of the nucleons whirling around inside the nucleus. Many radioactive nuclei also emit high energy photons, which then are called γ rays, in the range between 1 keV (or 0.2 fJ) and more than 20 MeV (or 3.3 pJ). The emission of γ rays by nuclei is similar to the emission of light by electrons in atoms. From the energy, the number and the lifetime of the excited states – they range from 1 ps to 300 d – researchers can deduce how the nucleons move inside the nucleus.

In short, the energies of the emitted and absorbed γ ray photons define the 'colour' of the nucleus. The γ ray spectrum can be used, like all colours, to distinguish nuclei from each other and to study their motion. In particular, the spectrum of the γ rays emitted by excited nuclei can be used to determine the chemical composition of a piece of matter. Some of these transition lines are so narrow that they can been used to study the change due to the chemical environment of the nucleus, to measure nuclear motion inside solids or to detect the gravitational Doppler effect.

The study of γ-rays also allows us to determine the *shape* of nuclei. Many nuclei are spherical; but many are prolate or oblate ellipsoids. Ellipsoids are favoured if the reduction in average electrostatic repulsion is larger than the increase in surface energy. All nuclei – except the lightest ones such as helium, lithium and beryllium – have a constant mass density at their centre, given by about 0.17 fermions per fm^3, and a skin thickness of about 2.4 fm, where their density decreases. Nuclei are thus small clouds, as illustrated in Figure 105.

We know that molecules can be of extremely involved shape. In contrast, nuclei are mostly spheres, ellipsoids or small variations of these. The reason is the short range, or better, the fast spatial decay of nuclear interactions. To get interesting shapes like in molecules, one needs, apart from nearest neighbour interactions, also next neighbour

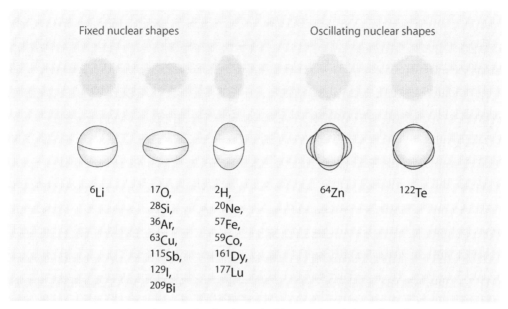

FIGURE 105 Various nuclear shapes – fixed: spherical, oblate, prolate (left) and oscillating (right), shown realistically as clouds (above) and simplified as geometric shapes (below).

interactions and next next neighbour interactions. The strong nuclear interaction is too short ranged to make this possible. Or does it? It might be that future studies will discover that some nuclei are of more unusual shape, such as smoothed pyramids. Some predictions have been made in this direction; however, the experiments have not been performed yet.

Ref. 155

The shape of nuclei does not have to be fixed; nuclei can also *oscillate* in shape. Such oscillations have been studied in great detail. The two simplest cases, the quadrupole and octupole oscillations, are shown in Figure 105. In addition, non-spherical nuclei can also rotate. Several rapidly spinning nuclei, with a spin of up to $60\hbar$ and more, are known. They usually slow down step by step, emitting a photon and reducing their angular momentum at each step. Recently it was even discovered that nuclei can also have bulges that rotate around a fixed core, a bit like the tides that rotate around the Earth.

Ref. 156

THE FOUR TYPES OF MOTION IN THE NUCLEAR DOMAIN

Nuclei are small because the nuclear interactions are short-ranged. Due to this short range, nuclear interactions play a role only in four types of motion:

— *scattering*,
— *bound motion*,
— *decay* and
— a combination of these three called *nuclear reactions*.

The history of nuclear physics has shown that the whole range of observed phenomena can be reduced to these four fundamental processes. Each process is a type of motion. And in each process, the main interest is the comparison of the start and the end situations; the intermediate situations are less interesting. Nuclear interactions thus lack the complex types of motion which characterize everyday life. That is also the main reason for the shortness of this chapter.

Scattering is performed in all accelerator experiments. Such experiments repeat for nuclei what we do when we look at an object. Eye observation, or seeing something, is a scattering experiment, as eye observation is the detection of scattered light. Scattering of X-rays was used to see atoms for the first time; scattering of high energy alpha particles was used to discover and study the nucleus, and later the scattering of electrons with even higher energy was used to discover and study the components of the proton.

Bound motion is the motion of protons and neutrons inside nuclei or the motion of quarks inside mesons and baryons. In particular, bound motion determines shape and changes of shape of compounds: hadrons and nuclei.

Decay is obviously the basis of radioactivity. Nuclear decay can be due to the electromagnetic, the strong or the weak nuclear interaction. Decay allows studying the conserved quantities of nuclear interactions.

Nuclear reactions are combinations of scattering, decay and possibly bound motion. Nuclear reactions are for nuclei what the touching of objects is in everyday life. Touching an object we can take it apart, break it, solder two objects together, throw it away, and much more. The same can be done with nuclei. In particular, nuclear reactions are responsible for the burning of the Sun and the other stars; they also tell the history of the nuclei inside our bodies.

Quantum theory showed that all four types of nuclear motion can be described in the same way. Each type of motion is due to *the exchange of virtual particles*. For example, scattering due to charge repulsion is due to exchange of virtual photons, the bound motion inside nuclei due to the strong nuclear interaction is due to exchange of virtual gluons, β decay is due to the exchange of virtual W bosons, and neutrino reactions are due to the exchange of virtual Z bosons. The rest of this chapter explains these mechanisms in more details.

NUCLEI REACT

Vol. IV, page 108

The first man thought to have made transuranic elements, the Italian genius Enrico Fermi, received the Nobel Prize for the discovery. Shortly afterwards, Otto Hahn and his collaborators Lise Meitner and Fritz Strassmann showed that Fermi was wrong, and that his prize was based on a mistake. Fermi was allowed to keep his prize, the Nobel committee gave Hahn and Strassmann the Nobel Prize as well, and to make the matter unclear to everybody and to women physicists in particular, the prize was not given to Lise Meitner. (After her death though, a new chemical element was named after her.)

When protons or neutrons are shot into nuclei, they usually remained stuck inside them, and usually lead to the transformation of an element into a heavier one. After having done this with all elements, Fermi used uranium; he found that bombarding it with neutrons, a new element appeared, and concluded that he had created a transuranic element. Alas, Hahn and his collaborators found that the element formed was well-known: it was barium, a nucleus with less than half the mass of uranium. Instead of remaining stuck as in the previous 91 elements, the neutrons had *split* the uranium nucleus. In short,

Fermi, Hahn, Meitner and Strassmann had observed reactions such as:

$$^{235}\text{U} + \text{n} \rightarrow {}^{143}\text{Ba} + {}^{90}\text{Kr} + 3n + 170\,\text{MeV} \;. \tag{52}$$

Meitner called the splitting process *nuclear fission*. The amount of energy liberated in fission is unusually large, millions of times larger than in a chemical interaction of an atom. In addition, several neutrons are emitted, which in turn can lead to the same process; fission can thus start a *chain reaction*. Later, and (of course) against the will of the team, the discovery would be used to make nuclear bombs.

Nuclear reactions are typically triggered by neutrons, protons, deuterons or γ particles. Apart from triggering fission, neutrons are used to transform lithium into tritium, which is used as (one type of) fuel in fusion reactors; and neutrons from (secondary) cosmic rays produce radiocarbon from the nitrogen in the atmosphere. Deuterons impinging on tritium produce helium in fusion reactors. Protons can trigger the transformation of lithium into beryllium. Photons can knock alpha particles or neutrons out of nuclei.

All nuclear reactions and decays are *transformations*. In each transformation, already the ancient Greek taught us to search, first of all, for conserved quantities. Besides the well-known cases of energy, momentum, electric charge and angular momentum conservation, the results of nuclear physics lead to several new conserved quantities. The behaviour is quite constrained. Quantum field theory implies that particles and antiparticles (commonly denoted by a bar) must behave in compatible ways. Both experiment and quantum field theory show for example that every reaction of the type

$$A + B \rightarrow C + D \tag{53}$$

implies that the reactions

$$A + \overline{C} \rightarrow \overline{B} + D \tag{54}$$

or

$$\overline{C} + \overline{D} \rightarrow \overline{A} + \overline{B} \tag{55}$$

or, if energy is sufficient,

$$A \rightarrow C + D + \overline{B} \;, \tag{56}$$

are also possible. Particles thus behave like conserved mathematical entities.

Experiments show that antineutrinos differ from neutrinos. In fact, all reactions confirm that the so-called *lepton number* is conserved in nature. The lepton number L is zero for nucleons or quarks, is 1 for the electron and the neutrino, and is −1 for the positron and the antineutrino.

In addition, all reactions conserve the so-called *baryon number*. The baryon number B for protons and neutrons is 1 (and 1/3 for quarks), and −1 for antiprotons and antineutrons (and thus −1/3 for antiquarks). So far, no process with baryon number violation has ever been observed. Baryon number conservation is one reason for the danger of radioactivity, fission and fusion.

FIGURE 106 The destruction of four nuclear reactors in 2011 in Fukushima, in Japan, which rendered life impossible at a distance of less than 30 km (courtesy Digital Globe).

BOMBS AND NUCLEAR REACTORS

Uranium fission is triggered by a neutron, liberates energy and produces several additional neutrons. Therefore, uranium fission can trigger a *chain reaction* that can lead either to an explosion or to a controlled generation of heat. Once upon a time, in the middle of the twentieth century, these processes were studied by quite a number of researchers. Most of them were interested in making weapons or in using nuclear energy, despite the high toll these activities place on the economy, on human health and on the environment.

Most stories around the development of nuclear weapons are almost incredibly absurd. The first such weapons were built during the second world war, with the help of the smartest physicists that could be found. Everything was ready, including the most complex physical models, several huge factories and an organization of incredible size. There was just one little problem: there was no uranium of sufficient quality. The mighty United States thus had to go around the world to shop for good uranium. They found it in the (then) Belgian colony of Congo, in central Africa. In short, without the support of Belgium, which sold the Congolese uranium to the USA, there would have been no nuclear bomb, no early war end and no superpower status.

Congo paid a high price for this important status. It was ruled by a long chain of military dictators up to this day. But the highest price was paid by the countries that actually built nuclear weapons. Some went bankrupt, others remained underdeveloped; even the richest countries have amassed huge debts and a large underprivileged population. There is *no* exception. The price of nuclear weapons has also been that some regions of our planet became uninhabitable, such as numerous islands, deserts, rivers, lakes and marine environments. But it could have been worse. When the most violent physicist ever, Edward Teller, made his first calculations about the hydrogen bomb, he predicted that the bomb would set the atmosphere into fire. Nobel Prize winner Hans Bethe* cor-

---

* Hans Bethe (b. 1906 Strasbourg, d. 2005) was one of the great physicists of the twentieth century, even

FIGURE 107 The explosion of a nuclear bomb: an involved method of killing many children in the country where it explodes and ruining the economic future of the children in the country that built it.

rected the mistake and showed that nothing of this sort would happen. Nevertheless, the military preferred to explode the hydrogen bomb in the Bikini atoll, the most distant place from their homeland they could find. The place is so radioactive that today it is Ref. 157 even dangerous simply to fly over that island!

It was then noticed that nuclear test explosions increased ambient radioactivity in the atmosphere all over the world. Of the produced radioactive elements, ^{3}H is absorbed by humans in drinking water, ^{14}C and ^{90}Sr through food, and ^{137}Cs in both ways. Fortunately, in the meantime, all countries have agreed to perform their nuclear tests underground.

Radioactivity is dangerous to humans, because it disrupts the processes inside living cells. Details on how radioactivity is measured and what effects on health it produces are provided below. Page 178

though he was head of the theory department that lead to the construction of the first atomic bombs. He worked on nuclear physics and astrophysics, helped Richard Feynman in developing quantum electrodynamics, and worked on solid state physics. When he got older and wiser, he became a strong advocate of arms control; he also was essential in persuading the world to stop atmospheric nuclear test explosions and saved many humans from cancer in doing so.

TABLE 13 Some radioactivity measurements.

MATERIAL	ACTIVITY IN BQ/KG
Air	$c.\ 10^{-2}$
Sea water	10^1
Human body	$c.\ 10^2$
Cow milk	max. 10^3
Pure ^{238}U metal	$c.\ 10^7$
Highly radioactive α emitters	$> 10^7$
Radiocarbon: ^{14}C (β emitter)	10^8
Highly radioactive β and γ emitters	$> 10^9$
Main nuclear fallout: ^{137}Cs, ^{90}Sr (α emitter)	$2 \cdot 10^9$
Polonium, one of the most radioactive materials (α)	10^{24}

Not only nuclear bombs, also peaceful nuclear reactors are dangerous. The reason was discovered in 1934 by Frédéric Joliot and his wife Irène, the daughter of Pierre and Marie Curie: *artificial radioactivity*. The Joliot–Curies discovered that materials irradiated by α rays become radioactive in turn. They found that α rays transformed aluminium into radioactive phosphorus:

$$^{27}_{13}\text{Al} + ^{4}_{2}\alpha \rightarrow ^{30}_{15}\text{P} + ^{30}_{15}\text{n}\ . \tag{57}$$

In fact, almost all materials become radioactive when irradiated with alpha particles, neutrons or γ rays. As a result, radioactivity itself can only be contained with difficulty. After a time span that depends on the material and the radiation, any box that contains radioactive material has itself become radioactive. The 'contagion' stops only for very small amounts of radioactive material.

The dangers of natural and artificial radioactivity are the reason for the high costs of nuclear reactors. After about thirty years of operation, reactors have to be dismantled. The radioactive pieces have to be stored in specially chosen, inaccessible places, and at the same time the workers' health must not be put in danger. The world over, many reactors now need to be dismantled. The companies performing the job sell the service at high price. All operate in a region not far from the border to criminal activity, and since radioactivity cannot be detected by the human senses, many companies cross that border.

In fact, one important nuclear reactor is (usually) not dangerous to humans: the Sun.
Page 183 We explore it shortly.

CURIOSITIES AND CHALLENGES ON NUCLEI AND RADIOACTIVITY

Nowadays, nuclear magnetic resonance is also used to check the quality of food. For example, modern machines can detect whether orange juice is contaminated with juice from other fruit and can check whether the fruit were ripe when pressed. Other machines can check whether wine was made from the correct grapes and how it aged.

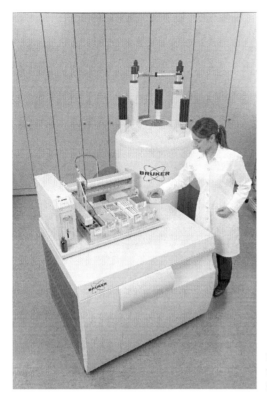

FIGURE 108 A machine to test fruit quality with the help of nuclear magnetic resonance (© Bruker).

∗ ∗

Magnetic resonance machines pose no danger; but they do have some biological effects, as Peter Mansfield, one of the inventors of the technique, explains. The first effect is due to the conductivity of blood. When blood in the aorta passes through a magnetic field, a voltage is induced. This effect has been measured and it might interfere with cardiac functioning at 7 T; usual machines have 1.5 T and pose no risk. The second effect is due to the switching of the magnetic field. Some people sense the switching in the thorax and in the shoulders. Not much is known about the details of such peripheral nerve stimulation yet.

Ref. 161

∗ ∗

The amount of radioactive radiation is called the *dose*. The unit for the radioactive dose is one *gray*: it is the amount of radioactivity that deposits the energy 1 J on 1 kg of matter: 1 Gy = 1 J/kg. A *sievert*, or 1 Sv, is the unit of radioactive dose *equivalent*; it is adjusted to humans by weighting each type of human tissue with a factor representing the impact of radiation deposition on it. Three to five sievert are a lethal dose to humans. In comparison, the natural radioactivity present inside human bodies leads to a dose of 0.2 mSv per year. An average X-ray image implies an irradiation of 1 mSv; a CAT scan 8 mSv. For other measurement examples, see Table 13.

Ref. 158

The *amount of radioactive material* is measured by the number of nuclear decays per second. One decay per second is called one *becquerel*, or 1 Bq. An adult human body

TABLE 14 Human exposure to radioactivity and the corresponding doses.

EXPOSURE	DOSE
Daily human exposure:	
Average exposure to cosmic radiation in Europe	
at sea level	0.3 mSv/a
at a height of 3 km	1.2 mSv/a
Average (and maximum) exposure from the soil, not counting radon effects	0.4 mSv/a (2 mSv/a)
Average (and maximum) inhalation of radon	1 mSv/a (100 mSv/a)
Average exposure due to internal radionuclides	0.3 mSv/a
natural content of ^{40}K in human muscles	10^{-4} Gy and 4500 Bq
natural content of Ra in human bones	$2 \cdot 10^{-5}$ Gy and 4000 Bq
natural content of ^{14}C in humans	10^{-5} Gy
Total average (and maximum) human exposure	2 mSv/a (100 mSv/a)
Common situations:	
Dental X-ray	*c.* 10 mSv equivalent dose
Lung X-ray	*c.* 0.5 mSv equivalent dose
Short one hour flight (see www.gsf.de/epcard)	*c.* 1 μSv
Transatlantic flight	*c.* 0.04 mSv
Maximum allowed dose at work	30 mSv/a
Smoking 60 cigarettes a day	26 to 120 mSv/a
Deadly exposures:	
Ionization	0.05 C/kg can be deadly
Dose	100 Gy=100 J/kg is deadly in 1 to 3 days
Equivalent dose	more than 3 Sv leads to death for 50% of untreated patients

typically contains 9 kBq, the European limit for food, in 2011, varies between 370 and 600 Bq/kg. The amount released by the Hiroshima bomb is estimated to have been between 4 PBq and 60 PBq, the amount released by the Chernobyl disaster was between 2 and 12 EBq, thus between 200 and 500 times larger. The numbers for the various Russian radioactive disasters in the 1960s and 1970 are similarly high. The release for the Fukushima reactor disaster in March 2011 is estimated to have been 370 to 630 PBq, which would put it at somewhere between 10 and 90 Hiroshima bombs.

The SI units for radioactivity are now common around the world; in the old days, 1 sievert was called 100 rem or 'Röntgen equivalent man'; the SI unit for dose, 1 gray, replaces what used to be called 100 rd or Rad. The SI unit for exposition, 1 C/kg, replaces the older unit 'röntgen', with the relation 1 R = $2.58 \cdot 10^{-4}$ C/kg. The SI unit becquerel replaces the curie (Ci), for which 1 Ci = 37 GBq.

* *

Not all γ-rays are due to radioactivity. In the year 2000, an Italian group discovered that thunderstorms also emit γ rays, of energies up to 10 MeV. The mechanisms are still being investigated; they seem to be related to the formation process of lightning.   Ref. 159

* *

Chain reactions are quite common in nature, and are not limited to the nuclear domain. *Fire* is a chemical chain reaction, as are exploding fireworks. In both cases, material needs heat to burn; this heat is supplied by a neighbouring region that is already burning.

* *

Radioactivity can be extremely dangerous to humans. The best example is plutonium. Only 1 μg of this α emitter inside the human body are sufficient to cause lung cancer. Another example is polonium. Polonium 210 is present in tobacco leaves that were grown with artificial fertilizers. In addition, tobacco leaves filter other radioactive substances from the air. Polonium, lead, potassium and the other radioactive nuclei found in tobacco are the main reason that *smoking produces cancer*. Table 14 shows that the dose is considerable and that it is by far the largest dose absorbed in everyday life.

* *

Why is nuclear power a dangerous endeavour? The best argument is Lake Karachay near Mayak, in the Urals in Russia. In less than a decade, the nuclear plants of the region have transformed it into the most radioactive place on Earth. In the 1970s, walking on the shore of the lake for an hour led to death on the shore. The radioactive material in the lake was distributed over large areas in several catastrophic explosions in the 1950s and 1960s, leading to widespread death and illness. Several of these accidents were comparable to the Chernobyl accident of 1986. The lake is now covered in concrete.

* *

All lead is slightly radioactive, because it contains the ^{210}Pb isotope, a β emitter. This lead isotope is produced by the uranium and thorium contained in the rock from where the lead is extracted. For sensitive experiments, such as for neutrino experiments, one needs radioactivity shields. The best shield material is lead, but obviously it has to be lead with a low radioactivity level. Since the isotope ^{210}Pb has a half-life of 22 years, one way to do it is to use *old* lead. In a precision neutrino experiment in the Gran Sasso in Italy, the research team uses lead mined during Roman times, thus 2000 years old, in order to reduce spurious signals.

* *

Not all nuclear reactors are human made. The occurrence of *natural* nuclear reactors have been predicted in 1956 by Paul Kuroda. In 1972, the first such example was found. In Oklo, in the African country of Gabon, there is a now famous geological formation where uranium is so common that two thousand million years ago a *natural* nuclear reactor has formed spontaneously – albeit a small one, with an estimated power generation of 100 kW. It has been burning for over 150 000 years, during the time when the uranium 235 percentage was 3% or more, as required for chain reaction. (Nowadays, the uranium

235 content on Earth is 0.7%.) The water of a nearby river was periodically heated to steam during an estimated 30 minutes; then the reactor cooled down again for an estimated 2.5 hours, since water is necessary to moderate the neutrons and sustain the chain reaction. The system has been studied in great detail, from its geological history up to the statements it makes about the constancy of the 'laws' of nature. The studies showed that 2000 million years ago the mechanisms were the same as those used today.

* *

Nuclear reactors exist in many sizes. The largest are used in power plants and can produce over 1000 MW in electrical power; the smallest are used in satellites, and usually produce around 10 kW. All work without refuelling for between one and thirty years.

* *

In contrast to massive particles, massless particles cannot decay at all. There is a simple reason for it: massless particles do not experience time, as their paths are 'null'. A particle
Challenge 125 s   that does not experience time cannot have a half-life. (Can you find another argument?)

* *

High energy radiation is dangerous to humans. In the 1950s, when nuclear tests were still made above ground by the large armies in the world, the generals overruled the orders of the medical doctors. They positioned many soldiers nearby to watch the explosion, and worse, even ordered them to walk to the explosion site as soon as possible after the explosion. One does not need to comment on the orders of these generals. Several of these unlucky soldiers made a strange observation: during the flash of the explosion,
Challenge 126 s   they were able to see the bones in their own hand and arms. How can this be?

* *

In 1958, six nuclear bombs were made to explode in the stratosphere by a vast group of criminals. A competing criminal group performed similar experiments in 1961, followed by even more explosions by both groups in 1962. (For reports and films, see en.wikipedia. org/wiki/High_altitude_nuclear_explosion.) As a result of most of these explosions, an *artificial aurora* was triggered the night following each of them. In addition, the electromagnetic pulse from the blasts destroyed satellites, destroyed electronics on Earth, disturbed radio communications, injured people on the surface of the Earth, caused problems with power plants, and distributed large amounts of radioactive material over the Earth – during at least 14 years following the blasts. The van Allen radiation belts around the Earth were strongly affected; it is expected that the lower van Allen belt will recover from the blasts only in a few hundred years. Fortunately for the human race, after 1962, this activity was stopped by international treaties.

* *

Nuclear bombs are terrible weapons. To experience their violence but also the criminal
Ref. 160   actions of many military people during the tests, have a look at the pictures of explosions. In the 1950 and 60s, nuclear tests were performed by generals who refused to listen to doctors and scientists. Generals ordered to explode these weapons in the air, making the complete atmosphere of the world radioactive, hurting all mankind in doing so; worse,

they even obliged soldiers to visit the radioactive explosion site a few minutes after the explosion, thus doing their best to let their own soldiers die from cancer and leukaemia. Generals are people to avoid.

∗ ∗

Several radioactive dating methods are used to date wine, and more are in development. A few are included in Table 12.

Page 168

∗ ∗

A few rare radioactive decay times can be changed by external influence. Electron capture, as observed in beryllium-7, is one of the rare examples were the decay time can change, by up to 1.5%, depending on the chemical environment. The decay time for the same isotope has also been found to change by a fraction of a per cent under pressures of 27 GPa. On the other hand, these effects are predicted (and measured) to be negligible for nuclei of larger mass. A few additional nuclides show similar, but smaller effects.

Ref. 150

The most interesting effect on nuclei is laser-induced fissioning of ^{238}U, which occurs for very high laser intensities.

Page 237

∗ ∗

Both γ radiation and neutron radiation can be used to image objects without destroying them. γ rays have been used to image the interior of the Tutankhamun mask. Neutron radiation, which penetrates metals as easily as other materials, has been used to image, even at film speed, the processes inside car engines.

∗ ∗

γ rays are used in Asia to irradiate potatoes, in order to prevent sprouting. In fact, over 30 countries allowed the food industry to irradiate food. For example, almost all spice in the world are treated with γ rays, to increase their shelf life. However, the consumer is never informed about such treatments.

∗ ∗

The non-radioactive isotopes ^{2}H – often written simply D – and ^{18}O can be used for measuring energy production in humans in an easy way. Give a person a glass of doubly labelled water to drink and collect his urine samples for a few weeks. Using a mass spectrometer one can determine his energy consumption. Why? Doubly labelled water ^{2}H$_2$^{18}O is processed by the body in three main ways. The oxygen isotope is expired as C^{18}O$_2$ or eliminated as H$_2$^{18}O; the hydrogen isotope is eliminated as ^{2}H$_2$O. Measurements on the urine allow one to determine carbon dioxide production, therefore to determine how much has food been metabolized, and thus to determine energy production.

Human energy consumption is usually given in joule per day. Measurements showed that high altitude climbers with 20 000 kJ/d and bicycle riders with up to 30 000 kJ/d are the most extreme sportsmen. Average humans produce 6 000 kJ/d.

∗ ∗

The percentage of the ^{18}O isotope in the water of the Earth's oceans can be used to deduce

Vol. I, page 313 where the water came from. This was told in the first volume of our adventure.

* *

Many nuclei oscillate in shape. The calculation of these shape oscillations is a research subject in itself. For example, when a spherical nucleus oscillates, it can do so in three mutually orthogonal axes. A spherical nucleus, when oscillating at small amplitudes, thus behaves like a three-dimensional harmonic oscillator. Interestingly, the symmetry of the three-dimensional harmonic oscillator is SU(3), the same symmetry that characterizes the strong nuclear interaction. However, the two symmetries are unrelated – at least fol-
Vol. VI, page 255 lowing present knowledge. A relation might appear in the future, though.

SUMMARY ON NUCLEI

Atomic nuclei are composed of protons and neutrons. Their diameter is between one and a few femtometres, and they have angular momentum. Their angular momentum, if larger than zero, allows us to produce magnetic resonance images. Nuclei can be spherical or ellipsoidal, they can be excited to higher energy states, and they can oscillate in shape. Nuclei have colours that are determined by their spectra. Nuclei can decay, can scatter, can break up and can react among each other. Nuclear reactions can be used to make bombs, power plants, generate biological mutations and to explore the human body. And as we will discover in the following, nuclear reactions are at the basis of the working of the Sun and of our own existence.

CHAPTER 6

THE SUN, THE STARS AND THE BIRTH OF MATTER

> " Lernen ist Vorfreude auf sich selbst.*
> Peter Sloterdijk "

NUCLEAR physics is the most violent part of physics. But despite this bad image, nuclear physics has also something fascinating to offer: by exploring nuclei, we learn to understand the Sun, the stars, the early universe, the birth of matter and our own history.

Ref. 162

Nuclei consist of protons and neutrons. Since protons are positively charged, they repel each other. Inside nuclei, protons must be bound by a force strong enough to keep them together against their electromagnetic repulsion. This is the *strong nuclear interaction*; it is needed to avoid that nuclei explode. The strong nuclear interaction is the strongest of the four interactions in nature; nevertheless, we do not experience it in everyday life, because its range is limited to distances of a few femtometres, i.e., to a few proton diameters. Despite this limitation, the strong interaction tells a good story about the burning of the Sun and about the flesh and blood we are made of.

THE SUN

At present, the Sun emits 385 YW of light. Where does this energy come from? If it came from burning coal, the Sun would stop burning after a few thousands of years. When radioactivity was discovered, researchers tested the possibility that this process was at the heart of the Sun's shining. However, even though radioactivity – or the process of fission that was discovered later – can produce more energy than chemical burning, the composition of the Sun – mostly hydrogen and helium – makes this impossible.

The origin of the energy radiated by the Sun was settled in 1939 by Hans Bethe: the Sun burns by *hydrogen fusion*. Fusion is the composition of a large nucleus from smaller ones. In the Sun, the central fusion reaction

Ref. 163

$$4\,^{1}\text{H} \rightarrow {}^{4}\text{He} + 2\,e^{+} + 2\,\nu + 4.4\,\text{pJ} \tag{58}$$

converts hydrogen nuclei into helium nuclei. The reaction is called the *hydrogen–hydrogen cycle* or *p–p cycle*. The hydrogen cycle is the result of a continuous cycle of

---

* 'Learning is anticipated joy about yourself.'

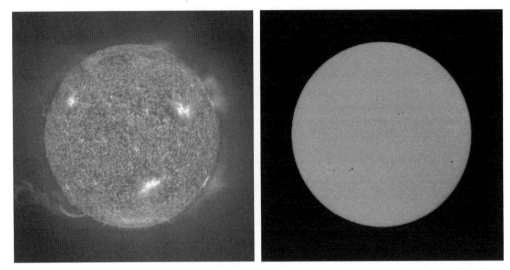

FIGURE 109 Photographs of the Sun at wavelengths of 30.4 nm (in the extreme ultraviolet, left) and around 677 nm (visible light, right, at a different date), by the SOHO mission (ESA and NASA).

three separate nuclear reactions:

$$p + p \rightarrow d + e^+ + v \quad \text{(a weak nuclear reaction)}$$
$$d + p \rightarrow {}^3He + \gamma \quad \text{(a strong nuclear reaction)}$$
$${}^3He + {}^3He \rightarrow \alpha + 2\,p + \gamma \,. \tag{59}$$

We can also write the p-p cycle as

$${}^1H + {}^1H \rightarrow {}^2H + e^+ + v \quad \text{(a weak nuclear reaction)}$$
$${}^2H + {}^1H \rightarrow {}^3He + \gamma \quad \text{(a strong nuclear reaction)}$$
$${}^3He + {}^3He \rightarrow {}^4He + 2\,{}^1H + \gamma \,. \tag{60}$$

In total, four protons are thus fused to one helium nucleus, or alpha particle; if we include the electrons, we can say that four hydrogen atoms are fused to one helium atom. The fusion process emits neutrinos and light with a total energy of 4.4 pJ (26.7 MeV). Most of the energy is emitted as light; around 10 % is carried away by neutrinos. This is the energy that makes the Sun shine.

The first of the three reactions of equation (59) is due to the weak nuclear interaction. This transmutation and the normal β decay have the same first-order Feynman diagram.

Challenge 127 e The weak interaction avoids that fusion happens too rapidly and ensures that the Sun will shine still for some time. Indeed, in the Sun, with a luminosity of 385 YW, there are Ref. 164 thus only 10^{38} fusions per second. This allows us to deduce that the Sun will last another handful of Ga (Gigayears) before it runs out of fuel.

The simplicity of the hydrogen-hydrogen cycle does not fully purvey the fascination of the process. On average, protons in the Sun's centre move with 600 km/s. Only if they hit each other precisely head-on can a nuclear reaction occur; in all other cases, the electro-

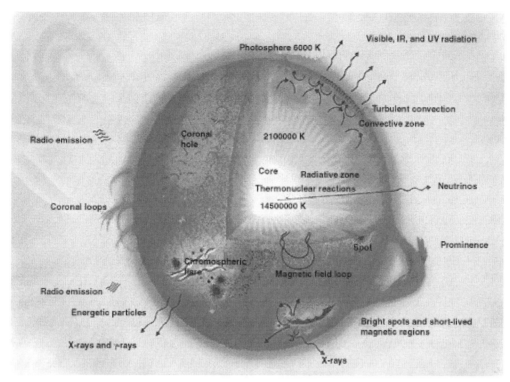

FIGURE 110 The interior of the Sun (courtesy NASA).

static repulsion between the protons keeps them apart. For an average proton, a head-on collision happens once every 7 thousand million years. Nevertheless, there are so many proton collisions in the Sun that every second four million tons of hydrogen are burned to helium.

The fusion reaction (59) takes place in the centre of the Sun, in the so-called *core*. Fortunately for us, the high energy γ photons generated in the Sun's centre are 'slowed' down by the outer layers of the Sun, namely the *radiaton zone*, the *convection zone* with its involved internal motion, and the so-called *photosphere*, the thin layer we actually see. (The last layer, the atmosphere,* is not visible during a day, but only during an eclipse.) In this slowing-down process, the γ photons are progressively converted to visible photons, mainly through scattering. Scattering takes time. Indeed, the sunlight of today was in fact generated at the time of the Neandertalers: a typical estimate is about 170 000 years Ref. 165 ago. In other words, the average effective speed of light inside the Sun is estimated to be around 300 km/year! After these one hundred and seventy thousand years, the photons take another 8.3 minutes to reach the Earth and to sustain the life of all plants and animals.

---

* The solar atmosphere consists of the *temperature minimum*, the *chromosphere*, the *transition region*, the *corona* and the *heliosphere*.

FIGURE 111 The evolution of a solar flare observed by the TRACE satellite (QuickTime film courtesy NASA).

MOTION IN AND ON THE SUN

Challenge 282 n

In its core, the Sun has a temperature of around 15 MK. At its surface, the temperature is around 5.8 kK. (Why is it cooler?) Since the Sun is cooler on its surface than in its centre, the Sun is not a homogeneous ball, but an inhomogeneous structure. The basic inhomogeneity is due to the induced convection processes. The convection, together with the rotation of the Sun around its axis, leads to fascinating structures shown in Figure 109: eruptions, including flares and coronal mass ejections, and solar spots.

In short, the Sun is not a static object. An impressive way to experience the violent processes it contains is to watch the film shown in Figure 111, which shows the evolution of a so-called *solar flare*.

When the Sun erupts, as shown in the lower left corner in Figure 109 or in Figure 112, matter is ejected far into space. When this matter reaches the Earth,* after being diluted by the journey, it affects our environment on Earth. Such *solar storms* can deplete the higher atmosphere and can thus possibly trigger usual Earth storms. Other effects of solar storms are the formation of auroras and the loss of orientation of birds during their migration; this happens during exceptionally strong solar storms, because the magnetic field of the Earth is disturbed in these situations. A famous effect of a solar storm was the loss of electricity in large parts of Canada in March of 1989. The flow of charged solar particles triggered large induced currents in the power lines, blew fuses and destroyed parts of the network, shutting down the power system. Millions of Canadians had no electricity, and in the most remote places it took two weeks to restore the electricity supply. Due to the coldness of the winter and a train accident resulting from the power loss, over 80 people died. In the meantime, the power network has been redesigned to withstand such events.

---

* It might even be that the planets affect the solar wind; the issue is not settled and is still under study.

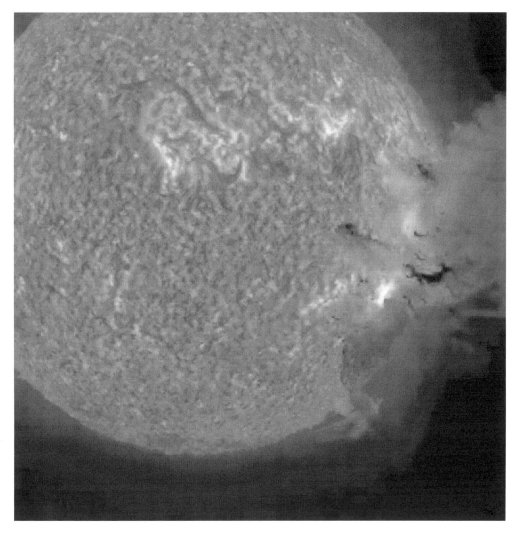

FIGURE 112 A spectacular coronal mass ejection observed on June 7, 2011 by the Solar Dynamic Observatory satellite (QuickTime film courtesy NASA).

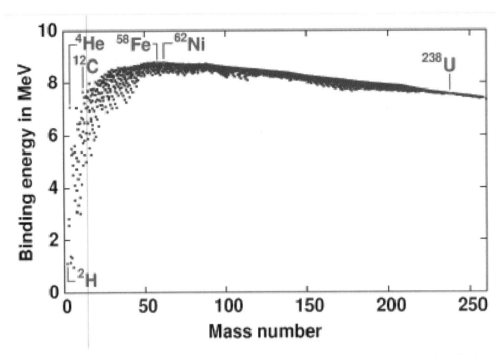

FIGURE 113 Measured values of the binding energy *per nucleon* in nuclei. The region on the left of the maximum, located at ^{58}Fe, is the region where *fusion* is possible; the right region is where *fission* is possible.

How can the Sun's surface have a temperature of 6 kK, whereas the Sun's *corona* – the thin gas emanating from and surrounding the Sun that is visible during a total solar eclipse – reaches two million Kelvin? In the latter part of the twentieth century it was shown, using satellites, that the magnetic field of the Sun is the cause; through the violent flows in the Sun's matter, magnetic energy is transferred to the corona in those places were flux tubes form knots, above the bright spots in the left of Figure 109 or above the dark spots in the right photograph. As a result, the particles of the corona are accelerated and heat the whole corona.

WHY DO THE STARS SHINE?

> Don't the stars shine beautifully? I am the only person in the world who knows why they do.
> Friedrich (Fritz) Houtermans

All stars shine because of fusion. When two light nuclei are fused to a heavier one, some energy is set free, as the average nucleon is bound more strongly. This energy gain is possible until nuclei of iron ^{56}Fe are produced. For nuclei beyond this nucleus, as shown in Figure 113, the binding energies per nucleon then decrease again; thus fusion is not energetically possible. It truns out that the heavier nuclei found on Earth and across the universe were formed through *neutron capture*. In short, nuclei below iron are made through fusion, nuclei above iron are made through neutron capture. And for the same reason, nuclei release energy through *fusion* when the result is lighter than iron, and release energy through *fission* when the starting point is above iron.

The different stars observed in the sky* can be distinguished by the type of fusion nuclear reaction that dominates them. Most stars, in particular young or light stars, run hydrogen fusion. But that is not all. There are several types of hydrogen fusion: the direct hydrogen–hydrogen (p–p) cycle, as found in the Sun and in many other stars, and the various CNO cycle(s) or Bethe-Weizsäcker cycle(s).

The hydrogen cycle was described above and can be summarized as Page 183

$$4\ {}^{1}\text{H} \rightarrow {}^{4}\text{He} + 2\,e^{+} + 2\,\nu + 4.4\,\text{pJ} \tag{61}$$

or, equivalently,

$$4\,\text{p} \rightarrow \alpha + 2\,e^{+} + 2\,\nu + 26.7\,\text{MeV} \,. \tag{62}$$

But this is not the only way for a star to burn. If a star has heavier elements inside it, the hydrogen fusion uses these elements as catalysts. This happens through the so-called Bethe-Weizsäcker cycle or CNO cycle, which runs as

$$
\begin{aligned}
{}^{12}\text{C} + {}^{1}\text{H} &\rightarrow {}^{13}\text{N} + \gamma \\
{}^{13}\text{N} &\rightarrow {}^{13}\text{C} + e^{+} + \nu \\
{}^{13}\text{C} + {}^{1}\text{H} &\rightarrow {}^{14}\text{N} + \gamma \\
{}^{14}\text{N} + {}^{1}\text{H} &\rightarrow {}^{15}\text{O} + \gamma \\
{}^{15}\text{O} &\rightarrow {}^{15}\text{N} + e^{+} + \nu \\
{}^{15}\text{N} + {}^{1}\text{H} &\rightarrow {}^{12}\text{C} + {}^{4}\text{He}
\end{aligned}
\tag{63}
$$

The end result of the cycle is the *same* as that of the hydrogen cycle, both in nuclei and in energy. The Bethe-Weizsäcker cycle is faster than hydrogen fusion, but requires higher temperatures, as the protons must overcome a higher energy barrier before reacting with carbon or nitrogen than when they react with another proton. (Why?) Inside the Sun, Challenge 129 s due to the comparatively low temperature of a few tens of million kelvin, the Bethe-Weizsäcker cycle (and its variations) is not as important as the hydrogen cycle.

The proton cycle and the Bethe-Weizsäcker cycle are not the only options for the burning of stars. Heavier and older stars than the Sun can also shine through other fusion reactions. In particular, when no hydrogen is available any more, stars run *helium burning*:

$$3\ {}^{4}\text{He} \rightarrow {}^{12}\text{C} \,. \tag{64}$$

This fusion reaction, also called the *triple-α process*, is of low probability, since it depends on three particles being at the same point in space at the same time. In addition, small amounts of carbon disappear rapidly via the reaction $\alpha + {}^{12}\text{C} \rightarrow {}^{16}\text{O}$. Nevertheless, since ${}^{8}\text{Be}$ is unstable, the reaction with 3 alpha particles is the only way for the universe to produce carbon. All these negative odds are countered only by one feature: carbon has an

---

* To find out which stars are in the sky above you at present, see the www.surveyor.in-berlin.de/himmel website.

excited state at 7.65 MeV, which is 0.3 MeV above the sum of the alpha particle masses; the excited state *resonantly enhances* the low probability of the three particle reaction. Only in this way the universe is able to produce the atoms necessary for apes, pigs and people. The prediction of this resonance by Fred Hoyle is one of the few predictions in physics made from the simple experimental observation that humans exist. The story Vol. III, page 282 has lead to a huge outflow of metaphysical speculations, most of which are unworthy of being even mentioned.

The studies of star burning processes also explain why the Sun and the stars do not collapse. In fact, the Sun and most stars are balls of hot gas, and the *gas pressure* due to the high temperature of its constituents prevents their concentration into a small volume. For other types of stars, the *radiation pressure* of the emitted photons prevents collapse; for still other stars, such as neutron stars, the role is taken by the *Pauli pressure*.

The nuclear reaction rates at the interior of a star are extremely sensitive to its temperature T. The carbon cycle reaction rate is proportional to between T^{13} for hot massive O stars and T^{20} for stars like the Sun. In red giants and supergiants, the triple-α reaction rate is proportional to T^{40}; these strong dependencies imply that stars usually shine with constancy over long times, often thousands and millions of years, because any change in temperature would be damped by a very efficient feedback mechanism. Of course, there are exceptions: variable stars get brighter and darker with periods of a few days; some stars change in brightness every few years. And even the Sun shows such effects. In the 1960s, it was discovered that the Sun pulsates with a frequency of 5 minutes. The amplitude is small, only 3 kilometres out of 1.4 million; nevertheless, it is measurable. In the meantime, helioseismologists have discovered numerous additional oscillations of the Sun, and in 1993, even on other stars. Such oscillations allow studying what is happening inside stars, even separately in each of the layers they consist of.

By the way, it is still not clear how much the radiation of the Sun changes over long time scales. There is an 11 year periodicity, the famous *solar cycle*, but the long term trend is still unknown. Precise measurements cover only the years from 1978 onwards, which makes only about 3 cycles. A possible variation of the intensity of the Suns, the so-called *solar constant* might have important consequences for climate research; however, the issue is still open.

WHY ARE FUSION REACTORS NOT COMMON YET?

Across the world, for over 50 years, a large number of physicists and engineers have tried to build fusion reactors. Fusion reactors try to copy the mechanism of energy release used by the Sun. The first machine that realized macroscopic energy production through fusion was, in 1991, the Joint European Torus* (JET for short) located in Culham in the United Kingdom. Despite this success, the produced power was still somewhat smaller than the power needed for heating.

Ref. 166   The idea of JET is to produce an extremely hot plasma that is as dense as possible. At high enough temperature and density, fusion takes place; the energy is released as a particle flux that is transformed (like in a fission reactor) into heat and then into electricity. To achieve ignition, JET used the fusion between deuterium and tritium, because this

---

* See www.jet.edfa.org.

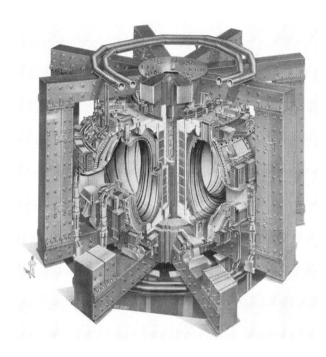

FIGURE 114 A simplified drawing of the Joint European Torus in operation at Culham, showing the large toroidal chamber and the magnets for the plasma confinement (© EFDA-JET).

reaction has the largest cross section and energy gain:

$$D + T \rightarrow He^4 + n + 17.6\,MeV\ . \tag{65}$$

Because tritium is radioactive, most research experiments are performed with the much less efficient deuterium–deuterium reactions, which have a lower cross section and a lower energy gain:

$$D + D \rightarrow T + H + 4\,MeV$$
$$D + D \rightarrow He^3 + n + 3.3\,MeV\ . \tag{66}$$

Fusion takes place when deuterium and tritium (or deuterium) collide at high energy. The high energy is necessary to overcome the electrostatic repulsion of the nuclei. In other words, the material has to be hot. To release energy from deuterium and tritium, one therefore first needs energy to heat it up. This is akin to the ignition of wood: in order to use wood as a fuel, one first has to heat it with a match.

Following the so-called *Lawson criterion*, rediscovered in 1957 by the English engineer John Lawson, after its discovery by Russian researchers, a fusion reaction releases energy only if the triple product of density n, reaction (or containment) time τ and temperature T exceeds a certain value. Nowadays this criterion is written as

Ref. 167

$$n\tau T > 3 \cdot 10^{28}\,s\,K/m^3\ . \tag{67}$$

In order to realize the Lawson criterion, JET uses temperatures of 100 to 200 MK, particle densities of 2 to $3 \cdot 10^{20}\,\mathrm{m}^{-3}$, and confinement times of 1 s. The temperature in JET is thus much higher than the 15 MK at the centre of the Sun, because the densities and the confinement times are much lower.

Matter at these temperatures is in form of a *plasma*: nuclei and electrons are completely separated. Obviously, it is impossible to pour a 100 MK plasma into a container: the walls would instantaneously evaporate. The only option is to make the plasma float in a vacuum, and to avoid that the plasma touches the container wall. The main challenge of fusion research in the past has been to find a way to keep a hot gas mixture of deuterium and tritium suspended in a chamber so that the gas never touches the chamber walls. The best way is to suspend the gas using a magnetic field. This works because in the fusion plasma, charges are separated, so that they react to magnetic fields. The most successful geometric arrangement was invented by the famous Russian physicists Igor Tamm and Andrei Sakharov: the *tokamak*. Of the numerous tokamaks around the world, JET is the largest and most successful. Its concrete realization is shown in Figure 114. JET manages to keep the plasma from touching the walls for about a second; then the situation becomes unstable: the plasma touches the wall and is absorbed there. After such a *disruption*, the cycle consisting of gas injection, plasma heating and fusion has to be restarted. As mentioned, JET has already achieved ignition, that is the state were more energy is released than is added for plasma heating. However, so far, no sustained commercial energy production is planned or possible, because JET has no attached electrical power generator.

The successor project, *ITER*, an international tokamak built with European, Japanese, US-American and Russian funding, aims to pave the way for commercial energy generation. Its linear reactor size will be twice that of JET; more importantly, ITER plans to achieve 30 s containment time. ITER will use superconducting magnets, so that it will have extremely cold matter at 4 K only a few metres from extremely hot matter at 100 MK. In other words, ITER will be a high point of engineering. The facility is being built in Cadarache in France. Due to its lack of economic sense, ITER has a good chance to be a modern version of the tower of Babylon; but maybe one day, it will start operation.

Like many large projects, fusion started with a dream: scientists spread the idea that fusion energy is safe, clean and inexhaustible. These three statements are still found on every fusion website across the world. In particular, it is stated that fusion reactors are not dangerous, produce much lower radioactive contamination than fission reactors, and use water as basic fuel. 'Solar fusion energy would be as clean, safe and limitless as the Sun.' In reality, the only reason that we do not feel the radioactivity of the Sun is that we are far away from it. Fusion reactors, like the Sun, are highly radioactive. The management of radioactive fusion reactors is much more complex than the management of radioactive fission reactors.

It is true that fusion fuels are almost inexhaustible: deuterium is extracted from water and the tritium – a short-lived radioactive element not found in nature in large quantities – is produced from *lithium*. The lithium must be enriched, but since material is not radioactive, this is not problematic. However, the production of tritium from lithium is a dirty process that produces large amounts of radioactivity. Fusion energy is thus inexhaustible, but not safe and clean.

In summary, of all technical projects ever started by mankind, fusion is by far the most challenging and ambitious. Whether fusion will ever be successful – or whether it ever should be successful – is another issue.

WHERE DO OUR ATOMS COME FROM?

People consist of electrons and various nuclei. Where did the nucleosynthesis take place?

About three minutes after the big bang, when temperature was around 0.1 MeV, protons and neutrons formed. About seven times as many protons as neutrons were formed, mainly due to their mass difference. Due to the high densities, the neutrons were captured, through the intermediate step of deuterium nuclei, in α particles. The process stopped around 20 minutes after the big bang, when temperatures became too low to allow fusion. After these seventeen minutes, the mass of the universe was split into 75% of hydrogen, 25% of helium, and traces of deuterium, lithium and beryllium. This process is called *primordial nucleosynthesis*. No heavier elements were formed, because the temperature fall prevented their accumulation in measureable quantities, and because there are stable nuclei with 5 or 8 nucleons.

Simulations of primordial nucleoynthesis agree well with the element abundances found in extremely distant, thus extremely old stars. The abundances are deduced from the spectra of these stars. In short, *hydrogen, helium, lithium and beryllium nuclei are formed shortly after ('during') the big bang.* These are the so-called *primordial elements.* Vol. II, page 227

All other nuclei are formed millions or billions of years after the big bang. In particular, *other light nuclei are formed in stars.* Young stars run hydrogen burning or helium burning; heavier and older stars run neon-burning or even silicon-burning. These latter processes require high temperatures and pressures, which are found only in stars with a mass at least eight times that of the Sun. All these fusion processes are limited by photodissociation and thus will only lead to nuclei up to ^{56}Fe. Ref. 168

Nuclei heavier than iron can only be made by neutron capture. There are *two* main neutron capture processes. The first process is the so-called *s-process* – for 'slow'. The process occurs inside stars, and gradually builds up heavy elements – including the most heavy stable neucleus, lead – from neutrons flying around. The second neutron capture processes is the rapid *r-process*. For a long time, its was unknown where it occured; for many decades, it was thought that it takes place in stellar explosions. Recent research points to *neutron star mergers* as the more likely place for the r-process. Such collisions emit material into space. The high neutron flux produces heavy elements. For example, it seems that most *gold* nuclei are synthesized in this way. The abundances of the heavy elements in the solar system can be measured with precision, a shown in Figure 116. These data points correspond well with what is expected from the material synthesized by neutron star mergers, the most likely candidate for the r-process at present, as illustrated by Figure 117. Ref. 170

In summary, the electron and protons in our body were made during the big bang; the lighter nuclei, such as carbon or oxygen, in stars; the heavier nuclei in neutron star mergers. But how did those nuclei get onto the Earth?

At a certain stage of their life, many stars explode. An exploding star is called a *supernova*. Such a supernova has an important effect: it distributes most of the matter of the star, such as carbon, nitrogen or oxygen, into space. This happens mostly in the form of neutral atoms. (Some elements are also synthesized during the explosion.) Exploding supernovae are thus essential for distributing material into space.

FIGURE 115 Two examples of how exploding stars shoot matter into interstellar space: the Crab nebula M1 and the Dumbbell nebula M27 (courtesy NASA and ESA, © Bill Snyder).

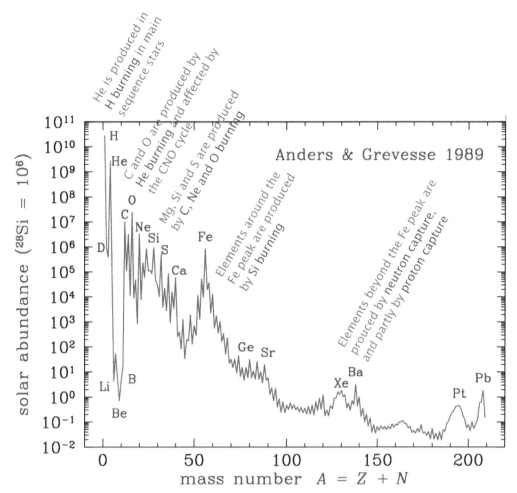

FIGURE 116 The measured nuclide abundances in the solar system and their main production processes (from Ref. 169).

The Sun is a second generation star – as so-called 'population I' star. and the solar system formed from the remnants of a supernova, as did, somewhat later, life on Earth. We all are made of recycled atoms.

We are recycled stardust. This is the short summary of the extended study by astrophysicists of all the types of stars found in the universe, including their birth, growth, mergers and explosions. The exploration of how stars evolve and then move in galaxies is a fascinating research field, and many aspects are still unknown.

Vol. II, page 196

CURIOSITIES ABOUT THE SUN AND THE STARS

What would happen if the Sun suddenly stopped shining? Obviously, temperatures would fall by several tens of degrees within a few hours. It would rain, and then all water would freeze. After four or five days, all animal life would stop. After a few weeks, the oceans would freeze; after a few months, air would liquefy. Fortunately, this will never

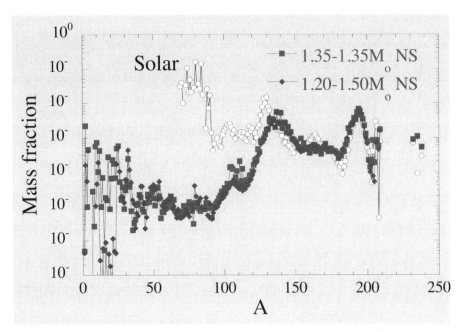

FIGURE 117 The comparison between measured nuclide abundances (dotted circles) in the solar system and the calculated values (red squares, blue diamonds) predicted by neutron star mergers (from reference Ref. 170).

happen.

* *

Not everything about the Sun is known. For example, the neutrino flux from the Sun oscillates with a period of 28.4 days. That is the same period with which the magnetic field of the Sun oscillates. The connections are still being studied.

* *

The Sun is a fusion reactor. But its effects are numerous. If the Sun were less brighter than it is, evolution would have taken a different course. We would not have eyelids, we would still have more hair, and would have a brighter skin. Can you find more examples?

Challenge 130 e

* *

Some stars shine like a police siren: their luminosity increases and decreases regularly. Such stars, called *Cepheids*, are important because their period depends on their average (absolute) brightness. Therefore, measuring their period and their brightness on Earth thus allows astronomers to determine their distance.

* *

By chance, the composition ratios between carbon, nitrogen and oxygen inside the Sun are the same as inside the human body.

* *

The first human-made hydrogen bomb explosion took place the Bikini atoll. Fortunately, none has ever been used on people.

But nature is much better at building bombs. The most powerful nuclear explosions take place on the surface of neutron stars in X-ray binaries. The matter falling into such a neutron star from the companion star, mostly hydrogen, will heat up until the temperature allows fusion. The resulting explosions can be observed in telescopes as light or X-ray flashes of about 10 s duration; the explosions are millions of times more powerful that those of human-made hydrogen bombs.

* *

Nucleosynthesis is mainly regulated by the strong interaction. However, if the electromagnetic interaction would be much stronger or much weaker, stars would either produce too little oxygen or too little carbon, and we would not exist. This famous argument is due to Fred Hoyle. Can you fill in the details?

Challenge 131 d

Summary on stars and nucleosynthesis

Stars and the Sun burn because of nuclear fusion. When stars have used up their nuclear fuel, they usually explode. In such an explosion, they distribute nuclei into space. Already in the distant past, such distributed nuclei recollected because of gravity and formed the Sun, the Earth and humans.

The energy liberated in nuclear fusion is due to the strong nuclear interaction. Nuclear processes also lead to nucleosynthesis. Nucleosynthesis during the big bang formed hydrogen and helium, nucleosynthesis in stars formed the light nuclei, and nucleosynthesis in neutron star mergers and supernovae formed the heavy nuclei.

CHAPTER 7

THE STRONG INTERACTION
– INSIDE NUCLEI AND NUCLEONS

Both radioactivity and medical images show that nuclei are composed systems. But quantum theory predicts even more: also protons and neutrons must be composed. There are two reasons: first, nucleons have a finite size, and second, their magnetic moments do not match the value predicted for point particles.

Ref. 171

The prediction of components inside protons was confirmed in the late 1960s when Kendall, Friedman and Taylor shot high energy electrons into hydrogen atoms. They found that a proton contains *three* constituents with spin 1/2. The experiment was able to 'see' the constituents through large angle scattering of electrons, in the same way that we see objects through large angle scattering of photons. These constituents correspond in number and (most) properties to the so-called *quarks* predicted in 1964 by George Zweig and also by Murray Gell-Mann.*

Ref. 172

Why are there three quarks inside a proton? And how do they interact? The answers are deep and fascinating.

THE FEEBLE SIDE OF THE STRONG INTERACTION

The mentioned deep inelastic scattering experiments show that the interaction keeping the protons together in a nucleus, which was first described by Yukawa Hideki,** is only

---

* The physicist George Zweig (b. 1937 Moscow) proposed the quark idea – he called them *aces* – in 1963, with more clarity than Gell-Mann. Zweig stressed the reality of aces, whereas Gell-Mann, in the beginning, did not believe in the existence of quarks. Zweig later moved on to a more difficult field: neurobiology.

Murray Gell-Mann (b. 1929 New York) received the Nobel Prize for physics in 1969. He is the originator of the term 'quark'. The term has two origins: officially, it is said to be taken from *Finnegans Wake*, a novel by James Joyce; in reality, Gell-Mann took it from a Yiddish and German term meaning 'lean soft cheese' and used figuratively in those languages to mean 'silly idea'.

Ref. 173

Gell-Mann was the central figure of particle physics in the 20th century; he introduced the concept of strangeness, the renormalization group, the flavour SU(3) symmetry and quantum chromodynamics itself. A disturbing story is that he took the idea, the data, the knowledge, the concepts and even the name of the V–A theory of the weak interaction from the bright physics student George Sudarshan and published it, together with Richard Feynman, as his own. The wrong attribution is still found in many textbooks.

Gell-Mann is also known for his constant battle with Feynman about who deserved to be called the most arrogant physicist of their university. A famous anecdote is the following. Newton's once used a common saying of his time in a letter to Hooke: 'If I have seen further than you and Descartes, it is by standing upon the shoulders of giants.' Gell-Mann is known for saying: 'If I have seen further than others, it is because I am surrounded by dwarfs.'

** Yukawa Hideki (b. 1907 Azabu, d. 1981 Kyoto), important Japanese physicist specialized in nuclear and particle physics. He founded the journal *Progress of Theoretical Physics* and together with his class mate

The proton

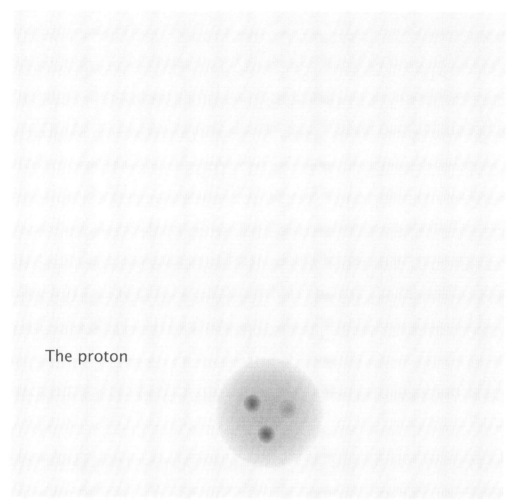

FIGURE 118 An illustration of deep inelastic electron scattering and of the resulting structure of the proton.

a feeble shadow of the interaction that keeps quarks together in a proton. Both interactions are called by the same name. The two cases correspond somewhat to the two cases of electromagnetism found in atomic matter. The clearest example is provided by neon atoms: the strongest and 'purest' aspect of electromagnetism is responsible for the attraction of the electrons to the neon nuclei; its feeble 'shadow', the Van-der-Waals interaction, is responsible for the attraction of neon atoms in liquid neon and for processes like its evaporation and condensation. Both attractions are electromagnetic, but the strengths differ markedly. Similarly, the strongest and 'purest' aspect of the strong interaction leads to the formation of the proton and the neutron through the binding of quarks; the feeble, 'shadow' aspect leads to the formation of nuclei and to α decay. Obviously, most information can be gathered by studying the strongest and 'purest' aspect.

Tomonaga Shin'ichiro, who also won the prize, he was an example to many scientists in Japan. Yukawa received the 1949 Nobel Prize for physics for his theory of mesons.

FIGURE 119 A typical experiment used to study the quark model: the Proton Synchroton at CERN in Geneva (© CERN).

BOUND MOTION, THE PARTICLE ZOO AND THE QUARK MODEL

Deep electron scattering showed that protons are made of interacting constituents. How can one study these constituents?

Physicists are simple people. To understand the constituents of matter, and of protons in particular, they had no better idea than to take all particles they could get hold of and to smash them into each other. Many researchers played this game for decades. Obviously, this is a facetious comment; in fact, quantum theory forbids any other method. Can you explain why?

Understanding the structure of particles by smashing them into each other is not simple. Imagine that you want to study how cars are built just by crashing them into each other. Before you get a list of all components, you must perform and study a non-negligible number of crashes. Most give the same result, and if you are looking for a particular part, you might have to wait for a long time. If the part is tightly attached to others, the crashes have to be especially energetic. In addition, the part most likely will be deformed. Compared to car crashes, quantum theory adds the possibility for debris to transform, to react, to bind and to get excited. Therefore the required diligence and patience is even greater for particle crashes than for car crashes. Despite these difficulties, for many decades, researchers have collected an ever increasing number of proton debris, also called *hadrons*. The list, a small part of which is given in Appendix B, is overwhelmingly long; the official full list, several hundred pages of fine print, is found at pdg.web. cern.ch and contains hundreds of hadrons. Hadrons come in two main types: integer spin hadrons are called *mesons*, half-integer spin hadrons are called *baryons*. The proton and the neutron themselves are thus baryons.

Then came the quark model. Using the ingenuity of many experimentalists and theoreticians, the quark model explained the whole meson and baryon catalogue as a con-

Ref. 174

Challenge 132 s

Page 317

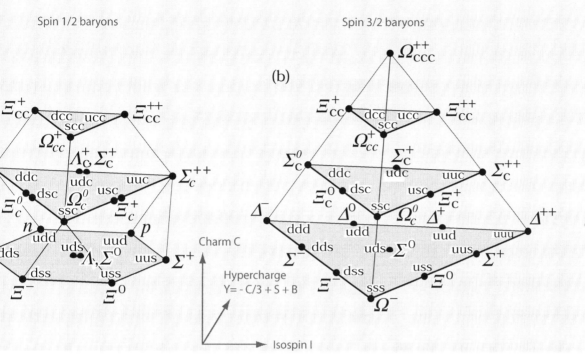

FIGURE 120 The family diagrams for the least massive baryons that can be built as qqq composites of the first four quark types (from Ref. 175).

sequence of only 6 types of bound quarks. Typically, a large part of the catalogue can be structured in graphs such as the ones given in Figure 121 and Figure 120. These graphs were the beginning of the end of high energy physics. The quark model explained all quantum numbers of the debris, and allowed understanding their mass ratios as well as their decays.

The quark model explained why debris come into two types: all *mesons* consist of a quark and an antiquark and thus have integer spin; all *baryons* consist of three quarks, and thus have half-integer spin. In particular, the proton and the neutron are seen as combinations of two quark types, called *up* (u) and *down* (d): the proton is a *uud* state, the neutron a *udd* state. The discovery of other hadrons lead to the addition of four additional types of quarks. The quark names are somewhat confusing: they are called *strange* (s), *charm* (c), *bottom* (b) – also called 'beauty' in the old days – and *top* (t) – called 'truth' in the past. The quark types are called *flavours*; in total, there are thus 6 quark flavours in nature.

All quarks have spin one half; they are fermions. Their electric charges are multiples of 1/3 of the electron charge. In addition, quarks carry a strong charge, called, again confusingly, *colour*. In contrast to electromagnetism, which has only positive, negative, and neutral charges, the strong interaction has red, blue, green quarks on one side, and anti-red, anti-blue and anti-green on the other. The neutral state is called 'white'. All baryons, including proton and neutrons, and all mesons are white, in the same way that all atoms are neutral.

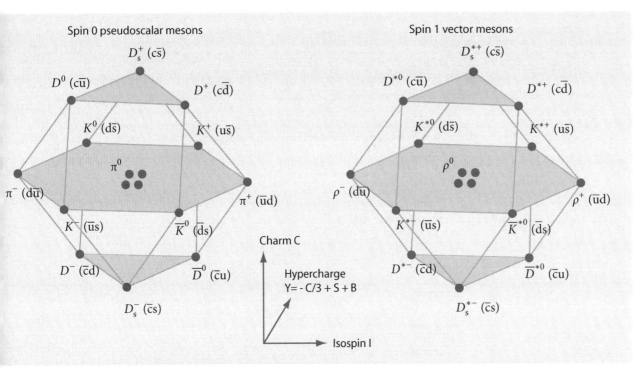

FIGURE 121 The family diagram for the least massive pseudoscalar and vector mesons that can be built as $q\bar{q}$ composites of the first four quark flavours.

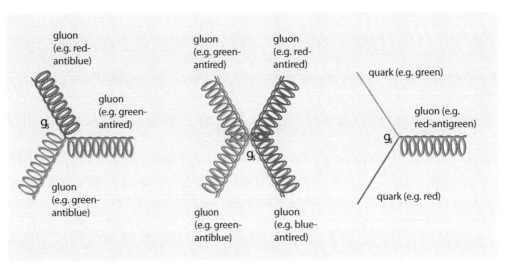

FIGURE 122 The essence of the QCD Lagrangian: the Feynman diagrams of the strong interaction.

THE ESSENCE OF QUANTUM CHROMODYNAMICS

The theory describing the bound states of quarks is called *quantum chromodynamics*, or

Ref. 176 QCD. It was formulated in its final form in 1973 by Fritzsch, Gell-Mann and Leutwyler. In the same way that in atoms, electrons and protons are held together by the exchange

TABLE 15 The quarks.

QUARK	MASS m (SEE TEXT)	SPIN J PARITY P	POSSIBLE COLOURS; POSSIBLE WEAK BE- HAVIOUR	CHARGE Q, ISOSPIN I, STRANGENESS S, CHARM C, BEAUTY B', TOPNESS T	LEPTON NUMBER L, BARYON NUMBER B
Down d	4.5 to 5.5 MeV/c^2	$\frac{1}{2}^+$	red, green, blue; singlet, doublet	$-\frac{1}{3}, -\frac{1}{2}, 0, 0, 0, 0$	$0, \frac{1}{3}$
Up u	1.8 to 3.0 MeV/c^2	$\frac{1}{2}^+$	red, green, blue; singlet, doublet	$+\frac{2}{3}, +\frac{1}{2}, 0, 0, 0, 0$	$0, \frac{1}{3}$
Strange s	95(5) MeV/c^2	$\frac{1}{2}^+$	red, green, blue; singlet, doublet	$-\frac{1}{3}, 0, -1, 0, 0, 0$	$0, \frac{1}{3}$
Charm c	1.275(25) GeV/c^2	$\frac{1}{2}^+$	red, green, blue; singlet, doublet	$+\frac{2}{3}, 0, 0, +1, 0, 0$	$0, \frac{1}{3}$
Bottom b	4.18(3) GeV/c^2	$\frac{1}{2}^+$	red, green, blue; singlet, doublet	$-\frac{1}{3}, 0, 0, 0, -1, 0$	$0, \frac{1}{3}$
Top t	173.5(1.4) GeV/c^2	$\frac{1}{2}^+$	red, green, blue; singlet, doublet	$+\frac{2}{3}, 0, 0, 0, 0, +1$	$0, \frac{1}{3}$

of virtual photons, in protons, quarks are held together by the exchange of virtual gluons. *Gluons* are the quanta of the strong interaction, and correspond to photons, the quanta of the electromagnetic interactions.

Quantum chromodynamics describes all motion due to the strong interaction with the three fundamental processes shown in Figure 122: two gluons can scatter, a gluon can emit or absorb another, and a quark can emit or absorb a gluon. In electrodynamics, only the last diagram is possible; in the strong interaction, the first two appear as well. Among others, the first two diagrams are responsible for the confinement of quarks, and thus for the lack of free quarks in nature. Ref. 177

QCD is a *gauge* theory: the fields of the strong interaction show gauge invariance under the Lie group SU(3). We recall that in the case of electrodynamics, the gauge group is U(1), and Abelian, or commutative. In contrast, SU(3) is non-Abelian; QCD is a non-Abelian gauge theory. Non-Abelian gauge theory was invented and popularized by Wolfgang Pauli. It is often incorrectly called *Yang–Mills theory* after the first two physicists who wrote down Pauli's ideas.

Due to the SU(3) gauge symmetry, there are 8 gluons; they are called red-antigreen, blue-antired, etc. Since SU(3) is non-Abelian, gluons interact among themselves, as shown in the first two processes in Figure 122. Out of the three combinations red-antired, blue-antiblue and green-antigreen, only two gluons are linearly independent, thus giving a total of $3^2 - 1 = 8$ gluons.

The coupling strength of the strong interaction, its three fundamental processes in Figure 122, together with its SU(3) gauge symmetry and the observed number of six quarks, completely determine the behaviour of the strong interaction. In particular, they completely determine its Lagrangian density.

THE LAGRANGIAN OF QUANTUM CHROMODYNAMICS*

The Lagrangian density of the strong interaction can be seen as a complicated formulation of the Feynman diagrams of Figure 122. Indeed, the Lagrangian density of quantum chromodynamics is

$$\mathcal{L}_{QCD} = -\frac{1}{4}F_{\mu\nu}^{(a)}F^{(a)\mu\nu} - c^2 \sum_q m_q \overline{\psi}_q^k \psi_{qk} + i\hbar c \sum_q \overline{\psi}_q^k \gamma^\mu (D_\mu)_{kl} \psi_q^l \qquad (68)$$

where the gluon field strength and the gauge covariant derivative are

$$F_{\mu\nu}^{(a)} = \partial_\mu A_\nu^a - \partial_\nu A_\mu^a + g_s f_{abc} A_\mu^b A_\nu^c$$

$$(D_\mu)_{kl} = \delta_{kl}\partial_\mu - i\frac{g_s}{2}\sum_a \lambda_{k,l}^a A_\mu^a .$$

Vol. I, page 243 We remember from the section on the principle of least action that Lagrangians are always sums of scalar products; this is clearly seen in expression (68). The index $a = 1 \ldots 8$ numbers the eight types of gluons and the index $k = 1, 2, 3$ numbers the three colours, all due to SU(3). The index $q = 1 \ldots 6$ numbers the six quark flavours. The fields $A_\mu^a(x)$ are the eight gluon fields, represented by the coiled lines in Figure 122. The fields $\psi_q^k(x)$ are those of the quarks of flavour q and colour k, represented by the straight line in the figure. The six times three quark fields, like those of any elementary fermion, are 4-component Dirac spinors with masses m_q.**

The Lagrangian (68) is that of a *local* field theory: observables are functions of position. In other words, QCD is similar to quantum electrodynamics and can be compared to experiment in the same way.

The first term of the Lagrangian (68) represents the kinetic energy of the radiation (the gluons), the second or mass term the kinetic energy of the matter particles (the quarks) and the third term the interaction between the two.

The mass term in the Lagrangian is the only term that spoils or *breaks flavour symmetry*, i.e., the symmetry under exchange of quark types. (In particle physics, this symmetry is also called *chiral symmetry*, for historical reasons.) Obviously, the mass term also breaks space-time conformal symmetry.

---

* This section can be skipped at first reading.

** In their simplest form, the matrices γ_μ can be written as

$$\gamma_0 = \begin{pmatrix} I & 0 \\ 0 & -I \end{pmatrix} \quad \text{and} \quad \gamma_n = \begin{pmatrix} 0 & \sigma^i \\ -\sigma^i & 0 \end{pmatrix} \quad \text{for } n = 1, 2, 3 \qquad (69)$$

Vol. IV, page 215   where the σ^i are the Pauli spin matrices.

The interaction term in the Lagrangian thus corresponds to the third diagram in Figure 122. The strength of the strong interaction is described by the *strong coupling constant* g_s. The constant is independent of flavour and colour, as observed in experiment. The Interaction term does not mix different quarks; as observed in experiments, flavour is conserved in the strong interaction, as is baryon number. The strong interaction also conserves spatial parity P and charge conjugation parity C. The strong interaction does not transform matter.

In QCD, the eight gluons are massless; also this property is taken from experiment. Therefore no gluon mass term appears in the Lagrangian. It is easy to see that massive gluons would spoil gauge invariance. As mentioned above, in contrast to electromagnetism, Challenge 133 ny where the gauge group U(1) is Abelian, the gauge group SU(3) of the strong interactions is non-Abelian. As a consequence, the colour field itself is charged, i.e., carries colour, and thus the index a appears on the fields A and F. As a result, gluons can interact with each other, in contrast to photons, which pass each other undisturbed. The first two diagrams of Figure 122 are thus reflected in the somewhat complicated definition of the field $F_{\mu\nu}^{(a)}$. In contrast to electrodynamics, the definition has an extra term that is *quadratic* in the fields A; it is described by the so-called structure constants f_{abc} and the interaction strength g_s. The numbers f_{abc} are the structure constants of the SU(3). Page 339

The behaviour of the gauge transformations and of the gluon field is described by the eight matrices $\lambda_{k,l}^a$. They are a fundamental, 3-dimensional representation of the *generators* of the SU(3) algebra and correspond to the eight gluon types. The matrices λ_a, $a = 1...8$, and the structure constants f_{abc} obey the relations

$$[\lambda_a, \lambda_b] = 2if_{abc}\lambda_c$$
$$\{\lambda_a, \lambda_b\} = 4/3\delta_{ab}I + 2d_{abc}\lambda_c \tag{70}$$

where I is the unit matrix. The structure constants f_{abc} of SU(3), which are odd under permutation of any pair of indices, and d_{abc}, which are even, have the values

abc	f_{abc}		abc	d_{abc}		abc	d_{abc}	
123	1		118	$1/\sqrt{3}$		355	1/2	
147	1/2		146	1/2		366	−1/2	
156	−1/2		157	1/2		377	−1/2	
246	1/2		228	$1/\sqrt{3}$		448	$−1/(2\sqrt{3})$	(71)
257	1/2		247	−1/2		558	$−1/(2\sqrt{3})$	
345	1/2		256	1/2		668	$−1/(2\sqrt{3})$	
367	−1/2		338	$1/\sqrt{3}$		778	$−1/(2\sqrt{3})$	
458	$\sqrt{3}/2$		344	1/2		888	$−1/\sqrt{3}$	
678	$\sqrt{3}/2$							

All other elements vanish. Physically, the structure constants of SU(3) describe the details of the interaction between quarks and gluons and of the interaction between the gluons themselves.

A fundamental 3-dimensional representation of the eight generators λ_a – corresponding to the eight gluon types – is given, for example, by the set of the *Gell-Mann matrices*

$$\lambda_1 = \begin{pmatrix} 0 & 1 & 0 \\ 1 & 0 & 0 \\ 0 & 0 & 0 \end{pmatrix} \lambda_2 = \begin{pmatrix} 0 & -i & 0 \\ i & 0 & 0 \\ 0 & 0 & 0 \end{pmatrix} \lambda_3 = \begin{pmatrix} 1 & 0 & 0 \\ 0 & -1 & 0 \\ 0 & 0 & 0 \end{pmatrix}$$

$$\lambda_4 = \begin{pmatrix} 0 & 0 & 1 \\ 0 & 0 & 0 \\ 1 & 0 & 0 \end{pmatrix} \lambda_5 = \begin{pmatrix} 0 & 0 & -i \\ 0 & 0 & 0 \\ i & 0 & 0 \end{pmatrix} \lambda_6 = \begin{pmatrix} 0 & 0 & 0 \\ 0 & 0 & 1 \\ 0 & 1 & 0 \end{pmatrix}$$

$$\lambda_7 = \begin{pmatrix} 0 & 0 & 0 \\ 0 & 0 & -i \\ 0 & i & 0 \end{pmatrix} \lambda_8 = \frac{1}{\sqrt{3}}\begin{pmatrix} 1 & 0 & 0 \\ 0 & 1 & 0 \\ 0 & 0 & -2 \end{pmatrix} . \tag{72}$$

There are eight matrices, one for each gluon type, with 3×3 elements, due to the 3 colours of the strong interaction. There is no ninth gluon, because that gluon would be colourless, or 'white'.

The Lagrangian is complete only when the 6 quark masses and the coupling constant g_s are included. These values, like the symmetry group SU(3), are not explained by QCD, of course.

Only quarks and gluons appear in the Lagrangian of QCD, because only quarks and gluons interact via the strong force. This can be also expressed by saying that only quarks and gluons carry colour; *colour* is the source of the strong force in the same way that electric charge is the source of the electromagnetic field. In the same way as electric charge, colour charge is *conserved* in all interactions. Electric charge comes in two types, positive and negative; in contrast, colour comes in three types, called *red*, *green* and *blue*. The neutral state, with no colour charge, is called *white*. Protons and neutrons, but also electrons or neutrinos, are thus 'white', thus *neutral* for the strong interaction.

In summary, the six quark types interact by exchanging eight gluon types. The interaction is described by the Feynman diagrams of Figure 122, or, equivalently, by the Lagrangian (68). Both descriptions follow from the requirements that the gauge group is SU(3) and that the masses and coupling constants are given. It was a huge amount of work to confirm that all experiments indeed agree with the QCD Lagrangian; various competing descriptions were discarded.

EXPERIMENTAL CONSEQUENCES OF THE QUARK MODEL

How can we pretend that quarks and gluons exist, even though they are never found alone? There are a number of arguments in favour.

∗ ∗

The quark model explains the non-vanishing magnetic moment of the neutron and explains the magnetic moments μ of the baryons. By describing the proton as a *uud* state and the neutron a *udd* state with no orbital angular momentum and using the precise wave functions, we get

Challenge 134 e

$$\mu_u = \tfrac{1}{5}(4\mu_p + \mu_n) \qquad \text{and} \qquad \mu_d = \tfrac{1}{5}(4\mu_n + \mu_p) . \tag{73}$$

Assuming that $m_u = m_d$ and that the quark magnetic moment is proportional to their charge, the quark model predicts a ratio of the magnetic moments of the proton and the neutron of

$$\frac{\mu_p}{\mu_n} = -\frac{3}{2} \, . \tag{74}$$

This prediction differs from measurements only by 3 %. Furthermore, using the same values for the magnetic moment of the quarks, magnetic moment values of over half a dozen of other baryons can be predicted. The results typically deviate from measurements only by around 10 %. In particular, the sign of the resulting baryon magnetic moment is always correctly calculated.

* *

The quark model describes all *quantum numbers* of mesons and baryons. P-parity, C-parity, and the absence of certain meson parities are all reproduced. The observed conservation of electric charge, baryon number, isospin, strangeness etc. is reproduced. Hadron family diagrams such as those shown in Figure 120 and in Figure 121 describe all existing hadron states (of lowest angular momentum) *completely*; the states not listed are not observed. The quark model thus produces a complete and correct classification of all hadrons as bound states of quarks.

* *

The quark model also explains the *mass spectrum* of hadrons. The best predictions are made by QCD lattice calculations. With months of computer time, researchers were able to reproduce the masses of proton and neutron to within a few per cent. Interestingly, Ref. 178 if one sets the *u* and *d* quark masses to zero, the resulting proton and neutron mass differ from experimental values only by 10 %. The mass of protons and neutrons is almost Ref. 179 completely due to the binding, not to the constituents. More details are given below. Page 212

* *

The number of colours of quarks must be taken into account to get correspondence of theory and calculation. For example, the measured decay time of the neutral pion is 83 as. The calculation without colour gives 750 as; if each quark is assumed to appear in 3 colours the value must be divided by 9, and then matches the measurement.

* *

In particle colliders, collisions of electrons and positrons sometimes lead to the production of hadrons. The calculated production rates also fit experiments only if quarks have three colours. In more detail, if one compares the ratio of muon–antimuon production and of hadron production, a simple estimate relates them to their charges: Challenge 135 s

$$R = \frac{\sum q_{\text{hadrons}}}{\sum q_{\text{muons}}} \tag{75}$$

Between 2 and 4 GeV, when only three quarks can appear, this argument thus predicts $R = 2$ if colours exist, or $R = 2/3$ if they don't. Experiments yield a value of $R = 2.2$, thus

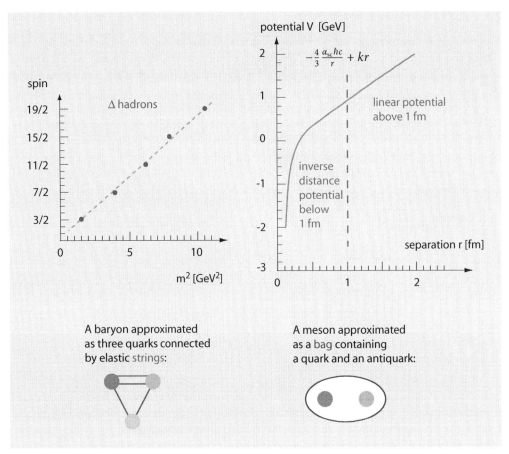

FIGURE 123 Top left: a Regge trajectory, or Chew–Frautschi plot, due to the confinement of quarks. Top right: the quark confinement potential. Bottom: two approximate ways to describe quark confinement: the string model and the bag model of hadrons.

confirming the number of colours. Many other such *branching ratios* can be calculated in this way. They agree with experiments only if the number of colours is three.

CONFINEMENT OF QUARKS – AND ELEPHANTS

Many of the observed hadrons are not part of the diagrams of Figure 120 and Figure 121; these additional hadrons can be explained as *rotational excitations* of the fundamental mesons from those diagrams. As shown by Tullio Regge in 1957, the idea of rotational excitations leads to quantitative predictions. Regge assumed that mesons and baryons are quarks connected by strings, like rubber bands – illustrated in Figure 123 and Figure 124 – and that the force or tension k between the quarks is thus constant over distance.

We assume that the strings, whose length we call $2r_0$, rotate around their centre of mass as rapidly as possible, as shown in Figure 124. Then we have

$$v(r) = c \frac{r}{r_0} \, . \tag{76}$$

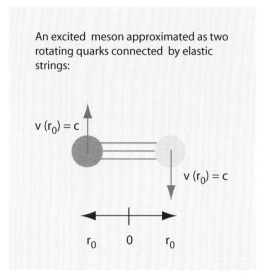

An excited meson approximated as two rotating quarks connected by elastic strings:

$v(r_0) = c$

$v(r_0) = c$

$r_0 \quad 0 \quad r_0$

FIGURE 124 Calculating masses of excited hadrons.

The quark masses are assumed negligible. For the total energy this implies the relation

$$E = c^2 m = 2 \int_0^{r_0} \frac{k}{\sqrt{1 - v(r)/c^2}} dr = kr_0 \pi \tag{77}$$

and for angular momentum the relation

$$J = \frac{2}{\hbar c^2} \int_0^{r_0} \frac{krv(r)}{\sqrt{1 - v(r)/c^2}} dr = \frac{kr_0^2}{2\hbar c} . \tag{78}$$

Including the spin of the quarks, we thus get

$$J = \alpha_0 + \alpha' m^2 \quad \text{where} \quad \alpha' = \frac{c^3}{2\pi k\hbar} . \tag{79}$$

Regge thus deduced a simple expression that relates the mass m of excited hadrons to their total spin J. For bizarre historical reasons, this relation is called a *Regge trajectory*.

The value of the constant α' is predicted to be independent of the quark–antiquark pairing. A few years later, as shown in Figure 123, such linear relations were found in experiments: the *Chew-Frautschi plots*. For example, the three lowest lying states of Δ are the spin 3/2 $\Delta(1232)$ with m^2 of 1.5 GeV^2, the spin 7/2 $\Delta(1950)$ with m^2 of 3.8 GeV^2, and the spin 11/2 $\Delta(2420)$ with m^2 of 5.9 GeV^2. The value of the constant α' is found experimentally to be around 0.93 GeV^{-2} for almost all mesons and baryons, whereas the value for α_0 varies from particle to particle. The quark string tension is thus found to be Ref. 177

$$k = 0.87 \, \text{GeV/fm} = 0.14 \, \text{MN} . \tag{80}$$

In other words, two quarks in a hadron attract each other with a force equal to the weight of two elephants: about 14 tons.

Experiments are thus clear: the observed Chew-Frautschi plots, as well as several other observations not discussed here, are best described by a quark–quark potential that grows, above 1 fm, *linearly* with distance. The slope of the linear potential, the force, has a value equal to the force with which the Earth attracts two elephants. As a result, quarks *never* appear as free particles: quarks are always *confined* in hadrons. This situation is in contrast with QED, where the force between charges goes to zero for large distances; electric charges are thus not confined, but can exist as free particles. At large distances, the electric potential *decreases* in the well-known way, with the inverse of the distance. In contrast, for the strong interaction, experiments lead to a quark potential given by

Ref. 174

$$V = -\frac{4}{3}\frac{\alpha_{sc}\hbar c}{r} + kr \qquad (81)$$

where k is the mentioned $0.87\,\text{GeV/fm}$, α_{sc} is 0.2, and $\hbar c$ is $0.1975\,\text{GeV/fm}$. The quark potential is illustrated in Figure 123.

Even though experiments are clear, theoreticians face a problem. So far, neither the quark-quark potential nor the quark bound states can be deduced from the QCD Lagrangian with a *simple* approximation method. Nevertheless, complicated non-perturbative calculations show that the QCD Lagrangian does predict a force between two coloured particles that levels off at a constant value (corresponding to a linearly increasing potential). These calculations show that the old empirical approximations of hadrons as quarks connected by strings or a quarks in bags, shown in Figure 123, can indeed be deduced from the QCD Lagrangian. However, the calculations are too complex to be summarized in a few lines. Independently, the constant force value has also been reproduced in computer calculations in which one simplifies space-time to a lattice and then approximates QCD by so-called *lattice QCD* or *lattice gauge theory*. Lattice calculations have further reproduced the masses of most mesons and baryons with reasonable accuracy. Using the most powerful computers available, these calculations have given predictions of the mass of the proton and other baryons within a few per cent. Discussing these complex and fascinating calculations lies outside the scope of this text, however.

Ref. 181

In fact, the challenge of explaining confinement in simple terms is so difficult that the brightest minds have been unable to solve it yet. This is not a surprise, as its solution probably requires the unification of the interactions and, most probably, also the unification with gravity. We therefore leave this issue for the last part of our adventure.

ASYMPTOTIC FREEDOM

QCD has another property that sets it apart form QED: the behaviour of its coupling with energy. In fact, there are three equivalent ways to describe the strong coupling strength. The first was is the quantity appearing in the QCD Lagrangian, g_s. The second way is often used to define the equivalent quantity $\alpha_s = g_s^2/4\pi$. Both α_s and g_s depend on the energy Q of the experiment. If they are known for one energy, they are known for all of them. Presently, the best experimental value is $\alpha_s(M_Z) = 0.1185 \pm 0.0010$.

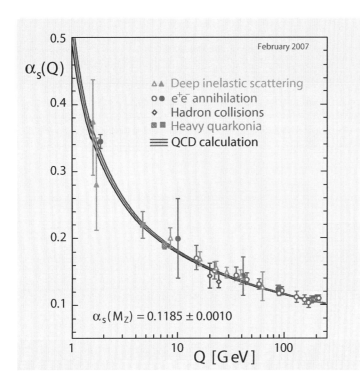

FIGURE 125 The measured and the calculated variation of the strong coupling with energy, showing the precision of the QCD Lagrangian and the asymptotic freedom of the strong interaction (© Siegfried Bethke, updated from Ref. 182).

The energy dependence of the strong coupling can be calculated with the standard renormalization procedures and is expected to be

Ref. 175, Ref. 177

$$\alpha_s(Q^2) = \frac{12\pi}{33 - 2n_f}\frac{1}{L}\left(1 - \frac{(918 - 114n_f)\ln L}{(33 - 2n_f)^2 L} + ...\right) \quad \text{where} \quad L = \ln\frac{Q^2}{\Lambda^2(n_f)} \quad (82)$$

where n_f is the number of quarks with mass below the energy scale Q, thus a number between 3 and 6. (The expression has been expanded to many additional terms with help of computer algebra.)

The third way to describe the strong coupling is thus the energy parameter $\Lambda(n_f)$. Experiments yield $\Lambda(3) = 230(60)$ GeV, $\Lambda(4) = 180(50)$ GeV and $\Lambda(5) = 120(30)$ GeV.

The accelerator experiments that measure the coupling are extremely involved, and hundreds of people across the world have worked for many years to gather the relevant data. The comparison of QCD and experiment, shown in Figure 125, does not show any contradiction between the two.

Figure 125 and expression (82) illustrate what is called *asymptotic freedom*: α_s decreases at high energies. In other words, at high energies quarks are *freed* from the strong interaction; they behave as free particles.* As a result of asymptotic freedom, in QCD, a perturbation expansion can be used only at energies much larger than Λ. Historically,

---

* Asymptotic freedom was discovered in 1972 by Gerard 't Hooft; since he had received the Nobel Prize already, the 2004 Prize was then given to the next people who highlighted it: David Gross, David Politzer and Frank Wilczek, who studied it extensively in 1973.

the discovery of asymptotic freedom was essential to establish QCD as a theory of the strong interaction.

Asymptotic freedom can be understood qualitatively if the situation is compared to QED. The electron coupling increases at small distances, because the screening due to the virtual electron-positron pairs has less and less effect. In QCD, the effective colour coupling also changes at small distances, due to the smaller number of virtual quark-antiquark pairs. However, the gluon properties lead to the opposite effect, an *antiscreening* that is even stronger: in total, the effective strong coupling decreases at small distances.

THE SIZES AND MASSES OF QUARKS

The size of quarks, like that of all elementary particles, is predicted to vanish by QCD, as in all quantum field theory. So far, no experiment has found any effect due to a finite quark size. Measurements show that quarks are surely smaller than 10^{-19} m. No size conjecture has been given by any hypothetical theory. Quarks are assumed point-like, or at most Planck-sized, in all descriptions so far.

We noted in several places that a neutral compound of charged particles is always less massive than its components. But if you look up the mass values for quarks in most tables, the masses of *u* and *d* quarks are only of the order of a few MeV/c^2, whereas the proton's mass is 938 MeV/c^2. What is the story here?

It turns out that the definition of the mass is more involved for quarks than for other particles. Quarks are never found as free particles, but only in bound states. As a result, the concept of quark mass depends on the theoretical framework one is using.

Due to *asymptotic freedom*, quarks behave almost like free particles only at high energies. The mass of such a 'free' quark is called the *current quark mass*; for the light quarks it is only a few MeV/c^2, as shown in Table 15.

At low energy, for example inside a proton, quarks are *not* free, but must carry along a large amount of energy due to the confinement process. As a result, bound quarks have a much larger effective, so-called *constituent quark mass*, which takes into account this confinement energy. To give an idea of the values, take a proton; the indeterminacy relation for a particle inside a sphere of radius 0.9 fm gives a momentum indeterminacy of around 190 MeV/c. In three dimensions this gives an energy of $\sqrt{3}$ times that value, or an effective, constituent quark mass of about 330 MeV/c^2. Three confined quarks are thus heavier than a proton, whose mass is 938 MeV/c^2; we can thus still say that a compound proton is less massive than its constituents.

In short, the mass of the proton and the neutron is (almost exclusively) the kinetic energy of the quarks inside them, as their rest mass is almost negligible. As Frank Wilczek says, some people put on weight even though they never eat anything heavy.

But also the small current quark mass values for the up, down, strange and charmed quarks that appear in the QCD Lagrangian are framework dependent. The values of Table 15 are those for a renormalization scale of 2 GeV. For half that energy, the mass values increase by 35%. The heavy quark masses are those used in the so-called $\overline{MS}$ scheme, a particular way to perform perturbation expansions.

The mass, shape and colour of protons

Frank Wilczek mentions that one of the main results of QCD, the theory of strong inter-   Ref. 184
actions, is to explain mass relations such as

$$m_{\text{proton}} \sim e^{-k/\alpha} m_{\text{Planck}} \quad \text{and} \quad k = 11/2\pi \, , \, \alpha_{\text{unif}} = 1/25 \, . \tag{83}$$

Here, the value of the coupling constant α_{unif} is taken at the grand unifying energy, a
factor of 1000 below the Planck energy. (See the section of grand unification below.) In   Page 246
other words, a general understanding of masses of bound states of the strong interaction,
such as the proton, requires almost purely a knowledge of the unification energy and the
coupling constant at that energy. The approximate value $\alpha_{\text{unif}} = 1/25$ is an extrapolation
from the low energy value, using experimental data. The proportionality factor k in expression (83) is not easy to calculate. It is usually determined on computers using lattice
QCD.

But the mass is not the only property of the proton. Being a cloud of quarks and gluons,
it also has a shape. Surprisingly, it took a long time before people started to become
interested in this aspect. The proton, being made of two up quarks and one down quark,
resembles a ionized H_2^+ molecule, where one electron forms a cloud around two protons.
Obviously, the H_2^+ molecule is elongated, or prolate.

Is the proton prolate? There is no spectroscopically measurable non-sphericty – or
quadrupole moment – of the proton. However, the proton has an *intrinsic* quadrupole
moment. The quadrupole moments of the proton and of the neutron are predicted to be
positive in all known calculation methods, implying an *prolate* shape. Recent measure-   Ref. 185
ments at Jefferson Laboratories confirm this prediction. A prolate shape is predicted for
all $J = 1/2$ baryons, in contrast to the oblate shape predicted for the $J = 3/2$ baryons.
The spin 0 pseudoscalar mesons are predicted to be prolate, whereas the spin 1 vector
mesons are expected to be oblate.

The shape of any molecule will depend on whether other molecules surround it. Recent research showed that similarly, both the size and the shape of the proton in nuclei
is slightly variable; both seem to depend on the nucleus in which the proton is built-in.   Ref. 186

Apart from shapes, molecules also have a colour. The colour of a molecule, like that
of any object, is due to the energy absorbed when it is irradiated. For example, the H_2^+
molecule can absorb certain light frequencies by changing to an excited state. Molecules
change mass when they absorb light; the excited state is heavier than the ground state. In
the same way, protons and neutrons can be excited. In fact, their excited states have been
studied in detail; a summary, also showing the limitation of the approach, is shown in   Ref. 175
Figure 126. Many excitations can be explained as excited quarks states, but many more
are predicted. The calculated masses agree with observations to within 10%. The quark
model and QCD thus structure and explain a large part of the baryon spectrum; but the
agreement is not yet perfect.

Obviously, in our everyday environment the energies necessary to excite nucleons do
not appear – in fact, they do not even appear inside the Sun – and these excited states
can be neglected. They only appear in particle accelerators and in cosmic rays. In a sense,
we can say that in our corner of the universe, the colour of protons usually is not visible.

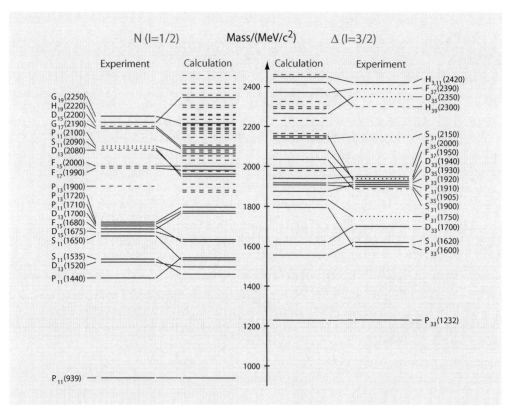

FIGURE 126 The mass spectrum of the excited states of the proton: experimental and calculated values (from Ref. 175).

TABLE 16 Correspondence between QCD and superconductivity.

QCD	SUPERCONDUCTIVITY
Quark	magnetic monopole
Colour force non-linearities	Electron–lattice interaction
Chromoelectric flux tube	magnetic flux tube
Gluon-gluon attraction	electron–electron attraction
Glueballs	Cooper pairs
Instability of bare vacuum	instability of bare Fermi surface
Discrete centre symmetry	continuous U(1) symmetry
High temperature breaks symmetry	low temperature breaks symmetry

CURIOSITIES ABOUT THE STRONG INTERACTION

In a well-known analogy, QCD can be compared to superconductivity. Table 16 gives an overview of the correspondence.

∗ ∗

The computer calculations necessary to extract particle data from the Lagrangian of

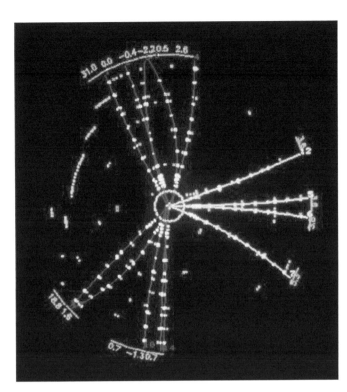

FIGURE 127 A three jet event observed at the PETRA collider in Hamburg in Germany. The event, triggered by an electron-positron collision, allowed detecting the decay of a gluon and measuring its spin (© DESY).

quantum chromodynamics are among the most complex calculations ever performed. They beat weather forecasts, fluid simulations and the like by orders of magnitude. Nobody knows whether this will be necessary also in the future: the race for a simple approximation method for finding solutions is still open.

* *

Even though gluons are massless, like photons and gravitons are, there is *no* colour radiation in nature. Gluons carry colour and couple to themselves; as a result, free gluons were predicted to directly decay into quark–antiquark pairs.

In 1979, the first clear decays of *gluons* have been observed at the PETRA particle collider in Hamburg. The occurrence of certain events, called *gluon jets*, are due to the decay of high-energy gluons into narrow beams of particles. Gluon jets appear in coplanar three-jet events. The observed rate and the other properties of these events confirmed the predictions of QCD. Experiments at PETRA also determined the spin $S = 1$ of the gluon and the running of the strong coupling constant. The hero of those times was the project manager Gustav-Adolf Voss, who completed the accelerator on budget and six months ahead of schedule.

Ref. 180

* *

Something similar to colour radiation, but still stranger, might have been found in 1997. It might be that a scalar meson with a mass of 1.5 GeV/c^2 is a *glueball*. This is a hypothetical meson composed of gluons only. Numerical results from lattice gauge theory seem to

Ref. 187 confirm the possibility of a glueball in that mass range. The existence of glueballs is hotly debated and still open.

* *

There is a growing consensus that most light scalar mesons below $1\,\text{GeV}/c^2$, are Ref. 188 *tetraquarks*. In 2003, experiments provided also candidates for heavier tetraquarks, namely the X(3872), Ds(2317) and Ds(2460). The coming years will show whether this interpretation is correct.

* *

Do particles made of five quarks, so-called *pentaquarks*, exist? So far, they seem to exist only in a few laboratories in Japan, whereas in other laboratories across the world they Ref. 189 are not seen. Most researchers do not believe the results any more.

* *

Whenever we look at a periodic table of the elements, we look at a manifestation of the strong interaction. The Lagrangian of the strong interaction describes the origin and properties of the presently known 115 elements.

Nevertheless one central aspect of nuclei is determined by the electromagnetic interaction. Why are there around 115 different elements? Because the electromagnetic coupling constant α is around $1/137.036$. In more detail, the answer is the following. If the charge of a nucleus were much higher than around 137, the electric field around nuclei would lead to spontaneous electron–positron pair generation; the generated electron would fall into the nucleus and transform one proton into a neutron, thus inhibiting a larger proton number. The finite number of the elements is thus due to the electromagnetic interaction.

* *

To know more about radioactivity, its effects, its dangers and what a government can do about it, see the English and German language site of the Federal Office for Radiation Protection at www.bfs.de.

* *

Challenge 136 s From the years 1990 onwards, it has regularly been claimed that extremely poor countries are building nuclear weapons. Why is this highly unlikely?

* *

Historically, nuclear reactions provided the first test of the relation $E = c^2\gamma m$. This was achieved in 1932 by Cockcroft and Walton. They showed that by shooting protons into lithium one gets the reaction

$$\begin{smallmatrix}7\\3\end{smallmatrix}\text{Li} + \begin{smallmatrix}1\\1\end{smallmatrix}\text{H} \rightarrow \begin{smallmatrix}8\\4\end{smallmatrix}\text{Be} \rightarrow \begin{smallmatrix}4\\2\end{smallmatrix}\text{He} + \begin{smallmatrix}4\\2\end{smallmatrix}\text{He} + 17\,\text{MeV} \; . \tag{84}$$

The measured energy on the right is exactly the same value that is derived from the differences in total mass of the nuclei on both sides.

* *

A large fraction of researchers say that QCD is defined by *two* parameters. Apart from the coupling constant, they count also the strong CP parameter. Indeed, it might be that the strong interaction violates CP invariance. This violation would be described by a second term in the Lagrangian; its strength would be described by a second parameter, a phase usually called θ_{CP}. However, many high-precision experiments have been performed to search for this effect, and no CP violation in the strong interaction has ever been detected.

A SUMMARY OF QCD AND ITS OPEN ISSUES

Quantum chromodynamics, the non-Abelian gauge theory based on the Lagrangian with SU(3) symmetry, describes the properties of gluons and quarks, the properties of the proton, the neutron and all other hadrons, the properties of atomic nuclei, the working of the stars and the origin of the atoms inside us and around us. Without the strong interaction, we would not have flesh and blood. And all these aspects of nature follow from a single number, the strong coupling constant, and the SU(3) gauge symmetry.

The strong interaction acts only on quarks and gluons. It conserves particle type, colour, electric charge, weak charge, spin, as well as C, P and T parity.

QCD and experiment agree wherever comparisons have been made. QCD is a perfect description of the strong interaction. The limitations of QCD are only conceptual. Like in all of quantum field theory, also in the case of QCD the mathematical form of the Lagrangian is almost uniquely defined by requiring renormalizability, Lorentz invariance and gauge invariance – SU(3) in this case. We say 'almost', because the Lagrangian, despite describing correctly all experiments, contains a few parameters that remain unexplained:

— The number, 6, and the masses m_q of the quarks are not explained by QCD.
— The coupling constant of the strong interaction g_s, or equivalently, α_s or Λ, is unexplained. QCD predicts its energy dependence, but not its absolute value.
— Experimentally, the strong interaction is found to be CP conserving. This is not obvious; the QCD Lagrangian assumes that any possible CP-violating term vanishes, even though there exist CP-violating Lagrangian terms that are Lorentz-invariant, gauge-invariant and renormalizable.
— The properties of space-time, in particular its Lorentz invariance, its continuity and the number of its dimensions are assumed from the outset and are obviously all unexplained in QCD.
— It is also not known how QCD has to be modified in strong gravity, thus in strongly curved space-time.

We will explore ways to overcome these limits in the last part of our adventure. Before we do that, we have a look at the other nuclear interaction observed in nature.

THE WEAK NUCLEAR INTERACTION AND THE HANDEDNESS OF NATURE

THE weirdest interaction in nature is the weak interaction. The weak interaction transforms elementary particles into each other, has radiation particles that have mass, violates parity and treats right and left differently. Fortunately, we do not experience the weak interaction in our everyday life, as its properties violate much of what we normally experience. This contrast makes the weak interaction the most fascinating of the four interactions in nature.

TRANSFORMATION OF ELEMENTARY PARTICLES

Radioactivity, in particular the so-called β decay, is a bizarre phenomenon. Experiments in the 1910s showed that when β sources emit electrons, atoms are *transformed* from one chemical element to another. For example, experiments such as those of Figure 128 show that tritium, an isotope of hydrogen, decays into helium as

$$_1^3\text{H} \rightarrow {_2^3}\text{He} + e^- + \overline{\nu}_e \ . \tag{85}$$

In fact, new elements appear in all β decays. In the 1930s it became clear that the transformation process is due to a neutron in the nucleus changing into a proton (and more):

$$n \rightarrow p + e^- + \overline{\nu}_e \ . \tag{86}$$

This reaction explains all β decays. In the 1960s, the quark model showed that β decay is in fact due to a down quark changing to an up quark:

$$d \rightarrow u + e^- + \overline{\nu}_e \ . \tag{87}$$

This reaction explains the transformation of a neutron – a *udd* state – into a proton – a *uud* state. In short, matter particles can transform into each other. We note that this transformation differs from what occurs in other nuclear processes. In fusion, fission or α decay, even though nuclei change, every neutron and every proton retains its nature. In β decay, elementary particles are not immutable. The dream of Democritus and Leucippus about immutable basic building blocks is definitely *not* realized in nature.

Experiments show that quark transformations cannot be achieved with the help of electromagnetic fields, nor with the help of gluon fields, nor with the help of gravitation. There must be another type of radiation in nature, and thus another, fourth interac-

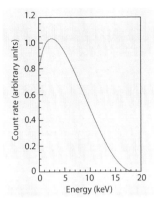
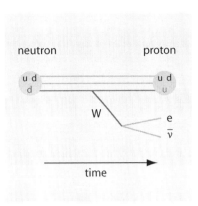

FIGURE 128 β decay (beta decay) in tritium: a modern, tritium-powered illuminated watch, the measured continuous energy spectrum of the emitted electrons from tritium, and the process occurring in the tritium nucleus (© www.traser.com, Katrin).

tion. Chemical transformations are also rare, otherwise we would not be running around with a constant chemical composition. The fourth interaction is therefore *weak*. Since the transformation processes were observed in the nucleus first, the interaction was named the *weak nuclear interaction*.

In β decay, the weak nuclear interaction transforms quarks into each other. In fact, the weak nuclear interaction can also transform leptons into each other, such as muons into electrons. But where does the energy released in β decay go to? Measurements in 1911 showed that the energy spectrum of the emitted electron is *continuous*. This is illustrated in Figure 128. How can this be? In 1930, Wolfgang Pauli had the courage and genius to explain this observation in with a daring hypothesis: the energy of the decay is split between the electron and a new, truly astonishing particle, the *neutrino* – more precisely, the electron anti-neutrino $\bar{v}_e$. In order to agree with data, the neutrino must be uncharged, cannot interact strongly, and must be of very low mass. As a result, neutrinos interact with ordinary matter only extremely rarely, and usually fly through the Earth without being affected. This property makes their detection very difficult, but not impossible; the first neutrino was finally detected in 1952. Later on, it was discovered that there are three types of neutrinos, now called the electron neutrino, the muon neutrino and the tau neutrino, each with its won antiparticle. For the experimental summary of these experimental efforts, see Table 17.

Page 220

THE WEAKNESS OF THE WEAK NUCLEAR INTERACTION

From the observation of β decay, and helped by the quark model, physicists quickly concluded that there must be an intermediate particle that carries the weak nuclear interaction, similar to the photon that carries the electromagnetic interaction. This 'weak radiation', in contrast to all other types of radiation, consists of *massive* particles.

> ▷ There are two types of weak radiation particles: the neutral Z boson with a mass of 91.2 GeV – that is the roughly mass of a silver atom – and the electrically charged W boson with a mass of 80.4 GeV.

TABLE 17 The leptons: the three neutrinos and the three charged leptons (antiparticles have opposite charge Q and parity P).

NEUTRINO	MASS m AND DECAY (SEE TEXT)	SPIN J PARITY P	COLOUR; POSSIBLE WEAK BE- HAVIOUR	CHARGE Q, ISOSPIN I, STRANGENESS S, CHARM C, BEAUTY B', TOPNESS T	LEPTON NUMBER L, BARYON NUMBER B
Electronneutrino ν_e	$< 2\,\mathrm{eV}/c^2$, oscillates	$\frac{1}{2}^+$	white; singlet, doublet	$0, 0, 0, 0, 0, 0$	$1, 0$
Muon neutrino ν_e	$< 2\,\mathrm{eV}/c^2$, oscillates	$\frac{1}{2}^+$	white; singlet, doublet	$0, 0, 0, 0, 0, 0$	$1, 0$
Tau neutrino ν_e	$< 2\,\mathrm{eV}/c^2$, oscillates	$\frac{1}{2}^+$	white; singlet, doublet	$0, 0, 0, 0, 0, 0$	$1, 0$
Electron e	$0.510\,998\,928(11)$ MeV/c^2, stable	$\frac{1}{2}^+$	white; singlet, doublet	$-1, 0, 0, 0, 0, 0$	$1, 0$
Muon μ	$105.658\,3715(35)$ MeV/c^2, $c.\ 99\%\ e\bar{\nu}_e\nu_\mu$	$\frac{1}{2}^+$	white; singlet, doublet	$-1, 0, 0, 0, 0, 0$	$1, 0$
Tau τ	$1.776\,82(16)$ GeV/c^2, $c.\ 17\%$ $\mu\bar{\nu}_\mu\nu_\tau$, $c.\ 18\%$ $e\bar{\nu}_e\nu_\tau$	$\frac{1}{2}^+$	white; singlet, doublet	$-1, 0, 0, 0, 0, 0$	$1, 0$

Together, the W and Z bosons are also called the *weak vector bosons*, or the *weak intermediate bosons*.

The masses of the weak vector bosons are so large that free weak radiation exists only for an extremely short time, about 0.1 ys; then the bosons decay. The large mass is the reason that the weak interaction is extremely short range and thus extremely weak. Indeed, any exchange of virtual carrier particles scales with the negative exponential of the intermediate particle's mass. A few additional properties are given in Table 18. In fact, the weak interaction is so weak that neutrinos, particles which interact *only* weakly, have a large probability to fly through the Sun without any interaction.

The existence of a massive charged intermediate vector boson, today called the W, was already deduced and predicted in the 1940s; but theoretical physicists did not accept the idea until the Dutch physicists Martin Veltman and Gerard 't Hooft proved that it was possible to have such a mass without having problems in the rest of the theory. For this proof they later received the Nobel Prize in Physics – after experiments confirmed their prediction.

The existence of an additional massive *neutral* intermediate vector boson, the Z boson, was predicted only much after the W boson, by Salam, Weinberg and Glashow. Experimentally, the Z boson was first observed as a *virtual* particle in 1973 at CERN in Geneva. The discovery was made by looking, one by one, at over 700 000 photographs made at

TABLE 18 The intermediate vector bosons of the weak interaction (the Z boson is its own antiparticle; the W boson has an antiparticle of opposite charge).

BOSON	MASS m	SPIN J	COLOUR; WEAK BE-HAVIOUR	CHARGE Q, ISOSPIN I, STRANGENESS S, CHARM C, BEAUTY B', TOPNESS T	LEPTON NUMBER L, BARYON NUMBER B
Z boson	91.1876(21) GeV/c^2	1	white; 'triplet'	0, 0, 0, 0, 0, 0	0, 0
W boson	80.398(25) GeV/c^2	1	white; 'triplet'	1, 0, 0, 0, 0, 0	0, 0

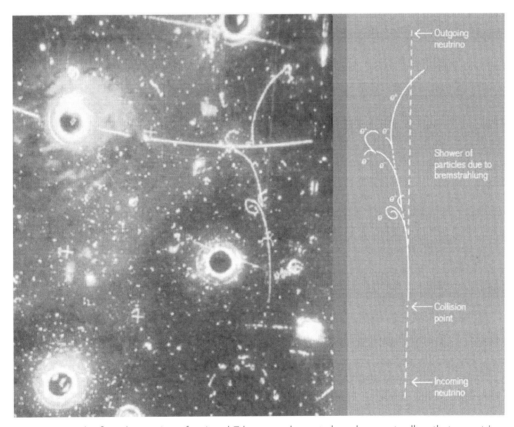

FIGURE 129 The first observation of a virtual Z boson: only neutral weak currents allow that a neutrino collides with an electron in the bubble chamber and leaves again (© CERN).

the Gargamelle bubble chamber. Only a few interesting pictures were found; the most famous one is shown in Figure 129.

In 1983, CERN groups produced and detected the first *real* W and Z bosons. This experiment was a five-year effort by thousands of people working together. The results are

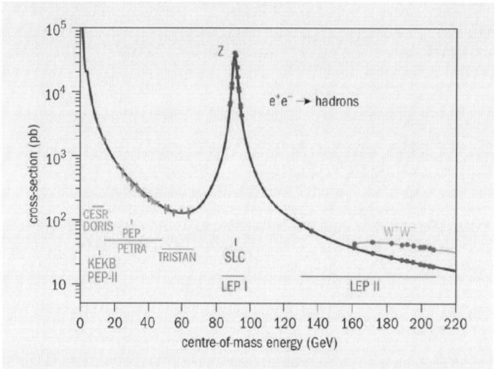

FIGURE 130 Top: the SPS, the proton–antiproton accelerator and collider at CERN, with 7 km circumference, that was used to make the first observations of real W and Z bosons. Bottom: the beautifully simple Z observation made with LEP, the successor machine (© CERN)

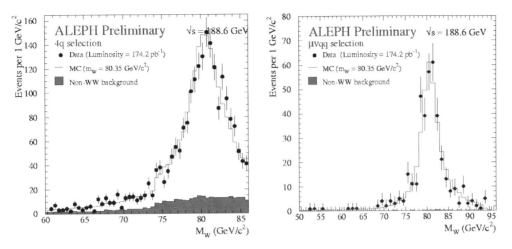

FIGURE 131 A measurement of the W boson mass at LEP (© CERN)

summarized in Table 18. The energetic manager of the project, Carlo Rubbia, whose bad temper made his secretaries leave, on average, after three weeks, and the chief technologist, Simon van der Meer, received the 1984 Nobel Prize in Physics for the discovery. This again confirmed the 'law' of nature that bosons are discovered in Europe and fermions in America. The simplest data that show the Z and W bosons is shown in Figure 130 and Figure 131; both results are deduced from the cross section of electron-positron collisions at LEP, a decade after the original discovery.

In the same way that photons are emitted by accelerated electric charges, W and Z bosons are emitted by accelerated weak charges. Due to the high mass of the W and Z bosons, the required accelerations are very high, so that they are only found in certain nuclear decays and in particle collisions. Nevertheless, the W and Z bosons are, apart from their mass and their weak charge, similar to the photon in most other aspects. In particular, the W and the Z are observed to be elementary. For example, the W gyromagnetic ratio has the value predicted for elementary particles.   Ref. 175

DISTINGUISHING LEFT FROM RIGHT

Another weird characteristic of the weak interaction is the non-conservation of parity P under spatial inversion. The weak interaction distinguishes between mirror systems; this is in contrast to everyday life, to gravitation, to electromagnetism, and to the strong interaction. The non-conservation of parity by the weak interaction had been predicted by 1956 by Lee Tsung-Dao and Yang Chen Ning in order to explain the ability of K^0 mesons to decay sometimes into 2 pions, which have even parity, and sometimes into 3 pions, which have odd parity.   Ref. 191

Lee and Yang suggested an experiment to Wu Chien-Shiung* The experiment she performed with her team is shown schematically in Figure 132. A few months after the

---

* Wu Chien-Shiung (b. 1912 Shanghai, d. 1997 New York) was called 'madame Wu' by her colleagues. She was a bright and driven physicist born in China. She worked also on nuclear weapons; later in life she was president of the American Physical Society.

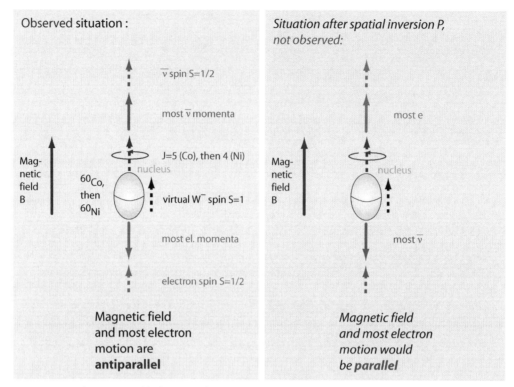

FIGURE 132 The measured behaviour of β decay, and its imagined, but unobserved behaviour under spatial inversion P (corresponding to a mirror reflection plus subsequent rotation by π around an axis perpendicular to the mirror plane).

first meetings with Lee and Yang, Wu and her team found that in the β decay of cobalt nuclei aligned along a magnetic field, the electrons are emitted mostly *against* the spin of the nuclei. In the experiment with inversed parity, the electrons would be emitted *along* the spin direction; however, this case is not observed. Parity is violated. This earned Lee and Yang a Nobel Prize in 1957.

Parity is thus violated in the weak interaction. Parity violation does not only occur in β decay. Parity violation has been found in muon decay and in every other weak process studied so far. In particular, when two electrons collide, those collisions that are mediated by the weak interaction behave differently in a mirror experiment. The number of experiments showing this increases from year to year. In 2004, two polarized beams of electrons – one left-handed and one right-handed – were shot at a matter target and the reflected electrons were counted. The difference was 0.175 parts per million – small, but measurable. The experiment also confirmed the predicted weak charge of −0.046 of the electron.

A beautiful consequence of parity violation is its influence on the *colour* of certain atoms. This prediction was made in 1974 by Bouchiat and Bouchiat. The weak interaction is triggered by the weak charge of electrons and nuclei; therefore, electrons in atoms do not exchange only virtual photons with the nucleus, but also virtual Z particles. The chance for this latter process is extremely small, around 10^{-11} times smaller than for ex-

Ref. 192

Ref. 193

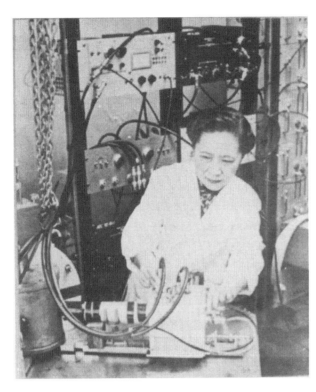

FIGURE 133 Wu Chien-Shiung (1912–1997) at her parity-violation experiment.

change of virtual photons. But since the weak interaction is not parity conserving, this process allows electron transitions which are impossible by purely electromagnetic effects. In 1984, measurements confirmed that certain optical transitions of caesium atoms that are impossible via the electromagnetic interaction, are allowed when the weak interaction is taken into account. Several groups have improved these results and have been able to confirm the calculations based on the weak interaction properties, including the weak charge of the nucleus, to within a few per cent.

Ref. 194
Ref. 195

The weak interaction thus allows one to distinguish left from right. Nature contains processes that differ from their mirror version. In short, particle physics has shown that nature is (weakly) left-handed.

The left-handedness of nature is to be taken literally. All experiments confirmed two central statements on the weak interaction that can be already guessed from Figure 132.

Challenge 137 e

▷ The weak interaction only couples to left-handed particles and to right-handed antiparticles. Parity is *maximally violated* in the weak interaction.

▷ All neutrinos observed so far are left-handed, and all antineutrinos are right-handed.

This result can only hold if neutrino masses vanish or are negligibly small. These two experimental results fix several aspects of the Lagrangian of the weak interaction.

DISTINGUISHING PARTICLES AND ANTIPARTICLES, CP VIOLATION

In the weak interaction, the observation that only right-handed particles and left-handed antiparticles are affected has an important consequence: it implies a violation of charge conjugation parity C. Observations of muons into electrons shows this most clearly: antimuon decay differs from muon decay. The weak interaction distinguishes particles from antiparticles.

Challenge 138 e

> ▷ Experiments show that C parity, like P parity, is *maximally violated* in the weak interaction.

Also this effect has been confirmed in all subsequent observations ever performed on the weak interaction.

But that is not all. In 1964, a now famous observation was made by Val Fitch and James Cronin in the decay of the neutral K mesons.

> ▷ The weak interaction also violates the combination of parity inversion with particle-antiparticle symmetry, the so-called CP invariance. In contrast to P violation and C violation, which are maximal, CP violation is a tiny effect.

The experiment, shown in Figure 134 earned them the Nobel Prize in 1980. CP violation has also been observed in neutral B mesons, in several different processes and reactions. The search for other manifestations of CP violation, such as in non-vanishing electric dipole moments of elementary particles, is an intense research field. The search is not simple because CP violation is a small effect in an already very weak interaction; this tends to makes experiments large and expensive.

Since the weak interaction violates CP invariance, it also violates motion (or time) reversal T.

> ▷ But like all gauge theories, the weak interaction does not violate the combined CPT symmetry: it is CPT invariant.

If CPT would be violated, the masses. lifetimes and magnetic moments of particles and antiparticles would differ. That is not observed.

WEAK CHARGE AND MIXINGS

All weak interaction processes can be described by the Feynman diagrams of Figure 135. But a few remarks are necessary. First of all, the W and Z act only on left-handed fermion and on right-handed anti-fermions. Secondly, the weak interaction conserves a so-called *weak charge* T_3, also called *weak isospin*. The three quarks u, c and t, as well as the three neutrinos, have weak isospin $T_3 = 1/2$; the other three quarks and the charged leptons have weak isospin $T_3 = -1/2$. In an idealized, SU(2)-symmetric world, the three vector bosons W^+, W^0, W^- would have weak isospin values $1, 0$ and -1 and be massless. However, a few aspects complicate the issue.

First of all, it turns out that the quarks appearing in Figure 135 are not those of the strong interaction: there is a slight difference, due to *quark mixing*. Secondly, also *neutri-*

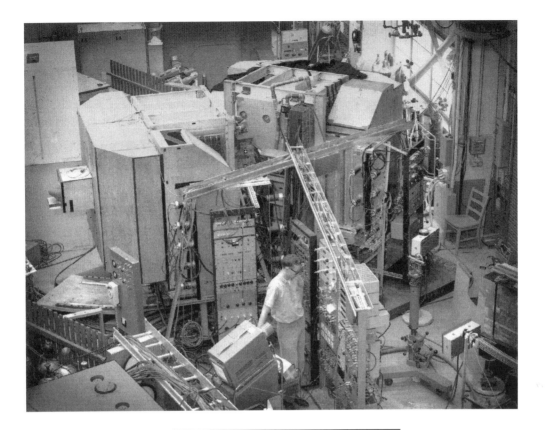

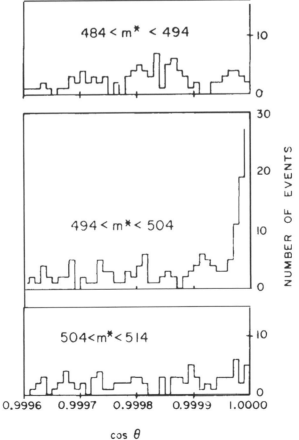

FIGURE 134 Top: the experimental set-up for measuring the behaviour of neutral K meson decay. Bottom: the measured angular dependence; the middle graph shows a peak at the right side that would not appear if CP symmetry would not be violated (© Brookhaven National Laboratory, Nobel Foundation).

FIGURE 135 The essence of the electroweak interaction Lagrangian.

nos mix. And thirdly, the vector bosons are massive and break the SU(2) symmetry of the imagined idealized world; the Lie group SU(2) is not an exact symmetry of the weak interaction, and the famous Higgs boson has mass.

1. Surprisingly, the weak interaction eigenstates of the quarks are not the same as the mass eigenstates. This discovery by Nicola Cabibbo is described by the so-called Cabibbo–Kobayashi–Maskawa or CKM mixing matrix. The matrix is defined by

$$\begin{pmatrix} d' \\ s' \\ b' \end{pmatrix} = (V_{ij}) \begin{pmatrix} d \\ s \\ b \end{pmatrix} .$$

(88)

where, by convention, the states of the +2/3 quarks (u, c, t) are unmixed. In its standard parametrization, the CKM matrix reads

$$V = \begin{pmatrix} c_{12}c_{13} & s_{12}c_{13} & s_{13}e^{-i\delta_{13}} \\ -s_{12}c_{23} - c_{12}s_{23}s_{13}e^{i\delta_{13}} & c_{12}c_{23} - s_{12}s_{23}s_{13}e^{i\delta_{13}} & s_{23}c_{13} \\ s_{12}s_{23} - c_{12}c_{23}s_{13}e^{i\delta_{13}} & -c_{12}s_{23} - s_{12}c_{23}s_{13}e^{i\delta_{13}} & c_{23}c_{13} \end{pmatrix}$$

(89)

where $c_{ij} = \cos\theta_{ij}$, $s_{ij} = \sin\theta_{ij}$ and i and j label the generation ($1 \leqslant i, j \leqslant 3$). In the limit $\theta_{23} = \theta_{13} = 0$, i.e., when only two generations mix, the only remaining parameter is the angle θ_{12}, called the *Cabibbo angle*, which was introduced when only the first two generations of fermions were known. The phase δ_{13}, lying between 0 and 2π, is different from zero in nature, and expresses the fact that CP invariance is violated in the case of the weak interactions. It appears in the third column and shows that CP violation is related to the existence of (at least) three generations.

The CP violating phase δ_{13} is usually expressed with the *Jarlskog invariant*, defined as $J = \sin\theta_{12}\sin\theta_{13}\sin\theta_{23}^2\cos\theta_{12}\cos\theta_{13}\cos\theta_{23}\sin\delta_{13}$. This expression is independent of the definition of the phase angles; it was discovered by Cecilia Jarlskog, an important

Ref. 196   Swedish particle physicist. Its measured value is $J = 2.96(20) \cdot 10^{-5}$.

The CKM mixing matrix is predicted to be unitary. The unitarity has been confirmed by all experiments so far. The 90 % confidence upper and lower limits for the *magnitude* Ref. 175 of the complex CKM matrix V are given by

$$|V| = \begin{pmatrix} 0.97428(15) & 0.2253(7) & 0.00347(16) \\ 0.2252(7) & 0.97345(16) & 0.0410(11) \\ 0.00862(26) & 0.0403(11) & 0.999152(45) \end{pmatrix} . \tag{90}$$

The values have been determined in dozens of experiments by thousands of physicists.

Also neutrinos mix, in the same way as the d, s and b quarks. The determination of the matrix elements is not as complete as for the quark case. This is an intense research field. Like for quarks, also for neutrinos the mass eigenstates and the flavour eigenstates differ. There is a dedicated neutrino mixing matrix, called the *Pontecorvo–Maki–Nakagawa–Sakata mixing matrix* or *PMNS mixing matrix*, with 4 angles for massive neutrinos (it would have 6 angles if neutrinos were massless). In 2012, the measured matrix values were

$$P = \begin{pmatrix} 0.82 & 0.55 & -0.15 + 0.038i \\ -0.36 + 0.020i & 0.70 + 0.013i & 0.61 \\ 0.44 + 0.026i & -0.45 + 0.017i & 0.77 \end{pmatrix} . \tag{91}$$

Many experiments are trying to measure these parameters with higher precision.

Symmetry breaking – and the lack of electroweak unification

The intermediate W and Z bosons are massive and their masses differ. Thus, the weak interaction does not show a SU(2) symmetry. In addition, electromagnetic and weak processes *mix*.

Beautiful research in the 1960s showed that the mixing of the electromagnetic and the weak interactions can be described by an *'electroweak' coupling constant g* and a *weak mixing angle θ_W*. The mixing angle describes the strength of the breaking of the SU(2) Ref. 175 symmetry.

It needs to be stressed that in contrast to what is usually said and written, the weak and the electromagnetic interactions do *not* unify. They have *never* been unified. Despite the incessant use of the term "*electroweak* unification" since several decades, the term is *wrong*. The electromagnetic and the weak interactions are two independent interactions, with two coupling constants, that *mix*. But they do not unify. Even though the Nobel Prize committee used the term 'unification', the relevant Nobel Prize winners confirm that the term is *not correct*.

▷ The electromagnetic and the weak interaction have *not* been unified. Their mixing has been elucidated.

The usual electromagnetic coupling constant e is related to the 'electroweak' coupling g and the mixing angle θ_w by

$$e = g \sin \theta_w , \tag{92}$$

which at low four-momentum transfers is the fine structure constant with the value

1/137.036. The electroweak coupling constant g also defines the historically defined *Fermi constant* G_F by

$$G_F = \frac{g^2 \sqrt{2}}{8 M_W^2} \,.$$

(93)

The broken SU(2) symmetry implies that in the *real* world, in contrast to the *ideal* SU(2) world, the intermediate vector bosons are

— the massless, neutral photon, given as $A = B \cos \theta_W + W^3 \sin \theta_W$;
— the massive neutral Z boson, given as $Z = -B \sin \theta_W + W^3 \cos \theta_W$;
— the massive charged W bosons, given as $W^\pm = (W^1 \mp i W^2)/\sqrt{2}$.

Together, the mixing of the electromagnetic and weak interactions as well as the breaking of the SU(2) symmetry imply that the electromagnetic coupling e, the weak coupling g and the intermediate boson masses by the impressive relation

$$\left(\frac{m_W}{m_Z}\right)^2 + \left(\frac{e}{g}\right)^2 = 1$$

(94)

The relation is well verified by experiments.

The mixing of the electromagnetic and weak interactions also suggests the existence a scalar, elementary *Higgs boson*. This prediction, from the year 1963, was made by Pe-

Ref. 197 ter Higgs and a number of other particles physicists, who borrowed ideas that Yoichiro Nambu and, above all, Philip Anderson introduced in solid state physics. The Higgs boson maintains the unitarity of longitudinal boson scattering at energies of a few TeV and influences the mass of all other elementary particles. In 2012, the Higgs boson has finally been observed in two large experiments at CERN.

THE LAGRANGIAN OF THE WEAK AND ELECTROMAGNETIC INTERACTIONS

If we combine the observed properties of the weak interaction mentioned above, namely its observed Feynman diagrams, its particle transforming ability, P and C violation, quark mixing, neutrino mixing and symmetry breaking, we arrive at the full Lagrangian density. It is given by:

$$
\begin{aligned}
\mathcal{L}_{\text{E\&W}} = \quad & \sum_k \overline{\psi}_k (i\not{\partial} - m_k - \tfrac{g m_k H}{2 m_W})\psi_k \\
& -e \sum_k q_k \overline{\psi}_k \gamma^\mu \psi_k A_\mu \\
& -\tfrac{g}{2\sqrt{2}} \sum_k \overline{\psi}_k \gamma^\mu (1 - \gamma^5)(T^+ W_\mu^+ + T^- W_\mu^-)\psi_k \\
& -\tfrac{g}{2\cos\theta_W} \sum_k \overline{\psi}_k \gamma^\mu (g_V^k - g_A^k \gamma^5)\psi_k Z_\mu \\
& -\tfrac{1}{4} F_{\mu\nu} F^{\mu\nu} \\
& -\tfrac{1}{2} W_{\mu\nu}^+ W^{-\mu\nu} - \tfrac{1}{4} Z_{\mu\nu} Z^{\mu\nu} \\
& +m_W^2 W^+ W^- + \tfrac{1}{2} m_Z^2 Z^2 \\
& -g W W A - g W W Z \\
& -\tfrac{g^2}{4}(W^4 + Z^4 + W^2 F^2 + Z^2 F^2) \\
& +\tfrac{1}{2}(\partial^\mu H)(\partial_\mu H) - \tfrac{1}{2} m_H^2 H^2 \\
& -\tfrac{g m_H^2}{4 m_W} H^3 - \tfrac{g^2 m_H^2}{32 m_W^2} H^4 \\
& +(g m_W H + \tfrac{g^2}{4} H^2)(W_\mu^+ W^{-\mu} + \tfrac{1}{2\cos^2\theta_w} Z_\mu Z^\mu)
\end{aligned}
$$

1. fermion mass terms
2. e.m. interaction
3. charged weak currents
4. neutral weak currents
5. electromagnetic field
6. weak W and Z fields
7. W and Z mass terms
8. cubic interaction
9. quartic interaction
10. Higgs boson mass
11. Higgs self-interaction
12. Higgs–W and Z int.

(95)

The terms in the Lagrangian are easily associated to the Feynman diagrams of Figure 135:

1. this term describes the inertia of every object around us, yields the motion of fermions, and represents the kinetic energy of the quarks and leptons, as it appears in the usual Dirac equation, modified by the so-called *Yukawa coupling* to the Higgs field H and possibly by a Majorana term for the neutrinos (not shown);

2. the second term describes the well-known interaction of matter and electromagnetic radiation, and explains practically all material properties and colours observed in daily life;

3. the term is the so-called *charged weak current interaction*, due to exchange of virtual W bosons, that is responsible for the β decay and for the fact that the Sun is shining;

4. this term is the *neutral weak current interaction*, the ‘$V - A$ theory’ of George Sudarshan, that explains the elastic scattering of neutrinos in matter;

5. this term represents the kinetic energy of photons and yields the evolution of the electromagnetic field in vacuum, thus the basic Maxwell equations;

6. this term represents the kinetic energy of the weak radiation field and gives the evolution of the intermediate W and Z bosons of the weak interaction;

7. this term is the kinetic energy of the vector bosons;

8. this term represents the triple vertex of the self-interaction of the vector boson;

9. this term represents the quadruple vertex of the self-interaction of the vector boson;

10. this term is the kinetic energy of Higgs boson;

11. this term is the self-interaction of the Higgs boson;

12. the last term is expected to represent the interaction of the vector bosons with the Higgs boson that restoring unitarity at high energies.

Let us look into the formal details. The quantities appearing in the Lagrangian are:

— The wave functions $\psi_k = (v'_k \ l^-_k)$ for leptons and $(u_k \ d'_k)$ for quarks are the left-handed fermion fields of the k-th fermion generation; every component is a spinor. The index $k = 1, 2, 3$ numbers the generation: the value 1 corresponds to (u d $v_e e^-$), the second generation is (c s $v_\mu \mu^-$) and the third (t b $v_\tau \tau^-$). The ψ_k transform as doublets under SU(2); the right handed fields are SU(2) singlets.

In the doublets, one has

$$d'_k = \sum_l V_{kl} d_l \ , \tag{96}$$

where V_{kl} is the Cabibbo–Kobayashi–Maskawa mixing matrix, d'_k are the quark flavour eigenstates and d_k are the quark mass eigenstates. A similar expression holds for the mixing of the neutrinos:

$$v'_k = \sum_l P_{kl} v_l \ , \tag{97}$$

where P_{kl} is the Pontecorvo–Maki–Nakagawa–Sakata mixing matrix, v'_l the neutrino flavour eigenstates and v_l the neutrino mass eigenstates.

— For radiation, A^μ and $F^{\mu\nu}$ is the field of the massless vector boson of the electromagnetic field, the *photon* γ.

$W^\pm_\mu$ are the massive charged gauge vector bosons of the weak interaction; the corresponding particles, W^+ and W^-, are each other's antiparticles.

Z_μ is the field of the massive neutral gauge vector boson of the weak interactions; the neutral vector boson itself is usually called Z^0.

— H is the field of the neutral scalar Higgs boson H^0, the only elementary scalar particle in the standard model. It is not yet discovered, so that all terms involving H still need final confirmation.

— Two charges appear, one for each interaction. The number q_k is the well-known *electric charge* of the particle ψ_k in units of the positron charge. The number $t_{3L}(k)$ is the *weak isospin*, or *weak charge*, of fermion k, whose value is $+1/2$ for u_k and v_k and is $-1/2$ for d_k and l_k. These two charges together define the so-called *vector coupling*

$$g^k_V = t_{3L}(k) - 2q_k \sin^2 \theta_W \tag{98}$$

and the axial coupling

$$g^k_A = t_{3L}(k) \ . \tag{99}$$

The combination $g^k_V - g^k_A$, or $V - A$ for short, expresses the maximal violation of P and C parity in the weak interaction.

— The operators T^+ and T^- are the weak isospin raising and lowering operators. Their action on a field is given e.g. by $T^+ l^-_k = v_k$ and $T^- u_k = d_k$.

We see that the Lagrangian indeed contains all the ideas developed above. The electroweak Lagrangian is essentially *unique*: it could not have a different mathematical form, because both the electromagnetic terms and the weak terms are fixed by the requirements of Lorentz invariance, U(1) and broken SU(2) gauge invariance, permutation symmetry and renormalizability.

The Lagrangian of the weak interaction has been checked and confirmed by thousands of experiments. Many experiments have been designed specifically to probe it to the highest precision possible. In all these cases, no contradictions between observation and theory has ever been found. Even though the last three terms of the Lagrangian are not fully confirmed, this is – most probably – the exact Lagrangian of the weak interaction.

Ref. 175

Curiosities about the weak interaction

The weak interaction, with its breaking of parity and its elusive neutrino, exerts a deep fascination on all those who have explored it. Let us explore this fascination a bit more.

Ref. 198

∗ ∗

The weak interaction is required to have an excess of matter over antimatter. Without the parity violation of the weak interactions, there would be no matter at all in the universe, because all matter and antimatter that appeared in the big bang would have annihilated. The weak interaction prevents the symmetry between matter and antimatter, which is required to have an excess of one over the other in the universe. In short, the parity violation of the weak interaction is a necessary condition for our own existence.

∗ ∗

The weak interaction is also responsible for the heat produced inside the Earth. This heat keeps the magma liquid. As a result, the weak interaction, despite its weakness, is responsible for most earthquakes, tsunamis and volcanic eruptions.

∗ ∗

The Lagrangian of the weak interaction was clarified by Steven Weinberg, Sheldon Glashow and Abdus Salam. They received the 1979 Nobel Prize in physics for their work.

Abdus Salam (b. 1926, Santokdas, d. 1996, Oxford) was an example to many scientists across the world and a deeply spiritual man. In his Nobel banquet speech he explained: "This, in effect, is the faith of all physicists: the deeper we seek, the more is our wonder excited, the more is the dazzlement for our gaze." Salam often connected his research to the spiritual aspects of Islam. Once he was asked in Pakistani television why he believed in unification of physics; he answered: "Because god is one!" When the parliament of Pakistan, in one of the great injustices of the twentieth century, declared Ahmadi Muslims to be non-Muslims and thus effectively started a religious persecution, Salam left Pakistan and never returned. The religious persecution continues to this day: on his tombstone in Pakistan, the word 'muslim' has been hammered away, and the internet is full of offensive comments about him by other muslims. Salam was an important science manager. With support of UNESCO, Salam founded the International Centre for Theoretical Physics and the Third World Academy of Sciences, both in Trieste, in Italy, and attracted there the best scientists of developing countries.

∗ ∗

β decay, due to the weak interaction, separates electrons and protons. Finally, in 2005,

FIGURE 136 A typical
underground
experiment to observe
neutrino oscillations:
the Sudbury Neutrino
Observatory in Canada
(© Sudbury Neutrino
Observatory)

people have proposed to use this effect to build long-life batteries that could be used in
satellites. Future will tell whether the proposals will be successful.

* *

Ref. 200   Every second around 10^{16} neutrinos fly through our body. They have five sources:

— *Solar neutrinos* arrive on Earth at $6 \cdot 10^{14}$ /m^{2}s, with an energy from 0 to 0.42 MeV;
   they are due to the p-p reaction in the sun; a tiny is due to the ^{8}B reaction and has
   energies up to 15 MeV.
— *Atmospheric neutrinos* are products of cosmic rays hitting the atmosphere, consist of
   2/3 of muon neutrinos and one third of electron neutrinos, and have energies mainly
   between 100 MeV and 5 GeV.

— *Earth neutrinos* from the radioactivity that keeps the Earth warm form a flux of Page 169
$6 \cdot 10^{10}$ /m²s.

— *Fossil neutrinos* from the big bang, with a temperature of 1.95 K are found in the universe with a density of 300 cm⁻³, corresponding to a flux of 10^{15} /m²s.

— *Man-made neutrinos* are produced in nuclear reactors (at 4 MeV) and as neutrino beams in accelerators, using pion and kaon decay. A standard nuclear plant produces $5 \cdot 10^{20}$ neutrinos per second. Neutrino beams are produced, for example, at the CERN in Geneva. They are routinely sent 700 km across the Earth to the Gran Sasso laboratory in central Italy, where they are detected. (In 2011, a famous measurement error led some people to believe, incorrectly, that these neutrinos travelled faster than light.)

Neutrinos are mainly created in the atmosphere by cosmic radiation, but also coming directly from the background radiation and from the centre of the Sun. Nevertheless, during our own life – around 3 thousand million seconds – we have only a 10 % chance that one of these neutrinos interacts with one of the $3 \cdot 10^{27}$ atoms of our body. The reason is, as usual, that the weak interaction is felt only over distances less than 10^{-17} m, about 1/100th of the diameter of a proton. The weak interaction is indeed weak.

— Already in 1957, the great physicist Bruno Pontecorvo imagined that travelling neutrinos could spontaneously change into their own antiparticles. Today, it is known experimentally that travelling neutrinos can change generation, and one speaks of *neutrino oscillations*. Such experiments are carried out in large deep underground caves; the most famous one is shown in Figure 136. In short, experiments show that the weak interaction mixes neutrino types in the same way that it mixes quark types.

— Only one type of particles interacts (almost) only weakly: neutrinos. Neutrinos carry no electric charge, no colour charge and almost no gravitational charge (mass). To get an impression of the weakness of the weak interaction, it is usually said that the probability of a neutrino to be absorbed by a lead screen of the thickness of one light-year is less than 50 %. The universe is thus essentially empty for neutrinos. Is there room for bound states of neutrinos circling masses? How large would such a bound state be? Can we imagine bound states, which would be called neutrinium, of neutrinos and antineutrinos circling each other? The answer depends on the mass of the neutrino. Bound states of massless particles do not exist. They could and would decay into two free massless particles.*

Since neutrinos are massive, a neutrino–antineutrino bound state is possible in principle. How large would it be? Does it have excited states? Can they ever be detected? These issues are still open. Challenge 139 ny

The weak interaction is so weak that a neutrino–antineutrino annihilation – which is only possible by producing a massive intermediate Z boson – has never been observed up to this day.

— The mixing of the weak and the electromagnetic interaction led to the prediction of the *Higgs boson*. The fascination of the Higgs boson is underlined by the fact that it is the only fundamental particle that bears the name of a physicist. By the way, the paper by Peter Higgs on the boson named after him is only 79 lines long, and has Ref. 199

---

* In particular, this is valid for photons bound by gravitation; this state is not possible.

only five equations.

* *

In the years 1993 and 1994 an intense marketing campaign was carried out across the United States of America by numerous particle physicists. They sought funding for the 'superconducting supercollider', a particle accelerator with a circumference of 80 km. This should have been the largest machine ever built, with a planned cost of more than twelve thousand million dollars, aiming at finding the Higgs boson before the Europeans would do so at a fraction of that cost. The central argument brought forward was the following: since the Higgs boson is the basis of particle masses, it was central to US science to know about it first. Apart from the issue of the relevance of the conclusion, the worst is that the premise is wrong.

Page 212

We have seen above that 99 % of the mass of protons, and thus of the universe, is due to quark confinement; this part of mass appears even if the quarks are approximated as massless. The Higgs boson is *not* responsible for the origin of mass itself; it just might

Ref. 184 shed some light on the issue. In particular, the Higgs boson does not allow calculating or understanding the mass of any particle. The whole campaign was a classic case of disinformation, and many people involved have shown their lack of honesty.* In the end, the project was stopped, mainly for financial reasons.

> Difficile est saturam non scribere.**
>   Juvenal, *Saturae* 1, 30.

* *

There is no generally accepted name for the quantum field theory of the weak interaction. The expression *quantum asthenodynamics* (QAD) – from the Greek word for 'weak' – has not been universally adopted.

* *

Do ruminating *cows* move their jaws equally often in clockwise and anticlockwise direction? In 1927, the theoretical physicists Pascual Jordan and Ralph de Laer Kronig pub-
Ref. 201 lished a study showing that in Denmark the two directions are almost equally distributed. The rumination direction of cows is thus not related to the weak interaction.

* *

Of course, the weak interaction is responsible for radioactive β decay, and thus for part of the radiation background that leads to mutations and thus to biological evolution.

* *

---

* We should not be hypocrites. The supercollider lie is negligible when compared to other lies. The biggest lie in the world is probably the one that states that to ensure its survival, the USA government need to spend more on the military than all other countries in the world combined. This lie is, every single year, around 40 times as big as the once-only supercollider lie. Many other governments devote even larger percentages of their gross national product to their own version of this lie. As a result, the defence spending lie is directly responsible for most of the poverty in all the countries that use it.

** 'It is hard not to be satirical.'

Due to the large toll it placed on society, research in nuclear physics, has almost disappeared from the planet, like poliomyelitis has. Like poliomyelitis, nuclear research is kept alive only in a few highly guarded laboratories around the world, mostly by questionable figures, in order to build dangerous weapons. Only a small number of experiments carried on by a few researchers are able to avoid this involvement and continue to advance the topic.

* *

Interesting aspects of nuclear physics appear when powerful lasers are used. In 1999, a British team led by Ken Ledingham observed laser induced uranium fission in ^{238}U nuclei. In the meantime, this has even be achieved with table-top lasers. The latest feat, in 2003, was the transmutation of ^{129}I to ^{128}I with a laser. This was achieved by focussing a 360 J laser pulse onto a gold foil; the ensuing plasma accelerates electrons to relativistic speed, which hit the gold and produce high energy γ rays that can be used for the transmutation.

Ref. 202

A SUMMARY OF THE WEAK INTERACTION

The weak interaction is described by a non-Abelian gauge theory based on a broken SU(2) gauge group for weak processes. The weak interaction mixes with the unbroken U(1) gauge group of the electrodynamic interaction. This description matches the observed properties of β decay, of particle transformations, of neutrinos and their mixing, of the massive intermediate W and Z bosons, of maximal parity violation, of the heat production inside the Earth, of several important reactions in the Sun and of the origin of matter in the universe. Even though the weak interaction is weak, it is a bit everywhere.

The weak interaction is described by a Lagrangian. After a century of intense research, the Lagrangian is known in all its details, including, since 2012, the Higgs boson. Theory and experiment agree whenever comparisons have been made.

All remaining limitations of the gauge theory of the weak interaction are only conceptual. Like in all of quantum field theory, also in the case of the weak interaction the mathematical form of the Lagrangian is almost uniquely defined by requiring renormalizability, Lorentz invariance, and (broken) gauge invariance – SU(2) in this case. We say again 'almost', as we did for the case of the strong interaction, because the Lagrangian of the weak and electromagnetic interactions contains a few parameters that remain unexplained:

Page 217

— The two coupling constants g and g' of the weak and the electromagnetic interaction are unexplained. (They define weak mixing angle $\theta_w = \arctan(g'/g)$.)
— The mass $M_Z = 91 \, \text{GeV}/c^2$ of the neutral Z boson is unexplained.
— The number $n = 3$ of generations is unexplained.
— The masses of the six leptons and the six quarks are unexplained.
— The four parameters of the Cabibbo–Kobayashi–Maskawa quark mixing matrix and the six parameters of the neutrino mixing matrix are unexplained, including the respective CP violating phases.
— The properties of space-time, in particular its Lorentz invariance, its continuity and the number of its dimensions are assumed from the outset and are obviously all un-

explained.

— It is also not known how the weak interaction behaves in strong gravity, thus in strongly curved space-time.

Before exploring how to overcome these limitations, we summarize all results found so far in the so-called *standard model of particle physics*.

CHAPTER 9

THE STANDARD MODEL OF PARTICLE PHYSICS – AS SEEN ON TELEVISION

THE expression *standard model of elementary particle physics* stands for the summary of all knowledge about the motion of quantum particles in nature. The standard model can be explained in four tables: the table of the elementary particles, the table of their properties, the table of possible Feynman diagrams and the table of fundamental constants.

The following table lists the known elementary particles found in nature.

TABLE 19 The elementary particles.

Radiation	electromagnetic	weak	strong
	$\boxed{\gamma}$	$\boxed{W^+}$, $\boxed{W^-}$ $\boxed{Z^0}$	$\boxed{g_1}$ ... $\boxed{g_8}$
	photon	weak bosons	gluons

Radiation particles are intermediate vector bosons, thus with spin 1. W^- is the antiparticle of W^+; the photons and the Z^0 are their own antiparticles. Only the $W^\pm$ and Z^0 are massive.

Matter	generation 1	generation 2	generation 3
Leptons	$\boxed{e}$ $\boxed{\nu_e}$	$\boxed{\mu}$ $\boxed{\nu_\mu}$	$\boxed{\tau}$ $\boxed{\nu_\tau}$
Quarks (each in three colours)	$\boxed{d}$ $\boxed{u}$	$\boxed{s}$ $\boxed{c}$	$\boxed{t}$ $\boxed{b}$

Matter particles are fermions with spin 1/2; all have a corresponding antiparticle. Leptons mix among themselves; so do quarks. All fermions are massive.

Vacuum state		
Higgs boson	$\boxed{H}$	Has spin 0 and a mass of 126 GeV.

The table has not changed much since the mid-1970s, except for the Higgs boson, which has been found in 2012. Assuming that the table is complete, it contains *all* constituents that make up all matter and all radiation in nature. Thus the table lists all constituents – the real 'uncuttables' or 'atoms', as the Greek called them – of material objects and beams of radiation. The elementary particles are the basis for materials science, geology, astronomy, engneering, chemistry, biology, medicine, the neurosciences and psychology. For this reason, the table regularly features in mass tabloids, on television and on the internet.

Ref. 175

The full list of elementary particles allows us to put together a full table of particle properties, shown in Table 20. It table lists *all* properties of the elementary particles. To save space, colour and weak isospin are not mentioned explicitly. Also the decay modes of the unstable particles are not given in detail; they are found in the standard references.

Vol. I, page 28

Table 20 is fascinating. It allows us to give a *complete* characterization of the intrinsic properties of *any composed moving entity*, be it an object or an image. At the beginning of our study of motion, we were looking for a complete list of the *intrinsic properties* of moving entities. Now we have it.

TABLE 20 Elementary particle properties.

PARTICLE	MASS m[a]	LIFETIME τ OR ENERGY WIDTH,[b] MAIN DECAY MODES	ISOSPIN I, SPIN J,[c] PARITY P, CHARGE PARITY C	CHARGE, ISOSPIN, STRANGE- NESS,[c] CHARM, BEAUTY, TOPNESS: $QISCBT$	LEPTON & BARYON NUM- BERS $L\,B$
Elementary radiation (bosons)					
photon γ	0 ($< 2 \cdot 10^{-54}$ kg)	stable	$I(J^{PC}) = 0, 1(1^{--})$	000000	0, 0
$W^{\pm}$	80.385(15) GeV/c^2	2.085(42) GeV 67.60(27)% hadrons, 32.40(27)% $l^+\nu$	$J = 1$	± 100000	0, 0
Z	91.1876(21) GeV/c^2	2.4952(23) GeV/c^2 69.91(6)% hadrons 10.0974(69)% l^+l^-	$J = 1$	000000	0, 0
gluon	0	stable	$I(J^P) = 0(1^-)$	000000	0, 0
Elementary matter (fermions): leptons					
electron e	9.109 382 91(40) $\cdot$ 10^{-31} kg = 81.871 0506(36) pJ/c^2 = 0.510 998 928(11) MeV/c^2 = 0.000 548 579 909 46(22) u gyromagnetic ratio $\mu_e/\mu_B = -1.001\ 159\ 652\ 180\ 76(27)$	$> 13 \cdot 10^{30}$ s	$J = \frac{1}{2}$	$-100\,000$	1, 0

PARTICLE	MASS m [a]	LIFETIME τ OR ENERGY WIDTH, [b] MAIN DECAY MODES	ISOSPIN I, SPIN J, [c] PARITY P, CHARGE PARITY C	CHARGE, ISOSPIN, STRANGE- NESS, [c] CHARM, BEAUTY, TOPNESS: $QISCBT$	LEPTON & BARYON NUM- BERS $L\,B$
	electric dipole moment [e] $d =< 0.87 \cdot 10^{-30} e$ m				
muon μ	0.188 353 109(16) yg	2.196 9811(22) µs	$J = \frac{1}{2}$	-100000	$1, 0$
		99 % $e^- \bar{\nu}_e \nu_\mu$			
	$= 105.658\ 3715(35)\ \mathrm{MeV}/c^2$ $= 0.113\ 428\ 9267(29)$ u				
	gyromagnetic ratio $\mu_\mu/(e\hbar/2m_\mu) = -1.001\ 165\ 9209(6)$				
	electric dipole moment $d = (-0.1 \pm 0.9) \cdot 10^{-21} e$ m				
tau τ	1.776 82(16) GeV/c^2	290.6(1.0) fs	$J = \frac{1}{2}$	-100000	$1, 0$
el. neutrino ν_e	< 2 eV/c^2		$J = \frac{1}{2}$		$1, 0$
muon neutrino ν_μ	< 2 eV/c^2		$J = \frac{1}{2}$		$1, 0$
tau neutrino ν_τ	< 2 eV/c^2		$J = \frac{1}{2}$		$1, 0$

Elementary matter (fermions): quarks [f]

PARTICLE	MASS	LIFETIME / DECAY	$I(J^P)$	$QISCBT$	$L\,B$
up u	1.8 to 3.0 MeV/c^2	see proton	$I(J^P) = \frac{1}{2}(\frac{1}{2}^+)$	$+\frac{2}{3}+\frac{1}{2}0000$	$0, \frac{1}{3}$
down d	4.5 to 5.5 MeV/c^2	see proton	$I(J^P) = \frac{1}{2}(\frac{1}{2}^+)$	$-\frac{1}{3}-\frac{1}{2}0000$	$0, \frac{1}{3}$
strange s	95(5) MeV/c^2		$I(J^P) = 0(\frac{1}{2}^+)$	$-\frac{1}{3}0-1000$	$0, \frac{1}{3}$
charm c	1.275(25) GeV/c^2		$I(J^P) = 0(\frac{1}{2}^+)$	$+\frac{2}{3}00+100$	$0, \frac{1}{3}$
bottom b	4.18(17) GeV/c^2	$\tau = 1.33(11)$ ps	$I(J^P) = 0(\frac{1}{2}^+)$	$-\frac{1}{3}000-10$	$0, \frac{1}{3}$
top t	173.5(1.4) GeV/c^2		$I(J^P) = 0(\frac{1}{2}^+)$	$+\frac{2}{3}0000+1$	$0, \frac{1}{3}$

Elementary boson

PARTICLE	MASS	LIFETIME	SPIN
Higgs H	126(1) GeV/c^2	not measured	$J = 0$

Notes:

a. See also the table of SI prefixes on page 301. About the eV/c^2 mass unit, see page 305.

b. The *energy width* Γ of a particle is related to its lifetime τ by the indeterminacy relation $\Gamma\tau = \hbar$. There is a difference between the *half-life* $t_{1/2}$ and the *lifetime* τ of a particle: they are related by $t_{1/2} = \tau \ln 2$, where $\ln 2 \approx 0.693\ 147\ 18$; the half-life is thus shorter than the lifetime. The unified *atomic mass unit* u is defined as 1/12 of the mass of a carbon 12 atom at rest and in its ground state. One has $1\ \mathrm{u} = \frac{1}{12}m(^{12}\mathrm{C}) = 1.660\ 5402(10)$ yg.

c. To keep the table short, the header does not explicitly mention *colour*, the charge of the strong interactions. This has to be added to the list of basic object properties. Quantum numbers containing the word 'parity' are multiplicative; all others are additive. Time parity T (not to be confused with topness T), better called motion inversion parity, is equal to CP. The isospin I (or I_Z) is defined only for up and down quarks and their composites, such as the proton and the neutron. In the literature one also sees references to the so-called *G-parity*, defined as $G = (-1)^{IC}$.

The table also does not mention the *weak charge* of the particles. The details on weak charge g, or, more precisely, on the *weak isospin*, a quantum number assigned to all left-handed fermions (and right-handed anti-fermions), but to no right-handed fermion (and no left-handed antifermion), are given in the section on the weak interactions.

Page 223

d. 'Beauty' is now commonly called *bottomness*; similarly, 'truth' is now commonly called *topness*. The signs of the quantum numbers S, I, C, B, T can be defined in different ways. In the standard assignment shown here, the sign of each of the non-vanishing quantum numbers is given by the sign of the charge of the corresponding quark.

Ref. 203, Ref. 204

e. The electron radius is observed to be less than 10^{-22} m. It is possible to store single electrons in traps for many months.

f. See page 212 for the precise definition and meaning of the quark masses.

The other aim that we formulated at the beginning of our adventure was to find the complete list of all *state properties*. This aim is also achieved, namely by the wave function and the field values due to the various bosons. Were it not for the possibility of space-time curvature, we would be at the end of our exploration.

The main ingredient of the standard model are the Lagrangians of the electromagnetic, the weak and the strong interactions. The combination of the Lagrangians, based on the U(1), SU(3) and broken SU(2) gauge groups, is possible only in one specific way. The Lagrangain can be summarized by the Feynman diagram of Figure 137.

To complete the standard model, we need the coupling constants of the three gauge interactions, the masses of all the particles, and the values of the mixing among quarks and among leptons. Together with all those constants of nature that define the SI system and the number of space-time dimensions, the following table therefore completes the standard model.

TABLE 21 Basic physical constants.

QUANTITY	SYMBOL	VALUE IN SI UNITS	UNCERT.[a]
Constants that define the SI measurement units			
Vacuum speed of light[c]	c	299 792 458 m/s	0
Vacuum permeability[c]	μ_0	$4\pi \cdot 10^{-7}$ H/m	0
		$= 1.256\,637\,061\,435\ldots$ µH/m0	
Vacuum permittivity[c]	$\varepsilon_0 = 1/\mu_0 c^2$	$8.854\,187\,817\,620\ldots$ pF/m	0
Original Planck constant	h	$6.626\,069\,57(52) \cdot 10^{-34}$ Js	$4.4 \cdot 10^{-8}$
Reduced Planck constant, quantum of action	$\hbar$	$1.054\,571\,726(47) \cdot 10^{-34}$ Js	$4.4 \cdot 10^{-8}$
Positron charge	e	$0.160\,217\,656\,5(35)$ aC	$2.2 \cdot 10^{-8}$
Boltzmann constant	k	$1.380\,6488(13) \cdot 10^{-23}$ J/K	$9.1 \cdot 10^{-7}$
Gravitational constant	G	$6.673\,84(80) \cdot 10^{-11}$ Nm^2/kg^2	$1.2 \cdot 10^{-4}$
Gravitational coupling constant $\kappa = 8\pi G/c^4$		$2.076\,50(25) \cdot 10^{-43}$ s^2/kg m	$1.2 \cdot 10^{-4}$
Fundamental constants (of unknown origin)			
Number of space-time dimensions		$3 + 1$	0[b]

TABLE 21 (Continued) Basic physical constants.

Quantity	Symbol	Value in SI units	Uncert.[a]
Fine-structure constant[d] or	$\alpha = \frac{e^2}{4\pi\varepsilon_0 \hbar c}$	$1/137.035\,999\,074(44)$	$3.2 \cdot 10^{-10}$
e.m. coupling constant	$= g_{em}(m_e^2 c^2)$	$= 0.007\,297\,352\,5698(24)$	$3.2 \cdot 10^{-10}$
Fermi coupling constant[d] or	$G_F/(\hbar c)^3$	$1.166\,364(5) \cdot 10^{-5}\,\text{GeV}^{-2}$	$4.3 \cdot 10^{-6}$
weak coupling constant	$\alpha_w(M_Z) = g_w^2/4\pi$	$1/30.1(3)$	$1 \cdot 10^{-2}$
Weak mixing angle	$\sin^2\theta_W(\overline{MS})$	$0.231\,24(24)$	$1.0 \cdot 10^{-3}$
	$\sin^2\theta_W$ (on shell)	$0.2224(19)$	$8.7 \cdot 10^{-3}$
	$= 1 - (m_W/m_Z)^2$		
Strong coupling constant[d]	$\alpha_s(M_Z) = g_s^2/4\pi$	$0.118(3)$	$25 \cdot 10^{-3}$
CKM quark mixing matrix	$\|V\|$	$\begin{pmatrix} 0.97428(15) & 0.2253(7) & 0.00347(16) \\ 0.2252(7) & 0.97345(16) & 0.0410(11) \\ 0.00862(26) & 0.0403(11) & 0.999152(45) \end{pmatrix}$	
Jarlskog invariant	J	$2.96(20) \cdot 10^{-5}$	
PMNS neutrino mixing m.	P	$\begin{pmatrix} 0.82 & 0.55 & -0.15 + 0.038i \\ -0.36 + 0.020i & 0.70 + 0.013i & 0.61 \\ 0.44 + 0.026i & -0.45 + 0.017i & 0.77 \end{pmatrix}$	

Particle masses: see previous table

a. Uncertainty: standard deviation of measurement errors.

b. Only measured from to 10^{-19} m to 10^{26} m.

c. Defining constant.

d. All coupling constants depend on the 4-momentum transfer, as explained in the section on renormalization. *Fine-structure constant* is the traditional name for the electromagnetic coupling constant g_{em} in the case of a 4-momentum transfer of $Q^2 = m_e^2 c^2$, which is the smallest one possible. At higher momentum transfers it has larger values, e.g., $g_{em}(Q^2 = M_W^2 c^2) \approx 1/128$. In contrast, the strong coupling constant has lover values at higher momentum transfers; e.g., $\alpha_s(34\,\text{GeV}) = 0.14(2)$.

Page 116

In short, with the three tables and the figure, the standard model describes *every* observation ever made in flat space-time. In particular, the standard model includes a minimum action, a maximum speed, electric charge quantization and the least action principle.

Summary and open questions

The standard model of particle physics clearly distinguishes *elementary* from *composed* particles. The standard model provides the full list of properties that characterizes a particle – and thus any moving object and image. These properties are: mass, spin, charge, colour, weak isospin, parity, charge parity, isospin, strangeness, charm, topness, beauty, lepton number and baryon number.

The standard model describes electromagnetic and nuclear interactions as gauge theories, i.e., as exchanges of virtual radiation particles. In particular, the standard model describes the three types of radiation that are observed in nature with full precision.

QED, describing the electromagnetic interaction

QCD, describing the strong nuclear interaction

QAD, describing the weak nuclear interaction
The diagrams use q' and l' to indicate quark and lepton mixing.

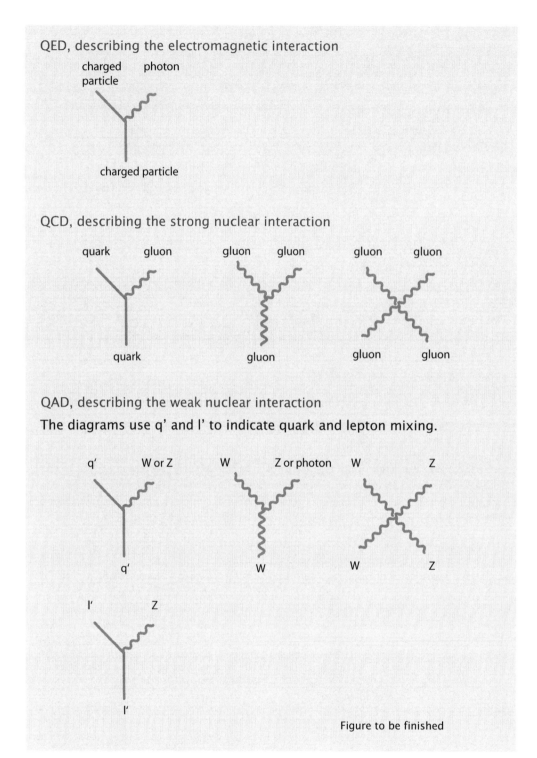

Figure to be finished

FIGURE 137 The Feynman diagrams of the standard model.

The standard model is based on quantization and conservation of electric charge, weak charge and colour, as well as on a smallest action value $\hbar$ and a maximum energy speed c. As a result, the standard model describes the structure of the atoms, their formation in the history of the universe, the properties of matter and the mechanisms of life. Despite the prospect of fame and riches, no deviation between the experiment and the standard model has been found.

In short, the standard model realizes the dream of Leucippus and Democritus, plus a bit more: we know the bricks that compose all of matter and radiation, and in addition we know how they move, interact and transform, in flat space-time, with *perfect* accuracy.

Despite this perfect accuracy, we also know what we still do *not* know:

— We do not know the origin of the coupling constants.
— We do not know why positrons and protons have the same charge.
— We do not know the origin of the masses of the particles.
— We do not know the origin of the mixing and CP violation parameters.
— We do not know the origin of the gauge groups.
— We do not know the origin of the three particle generations.
— We do not know whether the particle concept survives at high energy.
— We do not know what happens in curved space-time.

To study these issues, the simplest way is to explore nature at particle energies that are as high as possible. There are two methods: building large experiments and exploring hypothetical models. Both are useful.

CHAPTER 10

DREAMS OF UNIFICATION

> " Materie ist geronnenes Licht.*
>
> Albertus Magnus

Is there a common origin to the three particle interactions? We have seen in the preceding chapters that the Lagrangian densities of the three gauge interactions are determined almost uniquely by two types of requirements: to possess a certain gauge symmetry, and to be consistent with space-time, through Lorentz invariance and renormalizability. The search for *unification* of the interactions thus seems to require the identification of the one, unified symmetry of nature. (Do you Challenge 140 s agree?)

Between 1970 and 2010, several conjectures have fuelled the hope to achieve unification through higher symmetry. The most popular were grand unification, supersymmetry, conformal field theory, coupling constant duality and mathematical quantum field theory. We give a short summary of these efforts; we start with the first candidate, which is conceptually the simplest.

GRAND UNIFICATION

At all measured energies up to the year 2012, thus below about 2 TeV, there are no contradictions between the Lagrangian of the standard model and observation. On the other hand, the Lagrangian itself can be conjectured to be a low energy approximation to the unified theory. It should thus be possible – attention, this a belief – to find a unifying symmetry that *contains* the symmetries of the electroweak and strong interactions as subgroups. In this way, the three gauge interactions would be different aspects of a single, 'unified' interaction. We can then examine the physical properties that follow from this unifying symmetry and compare them with observation. This approach, called *grand unification*, attempts the unified description of all types of matter. All known elementary particles are seen as fields which appear in a Lagrangian determined by a single gauge symmetry group.

Like for each gauge theory described so far, also the grand unified Lagrangian is fixed by the symmetry group, the representation assignments for each particle, and the corresponding coupling constant. A general search for the grand symmetry group starts Ref. 205 with all those (semisimple) Lie groups which contain $U(1) \times SU(2) \times SU(3)$. The smallest

---

* 'Matter is coagulated light.' Albertus Magnus (b. *c.* 1192 Lauingen, d. 1280 Cologne) was the most important thinker of his time.

groups with these properties are SU(5), SO(10) and E(6); they are defined in Appendix C. For each of these candidate groups, the predicted consequences of the model must be studied and compared with experiment.   Ref. 206

COMPARING PREDICTIONS AND DATA

Grand unification models – also incorrectly called GUTs or *grand unified theories* – make several predictions that can be matched with experiment. First of all, any grand unified model predicts relations between the quantum numbers of quarks and those of leptons. In particular, grand unification explains why the electron charge is exactly the opposite of the proton charge.

Grand unification models predict a value for the weak mixing angle θ_w; this angle is not fixed by the standard model. The most frequently predicted value,   Ref. 205

$$\sin^2 \theta_{w,th} = 0.2 \tag{100}$$

is close to the measured value of

$$\sin^2 \theta_{w,ex} = 0.231(1) \,, \tag{101}$$

which is not a good match, but might be correct, in view of the approximations in the prediction.

All grand unified models predict the existence of *magnetic monopoles*, as was shown by Gerard 't Hooft. However, despite extensive searches, no such particles have been   Ref. 207 found yet. Monopoles are important even if there is only one of them in the whole universe: the existence of a single monopole would imply that electric charge is quantized. Ref. 208 If monopoles were found, grand unification would explain why electric charge appears in multiples of a smallest unit.

Grand unification predicts the existence of heavy intermediate vector bosons, called *X bosons*. Interactions involving these bosons do not conserve baryon or lepton number, but only the difference $B - L$ between baryon and lepton number. To be consistent with data, the X bosons must have a mass of the order of 10^{16} GeV. However, this mass is and always will be outside the range of experiments, so that the prediction cannot be tested directly.

Most spectacularly, the X bosons of grand unification imply that the *proton decays*. This prediction was first made by Pati and Abdus Salam in 1974. If protons decay, means that neither coal nor diamond* – nor any other material – would be for ever. Depending on the precise symmetry group, grand unification predicts that protons decay into pions, electrons, kaons or other particles. Obviously, we know 'in our bones' that the proton lifetime is rather high, otherwise we would die of leukaemia; in other words, the low level of cancer in the world already implies that the lifetime of the proton is larger than about 10^{16} years.

Detailed calculations for the proton lifetime τ_p using the gauge group SU(5) yield the

---

* As is well known, diamond is not stable, but metastable; thus diamonds are not for ever, but coal might be, as long as protons do not decay.

Ref. 205   expression

$$\tau_{\mathrm{p}} \approx \frac{1}{\alpha_G^2(M_X)} \frac{M_X^4}{M_{\mathrm{p}}^5} \approx 10^{31\pm1}\,\mathrm{a} \tag{102}$$

where the uncertainty is due to the uncertainty of the mass M_X of the gauge bosons involved and to the exact decay mechanism. Several large experiments have tried and are still trying to measure this lifetime. So far, the result is simple but clear. Not a single
Ref. 209   proton decay has ever been observed. The data can be summarized by

$$\tau(\mathrm{p} \to e^+\,\pi^0) > 5 \cdot 10^{33}\,\mathrm{a}$$
$$\tau(\mathrm{p} \to K^+\,\bar{\nu}) > 1.6 \cdot 10^{33}\,\mathrm{a}$$
$$\tau(\mathrm{n} \to e^+\,\pi^-) > 5 \cdot 10^{33}\,\mathrm{a}$$
$$\tau(\mathrm{n} \to K^0\,\bar{\nu}) > 1.7 \cdot 10^{32}\,\mathrm{a} \tag{103}$$

These values are higher than the prediction by SU(5) models. For the other gauge group candidates the situation is not settled yet.

THE STATE OF GRAND UNIFICATION

To settle the issue of grand unification definitively, one last prediction of grand unification remains to be checked: the unification of the coupling constants. Most estimates of the grand unification energy are near the Planck energy, the energy at which gravitation starts to play a role even between elementary particles. As grand unification does not take gravity into account, for a long time there was a doubt whether something was lacking in the approach. This doubt changed into certainty when the precision measurements
Ref. 210   of the coupling constants became available, in 1991, and were put into the diagram of Figure 138. The GUT prediction of the way the constants evolve with energy implies that the three constants do *not* meet at one energy. Simple grand unification by SU(5), SU(10) or E_6 is thus ruled out by experiment.

This state of affairs is changed if *supersymmetry* is taken into account. Supersymmetry is a conjecture on the way to take into account low-energy effects of gravitation in
Page 249   the particle world. Supersymmetry conjectures new elementary particles that change the curves at intermediate energies, so that they all meet at a grand unification energy of about 10^{16} GeV. (The line thicknesses in Figure 138 represent the experimental errors.) The inclusion of supersymmetry also puts the proton lifetime prediction back to a value higher (but not by much) than the present experimental bound and predicts the correct value of the mixing angle. With supersymmetry, one can thus retain the advantages of grand unification (charge quantization, one coupling constant) without being in contradiction with experiments.

Pure grand unification is thus in contradiction with experiments. This is not a surprise, as its goal, to unify the description of matter, *cannot* been achieved in this way. Indeed, the unifying gauge group must be introduced at the very beginning, because grand unification cannot deduce the gauge group from a general principle. Neither does grand unification tell us completely which elementary particles exist in nature. In other words,

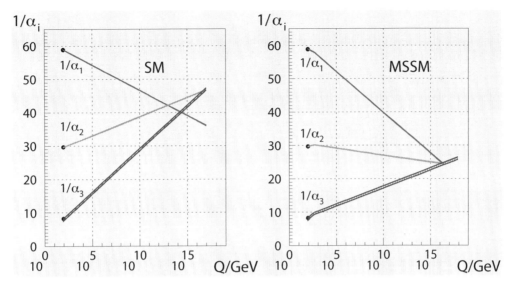

FIGURE 138 The behaviour of the three coupling constants with energy, for simple grand unification (left) and for the minimal supersymmetric model (right); the graph shows the constants $\alpha_1 = \frac{5}{3}\alpha_{em}/\cos^2\theta_W$ for the electromagnetic interaction (the factor 5/3 appears in GUTs), $\alpha_2 = \alpha_{em}/\sin^2\theta_W$ for the weak interaction and the strong coupling constant $\alpha_3 = \alpha_s$ (© W. de Boer).

grand unification only *shifts* the open questions of the standard model to the next level, while keeping most of the open questions unanswered. The name 'grand unification' was wrong from the beginning. We definitely need to continue our adventure.

SEARCHING FOR HIGHER SYMMETRIES

Since we want to reach the top of Motion Mountain, we go on. We have seen in the preceding sections that the main ingredients of the Lagrangian that describes motion are the symmetry properties. We recall that a Lagrangian is just the mathematical name for the concept that measures change. The discovery of the correct symmetry, together with mathematical consistency, usually restricts the possible choices for a Lagrangian down to a limited number, and to one in the best case. This Lagrangian then allows making experimental predictions.

The history of particle physics from 1920 to 1965 has shown that progress was always coupled to the discovery of *larger* symmetries, in the sense that the newly discovered symmetries always included the old ones as a subgroup. Therefore, in the twentieth century, researchers searched for the largest possible symmetry that is consistent with experiments on the one hand and with gauge theories on the other hand. Since grand unification is a failure, a better approach is to search directly for a symmetry that includes gravity.

SUPERSYMMETRY

In the search for possible symmetries of a Lagrangian describing a gauge theory and gravity, one way to proceed is to find general mathematical theorems which restrict the symmetries that a Lagrangian can possibly have.   Ref. 211

Ref. 212   A well-known theorem by Coleman and Mandula states that if the symmetry transformations transform fermions into fermions and bosons into bosons, no quantities other than the following can be conserved:

▪ the energy momentum tensor $T^{\mu\nu}$, a consequence of the *external* Poincaré space-time symmetry, and

▪ the internal quantum numbers, all scalars, associated with each gauge group generator – such as electric charge, colour, etc. – and consequences of the *internal* symmetries of the Lagrangian.

But, and here comes a way out, if transformations that *mix* fermions and bosons are considered, new conserved quantities become possible. This family of symmetries includes gravity and came to be known as *supersymmetry*. Its conserved quantities are not scalars but *spinors*. Therefore, *classical* supersymmetry does not exist; it is a purely quantum-mechanical symmetry. The study of supersymmetry has been a vast research field. For example, supersymmetry generalizes gauge theory to super-gauge theory. The possible super-gauge groups have been completely classified.

Supersymmetry can be extended to incorporate gravitation by changing it into a local gauge theory; in that case one speaks of *supergravity*. Supergravity is based on the idea that coordinates can be fermionic as well as bosonic. Supergravity thus makes specific statements on the behaviour of space-time at small distances. Supergravity predicts N additional conserved, spinorial charges. The number N lies between 1 and 8; each value leads to a different candidate Lagrangian. The simplest case is called $N = 1$ supergravity.

In short, supersymmetry is a conjecture to unify matter and radiation at low energies. Many researchers conjectured that supersymmetry, and in particular $N = 1$ supergravity, might be an approximation to reality.

Supersymmetric models make a number of predictions that can be tested by experiment.

— Supersymmetry predicts partners to the usual elementary particles, called *sparticles*. Fermions are predicted to have boson partners, and vice versa. For example, supersymmetry predicts a *photino* as fermionic partner of the photon, *gluinos* as partner of the gluons, a *selectron* as partner of the electron, etc. However, none of these particles have been observed yet.

— Supersymmetry allows for the unification of the coupling constants in a way compatible with the data, as shown already above.

Page 248

— Supersymmetry slows down the proton decay rates predicted by grand unified theories. The slowed-down rates are compatible with observation.

— Supersymmetry predicts electric dipole moments for the neutron and other elementary particles. The largest predicted values for the neutron, $10^{-30} e$ m, are in contradiction with observations; the smallest predictions have not yet been reached by experiment. In comparison, the values expected from the standard model are at most $10^{-33} e$ m. This is a vibrant experimental research field that can save tax payers from financing an additional large particle accelerator.

Up to the year 2012, there was no experimental evidence for supersymmetry. In particular, the Large Hadron Collider at CERN in Geneva has not found any hint of supersymmetry.

Is supersymmetry an ingredient of the unified theory? The safe answer is: this is unclear. The optimistic answer is: there is still a small chance that supersymmetry is discovered. The pessimistic answer is: supersymmetry is a belief system contradicting observations and made up to correct the failings of grand unified theories. The last volume of this adventure will tell which answer is correct.

OTHER ATTEMPTS

If supersymmetry is not successful, it might be that even higher symmetries are required for unification. Therefore, researchers have explored *quantum groups*, *non-commutative space-time*, *conformal symmetry*, *topological quantum field theory*, and other abstract symmetries. None of these approaches led to useful results; neither experimental predictions nor progress towards unification. But two other approaches deserve special mention: duality symmetries and extensions to renormalization.

DUALITIES – THE MOST INCREDIBLE SYMMETRIES OF NATURE

An important discovery of mathematical physics took place in 1977, when Claus Montonen and David Olive proved that the standard concept of symmetry could be expanded dramatically in a different and new way.

The standard class of symmetry transformations, which turns out to be only the first class, acts on fields. This class encompasses gauge symmetries, space-time symmetries, motion reversal, parities, flavour symmetries and supersymmetry.

The second, new class is quite different. If we take all coupling constants of nature, we can imagine that they are members of a continuous space of all possible coupling constants, called the *parameter space*.* Montonen and Olive showed that there are transformations in parameter space that leave nature invariant. These transformations thus form a new, second class of symmetries of nature.

In fact, we already encountered a member of this class: renormalization symmetry. But Olive and Montonen expanded the symmetry class considerably by the discovery of *electromagnetic duality*. Electromagnetic duality is a discrete symmetry exchanging Vol. III, page 86

$$e \leftrightarrow \frac{4\pi\hbar c}{e} \qquad (104)$$

where the right hand side turns out to be the unit of magnetic charge. Electro-magnetic duality thus relates the electric charge e and the magnetic charge m

$$Q_{el} = me \quad \text{and} \quad Q_{mag} = ng = 2\pi\hbar c/e \qquad (105)$$

and puts them on equal footing. In other words, the transformation exchanges

$$\alpha \leftrightarrow \frac{1}{\alpha} \quad \text{or} \quad \frac{1}{137.04} \leftrightarrow 137.04 \;, \qquad (106)$$

and thus exchanges weak and strong coupling. In other words, electromagnetic duality

---

* The space of solutions for all value of the parameters is called the *moduli space*.

relates a regime where particles make sense (the low coupling regime) with one where particles do not make sense (the strong coupling regime). It is the most mind-boggling symmetry ever conceived.

Dualities are among the deepest connections of physics. They contain $\hbar$ and are thus intrinsically quantum. They do not exist in classical physics and show that quantum theory is more fundamental than classical physics. More clearly stated, dualities are intrinsically non-classical symmetries. Dualities confirm that quantum theory stands on its own.

If one wants to understand the values of unexplained parameters such as coupling constants, an obvious thing to do is to study *all* possible symmetries in parameter space, thus all possible symmetries of the second class, possibly combining them with those of the first symmetry class. (Indeed, the combination of duality with supersymmetry is

Vol. VI, page 132

studied in string theory.)

These investigations showed that there are several types of dualities, which all are *non-*

Ref. 213

perturbative symmetries:
— S duality, the generalization of electromagnetic duality for all interactions;
— T duality, also called space-time duality, a mapping between small and large lengths and times following $l \leftrightarrow l_{\mathrm{Pl}}^2/l$;*
— infrared dualities.

Despite the fascination of the idea, research into dualities has not led to experimental predictions. However, the results highlighted a different way to approach quantum field

Vol. VI, page 132

theory. Dualities play an important role in string theory, which we will explore later on.

COLLECTIVE ASPECTS OF QUANTUM FIELD THEORY

For many decades, mathematicians asked physicists: What is the essence of quantum field theory? Despite intensive research, this question has yet to be answered precisely.

Half of the answer is given by the usual definition found in physics textbooks: QFT is the most general known way to describe quantum mechanically continuous systems with a *finite* number of types of quanta but with an *infinite* number of degrees of freedom. (Of course, this definition implies that the Lagrangian must be relativistically invariant and must be described by a gauge theory.) However, this half of the answer is already

Vol. VI, page 36

sufficient to spell trouble. We will show in the next part of our ascent that neither space-time nor physical systems are continuous; in particular, we will discover that nature does not have infinite numbers of degrees of freedom. In other words, *quantum field theory is an effective theory*; this is the modern way to say that it is approximate, or more bluntly, that it is wrong.

The second, still partly unknown half of the answer would specify which (mathematical) conditions a physical system, i.e., a Lagrangian, actually needs to realize in order to become a quantum field theory. Despite the work of many mathematicians, no complete list of conditions is known yet. But it is known that the list includes at least two conditions. First of all, a quantum field theory must be *renormalizable*. Secondly, a quantum

---

* Space-time duality, the transformation between large and small sizes, leads one to ask whether there is

Vol. IV, page 103
Page 267

an inside and an outside to particles. We encountered this question already in our study of gloves. We will encounter the issue again below, when we explore eversion and inversion. The issue will be fully clarified only in the last volume of our adventure.

field theory must be *asymptotically free*; in other words, the coupling must go to zero when the energy goes to infinity. This condition ensures that interactions are defined properly. Only a subset of renormalizable Lagrangians obey this condition.

In four dimensions, the *only* known quantum field theories with these two properties are the non-Abelian gauge theories. These Lagrangians have several general aspects which are not directly evident when we arrive at them through the usual way, i.e., by generalizing naive wave quantum mechanics. This standard approach, the historical one, emphasizes the *perturbative* aspects: we think of elementary fermions as field quanta and of interactions as exchanges of virtual bosons, to various orders of perturbation.

On the other hand, all field theory Lagrangians also show two other configurations, apart from particles, which play an important role. These mathematical solutions appear when a non-perturbative point of view is taken; they are *collective* configurations.

— Quantum field theories show solutions which are static and of finite energy, created by non-local field combinations, called *solitons*. In quantum field theories, solitons are usually magnetic monopoles and dyons; also the famous *skyrmions* are solitons. In this approach to quantum field theory, it is assumed that the actual equations of nature are non-linear at high energy. Like in liquids, one then expects stable, local- Vol. I, page 273 ized and propagating solutions, the solitons. These solitons could be related to the observed particles.

— Quantum field theories show self-dual or anti-self dual solutions, called *instantons*. Instantons play a role in QCD, and could also play a role in the fundamental Lagrangian of nature.

— Quantum field theory defines particles and interactions using perturbation expansions. Do particles exist non-perturbatively? When does the perturbation expansion break down? What happens in this case? Despite these pressing issues, no answer has ever been found.

All these fascinating topics have been explored in great detail by mathematical physicists. Ref. 214 Even though this research has deepened the understanding of gauge theories, none of the available results has helped to approach unification.

CURIOSITIES ABOUT UNIFICATION

From the 1970s onwards, it became popular to draw graphs such as the one of Figure 139. This approach for the final theory of motion was due to the experimental success of electroweak unification and to the success among theoreticians of grand unification.

Unfortunately, grand unification contradicts experiment. In fact, as explained above, not even the electromagnetic and the weak interactions have been unified. (It took about Page 229 a decade to convince people about the contrary, by introducing the term 'electroweak unification' instead of a term akin to 'electroweak compatibility' or 'electroweak mixing'.) In other words, the larger part of Figure 139 is not correct. The graph and the story behind it has led most researchers along the wrong path for several decades.

* *

For a good and up-to-date introduction into modern particle research, see the summer student lectures at CERN, which can be found at http://cdsweb.cern.ch/collection/SummerStudentLectures?ln=en.

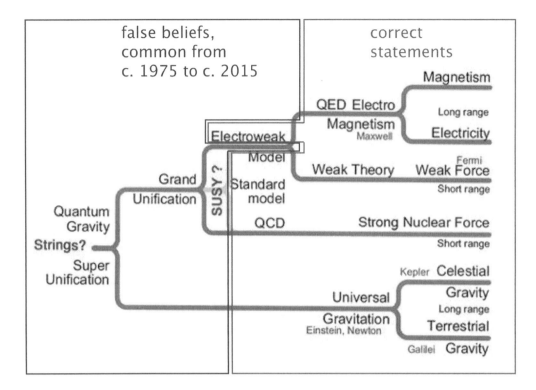

FIGURE 139 A typical graph on unification, as found in most publications on the topic, with its correct and incorrect parts (© CERN).

A SUMMARY ON MATHEMATICS AND HIGHER SYMMETRIES

Four decades of theoretical research has shown:

> ▷ Mathematical physics is *not* the way to search for unification.

All the searches for unification that were guided by *mathematical ideas* – by mathematical theorems or by mathematical generalizations – have failed. Mathematics is not helpful in this quest.

Four decades of of research have also shown a much more concrete result:

> ▷ The search for a *higher symmetry* in nature has failed.

Despite thousands of extremely smart people exploring many possible higher symmetries of nature, their efforts have not been successful.

It seems that researchers have fallen into the trap of music theory. Anybody who has learned to play an instrument has heard the statement that 'mathematics is at the basis of music'. Of course, this is nonsense; emotions are at the basis of music. But the incorrect statement about mathematics lurks in the head of every musician. Looking back to research in the twentieth century, it seems that the same has happened to physicists. From these failures we conclude:

> ▷ *Unification requires to extract an underlying* physical *principle.*

On the other hand, in the twentieth century, researchers have failed to find such a principle. This failure leads to two questions: Is there an alternative to the search for higher symmetry? And might researchers have relied on implicit incorrect assumptions about the structure of particles or of space-time? Before we explore these fascinating issues, we take a break to inspire us.

CHAPTER 11

BACTERIA, FLIES AND KNOTS

> “ La première et la plus belle qualité de la nature est le
> mouvement qui l'agite sans cesse ; mais ce
> mouvement n'est qu'une suite perpétuelle de crimes ;
> ce n'est que par des crimes qu'elle le conserve. ”
> Donatien de Sade, *Justine, ou les malheurs de la
> vertu*.*

Wobbly entities, in particular jellyfish or amoebas, open up a fresh vision of the world of motion, if we allow being led by the curiosity to study them in detail. We have missed many delightful insights by leaving them aside up to now. In fact, wobbly entities yield surprising connections between shape change and motion that will be of great use in the last part of our mountain ascent. Instead of continuing to look at the smaller and smaller, we now take a second look at everyday motion and its mathematical description.

To enjoy this chapter, we change a dear habit. So far, we always described any general example of motion as composed of the motion of *point particles*. This worked well in classical physics, in general relativity and in quantum theory; we based the approach on the silent assumption that during motion, each point of a complex system can be followed separately. We will soon discover that this assumption is *not* realized at smallest scales. Therefore the most useful description of motion of *extended* bodies uses methods that do not require that body parts be followed piece by piece. We explore these methods in this chapter; doing so is a lot of fun in its own right.

Vol. VI, page 108
If we describe elementary particles as extended entities – as we soon will have to – a particle moving through space is similar to a dolphin swimming through water, or to a bee flying through air, or to a vortex advancing in a liquid. Therefore we explore how this happens.

BUMBLEBEES AND OTHER MINIATURE FLYING SYSTEMS

If a butterfly passes by during our mountain ascent, we can stop a moment to appreciate a simple fact: a butterfly flies, and it is rather small. If we leave some cut fruit in the kitchen until it rots, we observe the even smaller fruit flies (*Drosophila melanogaster*),

---

* 'The primary and most beautiful of nature's qualities is motion, which agitates her at all times; but this motion is simply a perpetual consequence of crimes; she conserves it by means of crimes only.' Donatien Alphonse François de Sade (b. 1740 Paris, d. 1814 Charenton-Saint-Maurice) is the intense French writer from whom the term 'sadism' was deduced.

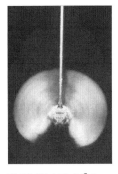
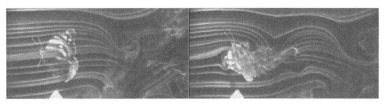

FIGURE 140 A flying fruit fly, tethered to a string (© Anton Unb)

FIGURE 141 Vortices around a butterfly wing (© Robert Srygley/Adrian Thomas).

just around one millimetre in size. Figure 140 shows a fruit fly in flight. If you have ever tried to build small model aeroplanes, or if you even only compare these insects to paper aeroplanes (possibly the smallest man-made flying thing you ever saw) you start to get a feeling for how well evolution optimized flying insects.

Compared to paper planes, insects also have engines, flapping wings, sensors, navigation systems, gyroscopic stabilizers, landing gear and of course all the features due to life, reproduction and metabolism, built into an incredibly small volume. Evolution really is an excellent engineering team. The most incredible flyers, such as the common house fly (*Musca domestica*), can change flying direction in only 30 ms, using the stabilizers that nature has given them by reshaping the original second pair of wings. Human engineers are getting more and more interested in the technical solutions evolution has developed; many engineers are trying to achieve similar miniaturization. The topic of miniature fly- Ref. 215 ing systems is extremely vast, so that we will pick out only a few examples.

How does a bumblebee (Bombus terrestris) fly? The lift mg generated by a *fixed* wing (as explained before) follows the empirical relation Vol. I, page 311

$$mg = f\, A\, v^2\, \rho \qquad (107)$$

where A is the surface of the wing, v is the speed of the wing in the fluid of density ρ. The factor f is a pure number, usually with a value between 0.2 and 0.4, that depends on the angle of the wing and its shape; here we use the average value 0.3. For a Boeing 747, the Ref. 216 surface is 511 m², the top speed at sea level is 250 m/s; at an altitude of 12 km the density Vol. I, page 37 of air is only a quarter of that on the ground, thus only 0.31 kg/m³. We deduce (correctly) that a Boeing 747 has a mass of about 300 ton. For bumblebees with a speed of 3 m/s and Challenge 141 e a wing surface of 1 cm², we get a lifted mass of about 35 mg, much less than the weight of the bee, namely about 1 g. The mismatch is even larger for fruit flies. In other words, an insect cannot fly if it keeps its wings *fixed*. It could not fly with fixed wings even if it had tiny propellers attached to them!

Due to the limitations of fixed wings at small dimensions, insects and small birds must *move* their wings, in contrast to aeroplanes. They must do so not only to take off or to gain height, but also to simply remain airborne in horizontal flight. In contrast, aeroplanes generate enough lift with *fixed* wings. Indeed, if you look at flying animals, such as the ones shown in Figure 142, you note that the larger they are, the less they need to move their wings (at cruising speed).

Challenge 142 s   Can you deduce from equation (107) that birds or insects can fly but people cannot? Conversely, the formula also (partly) explains why human-powered aeroplanes must be so large.*

But *how* do insects, small birds, flying fish or bats have to move their wings in order to fly? This is a tricky question and the answer has been uncovered only recently. The main point is that insect wings move in a way to produce eddies at the front edge which in turn thrust the insect upwards. Aerodynamic studies of butterflies – shown in Figure 141 – and studies of enlarged insect models moving in oil instead of in air are exploring the precise way insects make use of vortices. At the same time, more and more 'mechanical birds' and 'model aeroplanes' that use flapping wings for their propulsion are being built around the world. The field is literally in full swing.** Researchers are especially interested in understanding how vortices allow change of flight direction at the small dimensions typical for insects. Another aim is to reduce the size of flying machines.

Ref. 217
Ref. 218

The expression (107) for the lift of fixed wings also shows what is necessary for safe take-off and landing. The lift of all wings *decreases* for smaller speeds. Thus both animals and aeroplanes *increase* their wing surface in these occasions. Many birds also vigorously increase the flapping of wings in these situations. But even strongly flapping, enlarged wings often are insufficient for take-off. Many flying animals, such as swallows, therefore avoid landing completely. For flying animals which do take off from the ground, nature most commonly makes them hit the wings against each other, over their back, so that when the wings separate again, the low pressure between them provides the first lift. This method is used by insects and many birds, including pheasants. As bird watchers know, pheasants make a loud 'clap' when they take off. The clap is due to the low pressure region thus created.

Both wing use and wing construction depend on size. In fact, there are four types of wings in nature.

1. First of all, all large flying objects, such aeroplanes and large birds, fly using *fixed* wings, except during take-off and landing. This wing type is shown on the left-hand side of Figure 142.

2. Second, common size birds use *flapping* wings. (Hummingbirds can have over 50 wing beats per second.) These first two types of wings have a thickness of about 10 to 15 % of the wing depth. This wing type is shown in the centre of Figure 142.

---

* Another part of the explanation requires some aerodynamics, which we will not study here. Aerodynamics shows that the power consumption, and thus the resistance of a wing with given mass and given cruise speed, is inversely proportional to the square of the wingspan. Large wingspans with long slender wings are thus of advantage in (subsonic) flying, especially when energy is scarce.

One issue is mentioned here only in passing: why does an aircraft fly? The correct general answer is: *because it deflects air downwards.* How does an aeroplane achieve this? It can do so with the help of a tilted plank, a rotor, flapping wings, or a fixed wing. And when does a fixed wing deflect air downwards? First of all, the wing has to be tilted with respect to the air flow; in addition, the specific cross section of the wing can increase the downward flow. The relation between wing shape and downward flow is a central topic of applied aerodynamics.

** The website www.aniprop.de presents a typical research approach and the sites ovirc.free.fr and www.ornithopter.org give introductions into the way to build such systems for hobbyists.

FIGURE 142 Examples of the three larger wing types in nature, all optimized for rapid flows: turkey vulture (*Cathartes aura*), ruby-throated hummingbird (*Archilochus colubris*) and a dragonfly (© S.L. Brown, Pennsylvania Game Commission/Joe Kosack and nobodythere).

3. At smaller dimensions, a third wing type appears, the *membrane wing*. It is found in dragonflies and most everyday insects. At these scales, at Reynolds numbers of around 1000 and below, thin *membrane* wings are the most efficient. The *Reynolds number* measures the ratio between inertial and viscous effects in a fluid. It is defined as

$$R = \frac{lv\rho}{\eta} \qquad (108)$$

where l is a typical length of the system, v the speed, ρ the density and η the dynamic *viscosity* of the fluid.* A Reynolds number much larger than one is typical for rapid air flow and fast moving water. In fact, the Reynolds numbers specifies what is meant by a 'rapid' or 'fluid' flow on the one hand, and a 'slow' or 'viscous' flow on the other. This wing type is shown on the right-hand side of Figure 142. All the first three wing types, are all for *rapid* flows.

4. The fourth type of wings is found at the smallest possible dimensions, for insects smaller than one millimetre; their wings are not membranes at all. Typical are the cases of thrips and of parasitic wasps, which can be as small as 0.3 mm. All these small insects have wings which consist of a central *stalk* surrounded by hair. In fact, Figure 143 shows that some species of thrips have wings which look like miniature toilet brushes.

5. At even smaller dimensions, corresponding to Reynolds number below 10, nature does not use wings any more, though it still makes use of air transport. In principle, at the smallest Reynolds numbers gravity plays no role any more, and the process of flying merges with that of swimming. However, air currents are too strong compared

---

* The viscosity is the resistance to flow a fluid poses. It is defined by the force F necessary to move a layer of surface A with respect to a second, parallel one at distance d; in short, the (coefficient of) *dynamic viscosity* is defined as $\eta = dF/Av$. The unit is $1\,kg/s\,m$ or $1\,Pa\,s$ or $1\,N\,s/m^2$, once also called $10\,P$ or 10 poise. In other words, given a horizontal tube, the viscosity determines how strong the pump needs to be to pump the fluid through the tube at a given speed. The viscosity of air $20°C$ is $1.8 \times 10^{-5}\,kg/s\,m$ or $18\,\mu Pa\,s$ and increases with temperature. In contrast, the viscosity of liquids decreases with temperature. (Why?) The Challenge 143 ny viscosity of water at $0°C$ is $1.8\,mPa\,s$, at $20°C$ it is $1.0\,mPa\,s$ (or $1\,cP$), and at $40°C$ is $0.66\,mPa\,s$. Hydrogen has a viscosity smaller than $10\,\mu Pa\,s$, whereas honey has $25\,Pa\,s$ and pitch $30\,MPa\,s$.

Physicists also use a quantity v called the *kinematic viscosity*. It is defined with the help of the mass density of the fluid as $v = \eta/\rho$ and is measured in m^2/s, once called 10^4 stokes. The kinematic viscosity of water at $20°C$ is $1\,mm^2/s$ (or $1\,cSt$). One of the smallest values is that of acetone, with $0.3\,mm^2/s$; a larger one is glycerine, with $2000\,mm^2/s$. Gases range between $3\,mm^2/s$ and $100\,mm^2/s$.

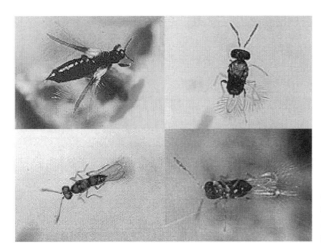

FIGURE 143 The wings of a few types of insects smaller than 1 mm (thrips, *Encarsia, Anagrus, Dicomorpha*) (HortNET).

with the speeds that such a tiny system could realize. No active navigation is then possible any more. At these small dimensions, which are important for the transport through air of spores and pollen, nature uses the air currents for passive transport, making use of special, but fixed shapes.

We summarize: active flying is only possible through shape change. Only two types of shape changes are possible for active flying: that of wings and that of propellers (or
Ref. 219
turbines). Engineers are studying with intensity how these shape changes have to take
Ref. 220
place in order to make flying most effective. Interestingly, a similar challenge is posed by swimming.

SWIMMING

Swimming is a fascinating phenomenon. The Greeks argued that the ability of fish to swim is a proof that water is made of atoms. If atoms would not exist, a fish could not advance through it. Indeed, swimming is an activity that shows that matter cannot be continuous. Studying swimming can thus be quite enlightening. But how exactly do fish swim?

Whenever dolphins, jellyfish, submarines or humans *swim*, they take water with their fins, body, propellers, hands or feet and push it backwards. Due to momentum conservation they then move forward.* In short, people swim in the same way that fireworks or rockets fly: by throwing matter behind them. This is macroscopic swimming. Does all swimming work in this way? In particular, do *small* organisms advancing through the molecules of a liquid use the same method? No. They use a different, *microscopic* way of swimming.

Small organisms such as bacteria do *not* have the capacity to propel or accelerate water against their surroundings. Indeed, the water remains attached around a microorganism without ever moving away from it. Physically speaking, in these cases of swimming the kinetic energy of the water is negligible. In order to swim, unicellular beings thus need to

---

Vol. I, page 85
* Fish could use propellers, as the arguments against wheels we collected at the beginning of our walk do not apply for swimming. But propellers with blood supply would be a weak point in the construction, and thus make fish vulnerable. Therefore, nature has not developed fish with propellers.

FIGURE 144 A swimming scallop (here from the genus *Chlamys*) (© Dave Colwell).

use other effects. In fact, their only possibility is to change their body shape in controlled ways. Seen from far away, the swimming of microorganisms thus resembles the motion of particles through vacuum: like microorganisms,* also particles have nothing to throw behind them.

A good way to distinguish macroscopic from microscopic swimming is provided by scallops. *Scallops* are molluscs up to a few cm in size; an example is shown in Figure 144. Scallops have a double shell connected by a hinge that they can open and close. If they close it *rapidly*, water is expelled and the mollusc is accelerated; the scallop then can glide for a while through the water. Then the scallop opens the shell again, this time *slowly*, and repeats the feat. When swimming, the larger scallops look like clockwork false teeth.

If we reduce the size of the scallop by a thousand times to the size of single cells we get a simple result: such a tiny scallop *cannot* swim. The lack of scalability of swimming methods is due to the changing ratio between inertial and dissipative effects at different scales. This ratio is measured by the Reynolds number. For the scallop the Reynolds number is about 100, which shows that when it swims, inertial effects are much more important than dissipative, viscous effects. For a bacterium the Reynolds number is much smaller than 1, so that inertial effects effectively play no role. There is no way to accelerate water *away* from a bacterial-sized scallop, and thus no way to glide. But this is not the only problem microorganism face when they want to swim.

A famous mathematical theorem states that no cell-sized being can move if the shape change is the same in the two halves of the motion (opening and closing). Such a shape change would simply make it move back and forward. Another mathematical theorem, the so-called *scallop theorem*, that states that no microscopic system can swim if it uses movable parts with only *one* degree of freedom. Thus it is impossible to move at cell dimensions using the method the scallop uses on centimetre scale.

Ref. 221

In order to swim, microorganisms thus need to use a more evolved, *two-dimensional motion* of their shape. Indeed, biologists found that all microorganisms use one of the following four swimming styles:

---

* There is an exception: gliding bacteria move by secreting slime, even though it is still unknown why this leads to motion.

1. Microorganisms of compact shape of diameter between 20 µm and about 20 mm, use *cilia*. Cilia are hundreds of little hairs on the surface of the organism. The organisms move the cilia in waves wandering around their surface, and these surface waves make the body advance through the fluid. All children watch with wonder *Paramecium*, the unicellular animal they find under the microscope when they explore the water in which some grass has been left for a few hours. *Paramecium*, which is between 100 µm and 300 µm in size, as well as many plankton species* use cilia for its motion. The cilia and their motion are clearly visible in the microscope. A similar swimming method is even used by some large animals; you might have seen similar waves on the borders of certain ink fish; even the motion of the manta (partially) belongs into this class. Ciliate motion is an efficient way to change the shape of a body making use of two dimensions and thus avoiding the scallop theorem.

   Ref. 222

2. Sperm and eukaryote microorganisms whose sizes are in the range between 1 µm and 50 µm swim using an (eukaryote) flagellum.** Flagella, Latin for 'small whips', work like flexible oars. Even though their motion sometimes appears to be just an oscillation, flagella get a kick only during one half of their motion, e.g. at every swing to the left. Flagella are indeed used by the cells like miniature oars. Some cells even twist their flagellum in a similar way that people rotate an arm. Some microorganisms, such as *Chlamydomonas*, even have two flagella which move in the same way as people move their legs when they perform the breast stroke. Most cells can also change the sense in which the flagellum is kicked, thus allowing them to move either forward or backward. Through their twisted oar motion, bacterial flagella avoid retracing the same path when going back and forward. As a result, the bacteria avoid the scallop theorem and manage to swim despite their small dimensions. The flexible oar motion they use is an example of a non-adiabatic mechanism; an important fraction of the energy is dissipated.

   Ref. 224
   Ref. 225

3. The smallest swimming organisms, bacteria with sizes between 0.2 µm and 5 µm, swim using *bacterial* flagella. These flagella, also called prokaryote flagella, are different from the ones just mentioned. Bacterial flagella move like turning corkscrews. They are used by the famous *Escherichia coli* bacterium and by all bacteria of the genus *Salmonella*. This type of motion is one of the prominent exceptions to the non-existence of wheels in nature; we mentioned it in the beginning of our walk. Corkscrew motion is an example of an adiabatic mechanism.

   Ref. 226

   A Coli bacterium typically has a handful of flagella, each about 30 nm thick and of corkscrew shape, with up to six turns; the turns have a 'wavelength' of 2.3 µm. Each flagellum is turned by a sophisticated rotation motor built into the cell, which the cell can control both in rotation direction and in angular velocity. For Coli bacteria, the range is between 0 and about 300 Hz.

   Vol. I, page 85
   Ref. 227

   A turning flagellum does not propel a bacterium like a propeller; as mentioned, the velocities involved are much too small, the Reynolds number being only about 10^{-4}. At these dimensions and velocities, the effect is better described by a corkscrew turning in honey or in cork: a turning corkscrew produces a motion against the material

---

* See the www.liv.ac.uk/ciliate website for an overview.

Ref. 223 ** The largest sperm, of 6 cm length, are produced by the 1.5 mm sized *Drosophila bifurca* fly, a relative of the famous *Drosophila melanogaster*.

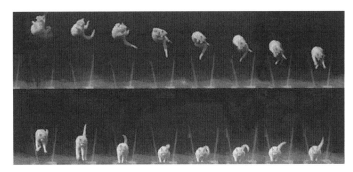

FIGURE 145 Cats can turn themselves, even with *no* initial angular momentum (photographs by Etienne-Jules Marey, 1894).

around it, in the direction of the corkscrew axis. The flagellum moves the bacterium in the same way that a corkscrew moves the turning hand with respect to the cork.

— One group of bacteria, the spirochaetes, move like a cork-screw through water. An example is *Rhodospirillum rubrum*, whose motion can be followed in the video on www.microbiologybytes.com/video/motility.com. These bacteria have an internal motor round an axial filament, that changes the cell shape in a non-symmetrical fashion and yield cork-screw motion. A different bacterium is *Spiroplasma*, a helical bacterium – not a spirochaete – that changes the cell shape, again in a non-symmetrical fashion, by propagating kink pairs along its body surface. Ref. 228

Note that still smaller bacteria do not swim at all. Indeed, each bacterium faces a minimum swimming speed requirement: it must outpace diffusion in the liquid it lives in. Ref. 229 Slow swimming capability makes no sense; numerous microorganisms therefore do not manage or do not try to swim at all. Some microorganisms are specialized to move along liquid–air interfaces. Others attach themselves to solid bodies they find in the liquid. Some of them are able to *move* along these solids. The amoeba is an example for a microorganism moving in this way. Also the smallest active motion mechanisms known, namely the motion of molecules in muscles and in cell membranes, work this way. Page 21

Let us summarize these observations in a different way. All known active motion, or self-propulsion, (in flat space) takes place in fluids – be it air or liquids. All active motion requires shape change. In order that shape change leads to motion, the environment, e.g. the water, must itself consist of moving components always pushing onto the swimming entity. The motion of the swimming entity can then be deduced from the particular shape change it performs. To test your intuition, you may try the following puzzle: is microscopic swimming possible in two spatial dimensions? In four? Challenge 144 s

ROTATION, FALLING CATS AND THE THEORY OF SHAPE CHANGE

At small dimensions, flying and swimming takes place through phase change. In the last decades, the description of shape change has changed from a fashionable piece of research to a topic whose results are both appealing and useful. There are many studies, both experimental and theoretical, about the exact way small systems move in water and air, about the achievable and achieved efficiency, and much more. The focus is on motion through translation.

But shape change can also lead to a *rotation* of a body. In this case, the ideas are not

FIGURE 146 Humans can turn themselves in mid air like cats: see the second, lateral rotation of Artem Silchenko, at the 2006 cliff diving world championship (© World High Diving Federation).

restricted to microscopic systems, but apply at all scales. In particular, the theory of shape change is useful in explaining how falling cats manage to fall always on their feet. Cats are not born with this ability; they have to learn it. But the feat has fascinated people for centuries, as shown in the ancient photograph given in Figure 145. In fact, cats confirm in three dimensions what we already knew for two dimensions: a deformable body can change its own orientation in space without outside help. Also humans can perform the feat: simply observe the second, lateral rotation of the diver in Figure 146. Astronauts in the space station and passengers of parabolic 'zero-gravity' flights regularly do the same, as do many artificial satellites sent into space.

In the 1980s, the work by Michael Berry, Wilczek and Zee as well as by Shapere and Wilczek made the point that all motion due to shape change is described by a gauge theory. The equivalence is shown in Table 22. A simple and beautiful example for these ideas has been given by Putterman and Raz and is shown in Figure 147. Imagine four spheres on perfect ice, all of the same mass and size, connected by four rods forming a parallelogram. Now imagine that this parallelogram can change length along one side, called a, and that it can also change the angle θ between the sides. Putterman and Raz call this the *square cat*. The figure shows that the square cat can change its own orientation on the ice while, obviously, keeping its centre of mass at rest. The figure also shows that this only works because the two motions that the cat can perform, the stretching and the

Vol. I, page 111

Ref. 230
Ref. 231

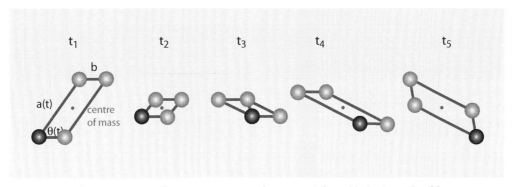

FIGURE 147 The square cat: in free space, or on perfect ice, a deformable body made of four masses that can change one body angle and one extension is able to rotate itself.

TABLE 22 The correspondence between shape change and gauge theory.

CONCEPT	SHAPE CHANGE	GAUGE THEORY
System	deformable body	matter–field combination
Gauge freedom	freedom of description of body orientation and position	freedom to define vector potential
Gauge-dependent quantity	shape's angular orientation and position	vector potential, phase
	orientation and position change along an open path	vector potential and phase change along open path
Gauge transformation	changes angular orientation and position	changes vector potential
Gauge-independent quantities	orientation and position after full stroke	phase difference on closed path, integral of vector potential along a closed path
	deformations	field strengths
Gauge group	e.g. possible rotations SO(3) or motions E(3)	U(1), SU(2), SU(3)

angle change, do *not* commute.

The rotation of the square cat occurs in *strokes*; large rotations are achieved by repeating strokes, similar to the situation of swimmers. If the square cat would be swimming in a liquid, the cat could thus rotate itself – though it could not advance.

When the cat rotates itself, each stroke results in a rotation angle that is independent of the speed of the stroke. (The same experience can be made when rotating oneself on an office chair by rotating the arm above the head: the chair rotation angle after arm turn is independent of the arm speed.) This leads to a puzzle: what is the largest angle that a cat can turn in one stroke?

Challenge 145 d

Rotation in strokes has a number of important implications. First of all, the number of strokes is a quantity that all observers agree upon: it is observer-invariant. Secondly, the orientation change after a *complete* stroke is also observer-invariant. Thirdly, the orientation change for *incomplete* strokes is observer-dependent: it depends on the way that

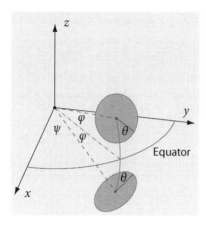

FIGURE 148 Swimming on a curved surface using two discs.

orientation is defined. For example, if orientation is defined by the direction of the body diagonal through the blue mass (see Figure 147), it changes in a certain way during a stroke. If the orientation is defined by the direction of the fixed bar attached to the blue mass, it changes in a different way during a stroke. Only when a full stroke is completed do the two values coincide. Mathematicians say that the choice of the definition and thus the value of the orientation is *gauge-dependent*, but that the value of the orientation change at a full stroke is *gauge-invariant*.

In summary, the square cat shows two interesting points. First, the orientation of a deformable body can change if the deformations it can perform are *non-commuting*. Secondly, such deformable bodies are described by *gauge theories*: certain aspects of the bodies are gauge-invariant, others are gauge-dependent. This summary leads to a question: can we use these ideas to increase our understanding of the gauge theories of the electromagnetic, weak and strong interaction? Shapere and Wilczek say no. We will find out definitively in the next volume. In fact, shape change bears even more surprises.

SWIMMING IN CURVED SPACE

In flat space it is not possible to produce translation through shape change. Only orientation changes are possible. Surprisingly, if space is *curved*, motion does become possible.
Ref. 232 A simple example was published in 2003 by Jack Wisdom. He found that cyclic changes in the shape of a body can lead to net translation, a rotation of the body, or both.

Indeed, we know from Galilean physics that on a frictionless surface we cannot move, but that we can change orientation. This is true only for a flat surface. On a curved surface, we can use the ability to turn and translate it into motion.

Take two massive discs that lie on the surface of a frictionless, spherical planet, as shown in Figure 148. Consider the following four steps: 1. the disc separation φ is increased by the angle $\Delta\varphi$, 2. the discs are rotated oppositely about their centres by the angle $\Delta\theta$, 3. their separation is decreased by $-\Delta\varphi$, and 4. they are rotated back by $-\Delta\theta$. Due to the conservation of angular momentum, the two-disc system changes its longi-
Challenge 146 ny tude $\Delta\psi$ as

$$\Delta\psi = \frac{1}{2}\gamma^2\Delta\theta\Delta\varphi \;, \tag{109}$$

where γ is the angular radius of the discs. This cycle can be repeated over and over. The cycle it allows a body, located on the surface of the Earth, to swim along the surface. Unfortunately, for a body of size of one metre, the motion for each swimming cycle is only around 10^{-27} m.

Wisdom showed that the same procedure also works in curved space, thus in the presence of gravitation. The mechanism thus allows a falling body to swim away from the path of free fall. Unfortunately, the achievable distances for everyday objects are negligibly small. Nevertheless, the effect exists.

In other words, there is a way to swim through curved space that looks similar to swimming at low Reynolds numbers, where swimming results of simple shape change. Does this tell us something about fundamental descriptions of motion? The last part of our ascent will tell.

Turning a sphere inside out

> A text should be like a lady's dress; long enough to cover the subject, yet short enough to keep it interesting.
>
> Anonymous

Exploring the theme of motion of wobbly entities, a famous example cannot be avoided. In 1957, the mathematician Stephen Smale proved that a sphere can be turned inside out. The discovery brought him the Fields medal in 1966, the highest prize for discoveries in mathematics. Mathematicians call his discovery the *eversion* of the sphere.   Ref. 233

To understand the result, we need to describe more clearly the rules of mathematical eversion. First of all, it is assumed that the sphere is made of a thin membrane which has the ability to stretch and bend without limits. Secondly, the membrane is assumed to be able to *intersect* itself. Of course, such a ghostly material does not exist in everyday life; but in mathematics, it can be imagined. A third rule requires that the membrane must be deformed in such a way that the membrane is not punctured, ripped nor creased; in short, everything must happen *smoothly* (or differentially, as mathematicians like to say).

Even though Smale proved that eversion is possible, the first way to actually perform it was discovered by the blind topologist Bernard Morin in 1961, based on ideas of Arnold Shapiro. After him, several additional methods have been discovered.   Ref. 234, Ref. 235

Several computer videos of sphere eversions are now available.* The most famous ones are *Outside in*, which shows an eversion due to William P. Thurston, and *The Optiverse*, which shows the most efficient method known so far, discovered by a team led by John Sullivan and shown in Figure 149.   Ref. 236

Why is sphere eversion of interest to physicists? If elementary particles were extended and at the same time were of spherical shape, eversion might be a particle symmetry. To see why, we summarize the effects of eversion on the whole surrounding space, not only

---

* Summaries of the videos can be seen at the website www.geom.umn.edu/docs/outreach/oi, which also has a good pedagogical introduction. Another simple eversion and explanation is given by Erik de Neve on his website www.usefuldreams.org/sphereev.htm. It is even possible to run the eversion film software at home; see the website www.cslub.uwaterloo.ca/~mjmcguff/eversion. Figure 149 is from the website new.math.uiuc.edu/optiverse.

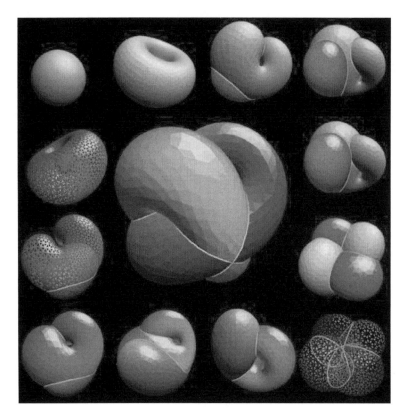

FIGURE 149 A way to turn a sphere inside out, with intermediate steps ordered clockwise (© John Sullivan).

on the sphere itself. The final effect of eversion is the transformation

$$(x, y, z) \rightarrow \frac{(x, y, -z)\, R^2}{r^2} \qquad (110)$$

where R is the radius of the sphere and r is the length of the coordinate vector (x, y, z), thus $r = \sqrt{x^2 + y^2 + z^2}$. Due to the minus sign in the z-coordinate, eversion differs from inversion, but not by too much. As we will find out shortly, a transformation similar to eversion, space-time duality, is a fundamental symmetry of nature.

Vol. VI, page 106

Clouds

Clouds are another important class of wobbly objects. The lack of a definite boundary makes them even more fascinating than amoebas, bacteria or falling cats. We can observe the varieties of clouds from any aeroplane.

Vol. III, page 178
The common cumulus or cumulonimbus in the sky, like all the other meteorological clouds, are vapour and water droplet clouds. Galaxies are clouds of stars. Stars are clouds of plasma. The atmosphere is a gas cloud. Atoms are clouds of electrons. Nuclei are clouds of protons and neutrons, which in turn are clouds of quarks. Comparing different cloud types is illuminating and fun.

Clouds of all types can be described by a shape and a size, even though in theory they have no bound. An *effective* shape and size can be defined by that region in which the cloud density is only, say, 1 % of the maximum density; slightly different procedures can also be used. All clouds are described by *probability densities* of the components making up the cloud. All clouds show *conservation* of the number of their constituents.

Whenever we see a cloud, we can ask why it does not collapse. Every cloud is an aggregate; all aggregates are kept from collapse in only *three* ways: through rotation, through pressure, or through the Pauli principle, i.e., the quantum of action. For example, galaxies are kept from collapsing by rotation. Most stars, the atmosphere and rain clouds are kept from collapsing by gas pressure. Neutron stars, the Earth, atomic nuclei, protons or the electron clouds of atoms are kept apart by the quantum of action.

Vol. I, page 225

A rain cloud is a method to keep several thousand tons of water suspended in the air. Can you explain what keeps it afloat, and what else keeps it from continuously diffusing into a thinner and thinner structure?

Challenge 147 s

Two rain clouds can merge. So can two atomic electron clouds. So can galaxies. But only atomic clouds are able to *cross* each other. We remember that a normal atom can be inside a Rydberg atom and leave it again without change. In contrast, rain clouds, stars, galaxies or other macroscopic clouds cannot cross each other. When their paths cross, they can only merge or be ripped into pieces. Due to this lack of crossing ability, only microscopic clouds can be counted. In the macroscopic cases, there is no real way to define a 'single' cloud in an accurate way. If we aim for full precision, we are unable to claim that there is more than one rain cloud, as there is no clear-cut boundary between them. Electronic clouds are different. True, in a piece of solid matter we can argue that there is only a single electronic cloud throughout the object; however, when the object is divided, the cloud is divided in a way that makes the original atomic clouds reappear. We thus can speak of 'single' electronic clouds.

Vol. IV, page 172

If one wants to be strict, galaxies, stars and rain clouds can be seen as made of localized particles. Their cloudiness is only apparent. Could the same be true for electron clouds? And what about space itself? Let us explore some aspects of these questions.

VORTICES AND THE SCHRÖDINGER EQUATION

Fluid dynamics is a topic with many interesting aspects. Take the *vortex* that can be observed in any deep, emptying bath tub: it is an extended, one-dimensional 'object', it is deformable, and it is observed to wriggle around. Larger vortices appear as tornadoes on Earth and on other planets, as *waterspouts*, and at the ends of wings or propellers of all kinds. Smaller, quantized vortices appear in superfluids. An example is shown in Figure 150; also the spectacular fire whirls and fire tornados observed every now and then are vortices.

Page 91

Vortices, also called *vortex tubes* or *vortex filaments*, are thus wobbly entities. Now, a beautiful result from the 1960s states that a vortex filament in a rotating liquid is described by the one-dimensional Schrödinger equation. Let us see how this is possible.

Ref. 237

Any deformable linear vortex, as illustrated in Figure 151, is described by a continuous set of position vectors $r(t, s)$ that depend on time t and on a single parameter s. The parameter s specifies the relative position along the vortex. At each point on the vortex, there is a unit tangent vector $e(t, s)$, a unit normal curvature vector $n(t, s)$ and a unit

FIGURE 150 A vortex in nature: a waterspout (© Zé Nogueira).

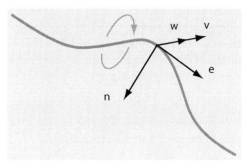

FIGURE 151 The mutually perpendicular tangent e, normal n, torsion w and velocity v of a vortex in a rotating fluid.

torsion vector $w(t, s)$. The three vectors, shown in Figure 151, are defined as usual as

$$e = \frac{\partial r}{\partial s} \, ,$$
$$\kappa n = \frac{\partial e}{\partial s} \, ,$$
$$\tau w = -\frac{\partial(e \times n)}{\partial s} \, , \tag{111}$$

where κ specifies the value of the curvature and τ specifies the value of the torsion. In general, both numbers depend on time and on the position along the line.

In the simplest possible case the rotating environment induces a local velocity v for the vortex that is proportional to the curvature κ, perpendicular to the tangent vector e and perpendicular to the normal curvature vector n:

$$v = \eta\kappa(e \times n)\,, \tag{112}$$

where η is the so-called *coefficient of local self-induction* that describes the coupling be- Ref. 237
tween the liquid and the vortex motion. This is the evolution equation of the vortex.

We now assume that the vortex is deformed only *slightly* from the straight configuration. Technically, we are thus in the *linear* regime. For such a linear vortex, directed along the x-axis, we can write

$$r = (x, y(x, t), z(x, t))\,. \tag{113}$$

Slight deformations imply $\partial s \approx \partial x$ and therefore

$$e = (1, \frac{\partial y}{\partial x}, \frac{\partial z}{\partial x}) \approx (1, 0, 0)\,,$$

$$\kappa n \approx (0, \frac{\partial^2 y}{\partial x^2}, \frac{\partial^2 z}{\partial x^2})\,, \text{ and}$$

$$v = (0, \frac{\partial y}{\partial t}, \frac{\partial z}{\partial t})\,. \tag{114}$$

We can thus rewrite the evolution equation (112) as

$$(0, \frac{\partial y}{\partial t}, \frac{\partial z}{\partial t}) = \eta\,(0, -\frac{\partial^2 z}{\partial x^2}, \frac{\partial^2 y}{\partial x^2})\,. \tag{115}$$

This equation is well known; if we drop the first coordinate and introduce complex numbers by setting $\Phi = y + iz$, we can rewrite it as

$$\frac{\partial\Phi}{\partial t} = i\eta\frac{\partial^2\Phi}{\partial x^2}\,. \tag{116}$$

This is the one-dimensional Schrödinger equation for the evolution of a free wave function! The complex function Φ specifies the transverse deformation of the vortex. In other Ref. 238
words, we can say that the Schrödinger equation in one dimension describes the evolution of the deformation for an almost linear vortex in a rotating liquid. We note that there is no constant $\hbar$ in the equation, as we are exploring a *classical* system.

Schrödinger's equation is linear in Φ. Therefore the fundamental solution is

$$\Phi(x, y, z, t) = a\,e^{i(\tau x - \omega t)} \quad \text{with} \quad \omega = \eta\tau^2 \quad \text{and} \quad \kappa = a\tau^2\,, \tag{117}$$

The amplitude a and the wavelength or pitch $b = 1/\tau$ can be freely chosen, as long as

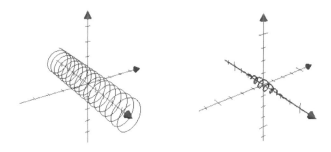

FIGURE 152 Motion of a vortex: the fundamental helical solution and a moving helical 'wave packet'.

the approximation of small deviation is fulfilled; this condition translates as $a \ll b$.* In the present interpretation, the fundamental solution corresponds to a vortex line that is deformed into a *helix*, as shown in Figure 152. The angular speed ω is the rotation speed around the axis of the helix.

Challenge 148 ny A helix moves along the axis with a speed given by

$$v_{\text{helix along axis}} = 2\eta\tau \; . \tag{118}$$

In other words, for extended entities following evolution equation (112), rotation and translation are coupled.** The momentum p can be defined using $\partial\Phi/\partial x$, leading to

$$p = \tau = \frac{1}{b} \; . \tag{119}$$

Momentum is thus inversely proportional to the helix wavelength or pitch, as expected. The energy E is defined using $\partial\Phi/\partial t$, leading to

$$E = \eta\tau^2 = \frac{\eta}{b^2} \; . \tag{120}$$

Energy and momentum are connected by

$$E = \frac{p^2}{2\mu} \quad \text{where} \quad \mu = \frac{1}{2\eta} \; . \tag{121}$$

---

* The curvature is given by $\kappa = a/b^2$, the torsion by $\tau = 1/b$. Instead of $a \ll b$ one can thus also write $\kappa \ll \tau$.
Challenge 149 ny ** A wave packet moves along the axis with a speed given by $v_{\text{packet}} = 2\eta\tau_0$, where τ_0 is the torsion of the helix of central wavelength.

In other words, a vortex with a coefficient η – describing the coupling between environment and vortex – is thus described by a number μ that behaves like an effective mass. We can also define the (real) quantity $|\Phi| = a$; it describes the *amplitude* of the deformation.

Page 271

In the Schrödinger equation (116), the second derivative implies that the deformation 'wave packet' has tendency to spread out over space. Can you confirm that the wavelength–frequency relation for a vortex wave group leads to something like the indeterminacy relation (however, without a $\hbar$ appearing explicitly)?

Challenge 150 ny

In summary, the complex amplitude Φ for a linear vortex in a rotating liquid behaves like the one-dimensional wave function of a non-relativistic free particle. In addition, we found a suggestion for the reason that complex numbers appear in the Schrödinger equation of quantum theory: they could be due to the intrinsic rotation of an underlying substrate. Is this suggestion correct? We will find out in the last part of our adventure.

Vol. VI, page 160

FLUID SPACE-TIME

General relativity shows that space can move and oscillate: space is a wobbly entity. Is space more similar to clouds, to fluids, or to solids?

An intriguing approach to space-time as a fluid was published in 1995 by Ted Jacobson. He explored what happens if space-time, instead of assumed to be continuous, is assumed to be the statistical average of numerous components moving in a disordered fashion.

Ref. 239

The standard description of general relativity describes space-time as an entity similar to a flexible mattress. Jacobson studied what happens if the mattress is assumed to be made of a fluid. A fluid is a collection of (undefined) components moving randomly and described by a temperature varying from place to place.

Vol. II, page 134

Jacobson started from the Fulling–Davies–Unruh effect and assumed that the local fluid temperature is given by a multiple of the local gravitational acceleration. He also used the proportionality – correct on horizons – between area and entropy. Since the energy flowing through a horizon can be called heat, one can thus translate the expression $\delta Q = T\delta S$ into the expression $\delta E = a\delta A(c^2/4G)$, which describes the behaviour of space-time at horizons. As we have seen, this expression is fully equivalent to general relativity.

Page 114

Vol. VI, page 30

In other words, imagining space-time as a fluid is a powerful analogy that allows deducing general relativity. Does this mean that space-time actually *is* similar to a fluid? So far, the analogy is not sufficient to answer the question and we have to wait for the last part of our adventure to settle it. In fact, just to confuse us a bit more, there is an old argument for the opposite statement.

DISLOCATIONS AND SOLID SPACE-TIME

General relativity tells us that space behaves like a deformable mattress; space thus behaves like a solid. There is a second argument that underlines this point and that exerts a continuing fascination. This argument is connected to a famous property of the motion of dislocations.

Dislocations are one-dimensional construction faults in crystals, as shown in Figure 153. A general dislocation is a mixture of the two pure dislocation types: *edge dislocations* and *screw dislocations*. Both are shown in Figure 153.

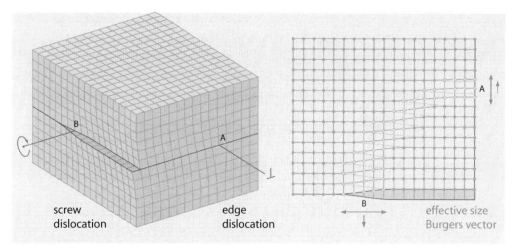

FIGURE 153 The two pure dislocation types, edge and screw dislocations, seen from the outside of a cubic crystal (left) and the mixed dislocation – a quarter of a dislocation loop – joining them in a horizontal section of the same crystal (right) (© Ulrich Kolberg).

Challenge 151 e

If one explores how the atoms involved in dislocations can rearrange themselves, one finds that edge dislocations can only move perpendicularly to the added plane. In contrast, screw dislocations can move in all directions.* An important case of general, mixed dislocations, i.e., of mixtures of edge and screw dislocations, are closed *dislocation rings*. On such a dislocation ring, the degree of mixture changes continuously from place to place.

Any dislocation is described by its *strength* and by its effective *size*; they are shown, respectively, in red and blue in Figure 153. The *strength* of a dislocation is measured by the so-called *Burgers vector*; it measures the misfits of the crystal around the dislocation. More precisely, the Burgers vector specifies by how much a section of perfect crystal needs to be displaced, after it has been cut open, to produce the dislocation. Obviously, the strength of a dislocation is quantized in multiples of a minimal Burgers vector. In fact, dislocations with large Burgers vectors can be seen as composed of dislocations of minimal Burgers vector, so that one usually studies only the latter.

The size or *width* of a dislocation is measured by an *effective width w*. Also the width is a multiple of the lattice vector. The width measures the size of the deformed region of the crystal around the dislocation. Obviously, the size of the dislocation depends on the elastic properties of the crystal, can take continuous values and is direction-dependent. The width is thus related to the energy content of a dislocation.

A general dislocation can move, though only in directions which are both perpendicular to its own orientation and to its Burgers vector. Screw dislocations are simpler: they can move in any direction. Now, the motion of screw dislocations has a peculiar property. We call c the speed of sound in a pure (say, cubic) crystal. As Frenkel and Kontorowa

---

* See the uet.edu.pk/dmems/edge_dislocation.htm, uet.edu.pk/dmems/screw_dislocation.htm and uet.edu.pk/dmems/mixed_dislocation.htm web pages to watch a moving dislocation.

found in 1938, when a screw dislocation moves with velocity v, its width w changes as   Ref. 240

$$w = \frac{w_0}{\sqrt{1 - v^2/c^2}} \; .$$ (122)

In addition, the energy of the moving dislocation obeys

$$E = \frac{E_0}{\sqrt{1 - v^2/c^2}} \; .$$ (123)

A screw dislocation thus cannot move faster than the speed of sound c in a crystal and its width shows a speed-dependent contraction. (Edge dislocations have similar, but more complex behaviour.) The motion of screw dislocations in solids is thus described by the same effects and formulae that describe the motion of bodies in special relativity; the speed of sound is the limit speed for dislocations in the same way that the speed of light is the limit speed for objects.

Does this mean that elementary particles are dislocations of space or even of space-time, maybe even dislocation rings? The speculation is appealing, even though it supposes that space-time is a solid crystal, and thus contradicts the model of space or space-time as a fluid. Worse, we will soon encounter other reasons to reject modelling space-   Vol. VI, page 70
time as a lattice; maybe you can find a few arguments already by yourself. Still, expres-   Challenge 152 s
sions (122) and (123) for dislocations continue to fascinate.

At this point, we are confused. Space-time seems to be solid and liquid at the same time. Despite this contrast, the discussion somehow gives the impression that there is something waiting to be discovered. But what? We will find out in the last part of our adventure.

POLYMERS

The study of polymers is both economically important and theoretically fascinating. Poly-   Ref. 241
mers are materials built of long and flexible macromolecules that are sequences of many ('poly' in Greek) similar monomers. These macromolecules are thus wobbly entities.

Polymers form *solids*, like rubber or plexiglas, *melts*, like those used to cure teeth, and many kinds of *solutions*, like glues, paints, eggs, or people. Polymer gases are of lesser importance.

All the material properties of polymers, such as their elasticity, their viscosity, their electric conductivity or their unsharp melting point, depend on the number of monomers and the topology of their constituent molecules. In many cases, this dependence can be calculated. Let us explore an example.

If L is the contour length of a free, ideal, unbranched polymer molecule, the *average* end-to-end distance R is proportional to the square root of the length L:

$$R = \sqrt{Ll} \sim \sqrt{L} \quad \text{or} \quad R \sim \sqrt{N}$$ (124)

where N is the number of monomers and l is an effective monomer length describing the

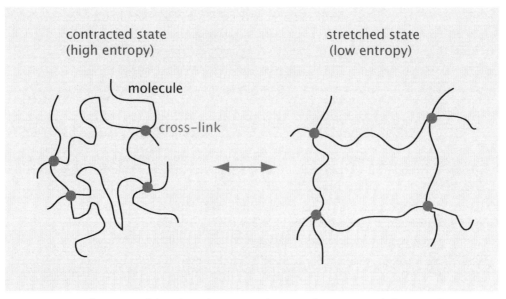

FIGURE 154 An illustration of the relation between polymer configurations and elasticity. The molecules in the stretched situation have fewer possible shape configurations and thus lower entropy; therfore, the material tends back to the contracted situation.

scale at which the polymer molecule is effectively stiff. R is usually much smaller than L; this means that free, ideal polymer molecules are usually in a *coiled* state.

Obviously, the end-to-end distance R varies from molecule to molecule, and follows a Gaussian distribution for the probability P of a end-to-end distance R:

$$P(R) \sim e^{\frac{-3R^2}{2Nl^2}} \ . \tag{125}$$

The average end-to-end distance mentioned above is the root-mean-square of this distribution. Non-ideal polymers are polymers which have, like non-ideal gases, interactions with neighbouring molecules or with solvents. In practice, polymers follow the ideal behaviour quite rarely: polymers are ideal only in certain solvents and in melts.

If a polymer is *stretched*, the molecules must rearrange. This changes their entropy and produces an elastic force f that tries to inhibit the stretching. For an ideal polymer, the force is not due to molecular interactions, but is entropic in nature. Therefore the force can be deduced from the *free energy*

$$F \sim -T \ln P(R) \tag{126}$$

of the polymer: the force is then simply given as $f = \partial F(R)/\partial R$. For an ideal polymer, using its probability distribution, the force turns out to be proportional to the stretched length. Thus the spring constant k can be introduced, given by

Challenge 153 e

$$k = \frac{f}{R} = \frac{3T}{Ll} \ . \tag{127}$$

We thus deduced a material property, the spring constant k, from the simple idea that polymers are made of long, flexible molecules. The proportionality to temperature T is a result of the entropic nature of the force; the dependence on L shows that longer molecules are more easy to stretch. For a real, non-ideal polymer, the calculation is more complex, but the procedure is the same. Indeed, this is the mechanism at the basis of the elasticity of rubber.

Using the free energy of polymer conformations, we can calculate the material properties of macromolecules in many other situations, such as their reaction to compression, their volume change in the melt, their interactions in solutions, the effect of branched molecules, etc. This is a vast field of knowledge on its own, which we do not pursue here. Modern research topics include the study of knotted polymers and the study of polymer mixtures. Extensive computer calculations and experiments are regularly compared.

Do polymers have some relation to the structure of physical space? The issue is open. It is sure, however, that polymers are often knotted and linked.

KNOTS AND LINKS

> Don't touch this, or I shall tie your fingers into knots!
>
> (Nasty, but surprisingly efficient child education technique.)

Knots and their generalization are central to the study of wobbly object motion. A *(mathematical) knot* is a closed piece of rubber string, i.e., a string whose ends have been glued together, which cannot be deformed into a circle or a simple loop. The simple loop is also called the *trivial knot*.

Knots are of importance in the context of this chapter as they visualize the limitations of the motion of wobbly entities. In addition, we will discover other reasons to study knots later on. In this section, we just have a bit of fun.*   Ref. 242

In 1949, Schubert proved that every knot can be decomposed in a unique way as sum of prime knots. Knots thus behave similarly to integers.   Ref. 243

If prime knots are ordered by their crossing numbers, as shown in Figure 155, the trivial knot (0_1) is followed by the trefoil knot (3_1) and by the figure-eight knot (4_1). The figure only shows *prime knots*, i.e., knots that cannot be decomposed into two knots that are connected by two parallel strands. In addition, the figure only shows one of the often possible two mirror versions.

Together with the search for invariants, the tabulation of knots – a result of their classification – is a modern mathematical sport. Flat knot diagrams are usually ordered by the minimal number of crossings as done in Figure 155. There is 1 knot with zero, 1 with three and 1 with four crossings (not counting mirror knots); there are 2 knots with five   Ref. 243 and 3 with six crossings, 7 knots with seven, 21 knots with eight, 41 with nine, 165 with ten, 552 with eleven, 2176 with twelve, 9988 with thirteen, 46 972 with fourteen, 253 293 with fifteen and 1 388 705 knots with sixteen crossings.

The mirror image of a knot usually, but not always, is different from the original. If

---

* Beautiful illustrations and detailed information about knots can be found on the Knot Atlas website at katlas.math.toronto.edu and at the KnotPlot website at www.knotplot.com.

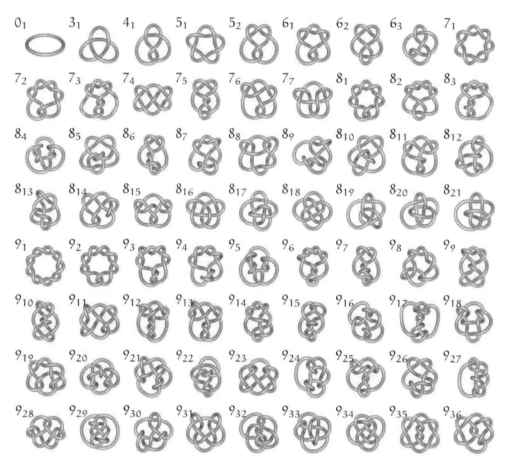

FIGURE 155 The knot diagrams for the simplest prime knots (© Robert Scharein).

you want a challenge, try to show that the trefoil knot, the knot with three crossings, is different from its mirror image. The first mathematical proof was by Max Dehn in 1914.

Antiknots do not exist. An *antiknot* would be a knot on a rope that cancels out the corresponding knot when the two are made to meet along the rope. It is easy to prove that this is impossible. We take an infinite sequence of knots and antiknots on a string, $K − K + K − K + K − K$.... On the one hand, we could make them disappear in this way $K − K + K − K + K − K... = (K − K) + (K − K) + (K − K)... = 0$. On the other hand, we could do the same thing using $K − K + K − K + K − K... = K(−K + K) + (−K + K) + (−K + K)... = K$. The only knot K with an antiknot is thus the unknot $K = 0$.[*]

How do we describe such a knot through the telephone? Mathematicians have spent a lot of time to figure out smart ways to achieve it. The obvious way is to flatten the knot onto a plane and to list the position and the type (below or above) of the crossings. (See Figure 156.) But what is the *simplest* way to describe knots by the telephone? The task is not completely finished, but the end is in sight. Mathematicians do not talk about 'telephone messages', they talk about *knot invariants*, i.e., about quantities that do not depend

Ref. 244

---

[*] This proof does *not* work when performed with numbers; we would be able to deduce $1 = 0$ by setting $K=1$. Why is this proof valid with knots but not with numbers?

Challenge 154 s

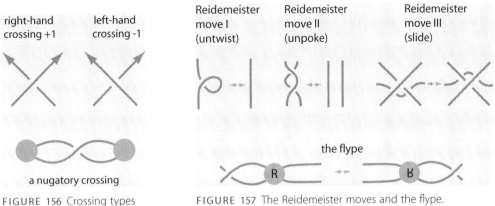

FIGURE 156 Crossing types in knots.

FIGURE 157 The Reidemeister moves and the flype.

FIGURE 158 A tight open overhand knot and a tight open figure-eight knot (© Piotr Pieranski)

on the precise shape of the knot. At present, the best description of knots use polynomial invariants. Most of them are based on a discovery by Vaughan Jones in 1984. However, though the Jones polynomial allows us to uniquely describe most simple knots, it fails to do so for more complex ones. But the Jones polynomial finally allowed to prove that a diagram which is alternating and eliminates nugatory crossings (i.e. if it is 'reduced') is indeed one which has minimal number of crossings. The polynomial also allows showing that any two reduced alternating diagrams are related by a sequence of flypes.

In short, the simplest way to describe a knot through the telephone is to give its Kauffman polynomial, together with a few other polynomials.

Since knots are stable in time, a knotted line in three dimensions is equivalent to a knotted surface in space-time. When thinking in higher dimensions, we need to be careful. Every knot (or knotted line) can be untied in four or more dimensions. However, there is no surface *embedded* in four dimensions which has as $t = 0$ slice a knot, and as $t = 1$ slice the circle. Such a surface embedding needs at least five dimensions.

In higher dimensions, knots are thus possible *only* if n-spheres are tied instead of circles; for example, as just said, 2-spheres can be tied into knots in 4 dimensions, 3-spheres in 5 dimensions and so forth.

THE HARDEST OPEN PROBLEMS THAT YOU CAN TELL YOUR GRANDMOTHER

Even though mathematicians have achieved good progress in the classification of knots, surprisingly, they know next to nothing about the *shapes* of knots. Here are a few problems that are still open today:

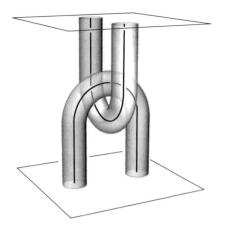 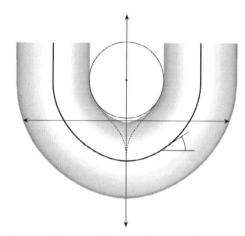

FIGURE 159 The ropelength problem for the simple clasp, and the candidate configuration that probably minimizes ropelength, leaving a gap between the two ropes (© Jason Cantarella).

Challenge 155 r

— This is the simplest unsolved knot problem: Imagine an ideally wobbly rope, that is, a rope that has the same radius everywhere, but whose curvature can be changed as one prefers. Tie a trefoil knot into the rope. By how much do the ends of the rope get nearer? In 2006, there are only numerical estimates for the answer: about 10.1 radiuses. There is no formula yielding the number 10.1. Alternatively, solve the following problem: what is the rope length of a closed trefoil knot? Also in this case, only numerical values are known – about 16.33 radiuses – but no exact formula. The same is valid for any other knot, of course.

Ref. 246

— For mathematical knots, i.e., *closed* knots, the problem is equally unsolved. For example: the ropelength of the tight trefoil knot is known to be around 16.33 diameters, and that of the figure-eight knot about 21.04 diameters. For beautiful visualizations of the tightening process, see the animations on the website www.jasoncantarella.com/ movs. But what is the formula giving the ropelength values? Nobody knows, because the precise shape of the trefoil knot – or of any other knot – is unknown. Lou Kauffman has a simple comment for the situation: 'It is a scandal of mathematics!'

Ref. 247

— Mathematicians also study more general structures than knots. *Links* are the generalization of knots to several closed strands. *Braids* and *long links* are the generalization of links to *open* strands. Now comes the next surprise, illustrated in Figure 159. Even for two ropes that form a simple *clasp*, i.e., two linked letters 'U', the ropelength problem is unsolved – and there is not even a knot involved! In fact, in 2004, Jason Cantarella and his colleagues have presented a candidate for the shape that minimizes ropelength. Astonishingly, the candidate configuration leaves a small gap between the two ropes, as shown in Figure 159.

In short, the shape of knots is a research topic that has barely taken off. Therefore we have to leave these questions for a future occasion.

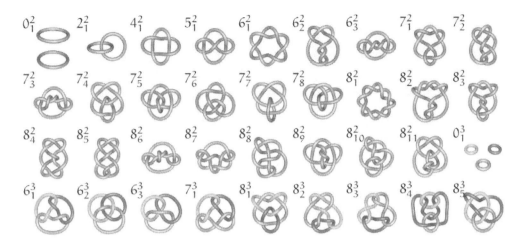

FIGURE 160 The diagrams for the simplest links with two and three components (© Robert Scharein).

FIGURE 161 A hagfish tied into a knot (© Christine Ortlepp).

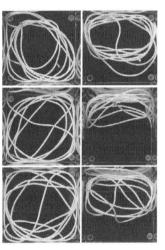

FIGURE 162 How apparent order for long rope coils (left) changes over time when shaking the container (right) (© 2007 PNAS).

CURIOSITIES AND FUN CHALLENGES ON KNOTS AND WOBBLY ENTITIES

Knots appear rarely in nature. For example, tree branches or roots do not seem to grow many knots during the lifetime of a plant. How do plants avoid this? In other words, why are there no knotted bananas of knotted flower stems in nature?

Challenge 156 r

∗ ∗

Also links can be classified. The simplest links, i.e., the links for which the simplest configuration has the smallest number of crossings, are shown in Figure 160.

∗ ∗

Ref. 249 A famous type of eel, the *knot fish Myxine glutinosa*, also called hagfish or slime eel, is able to make a knot in his body and move this knot from head to tail. Figure 161 shows an example. The hagfish uses this motion to cover its body with a slime that prevents predators from grabbing it; it also uses this motion to escape the grip of predators, to get rid of the slime after the danger is over, and to push against a prey it is biting in order to extract a piece of meat. All studied knot fish form only left handed trefoil knots, by the way; this is another example of chirality in nature.

∗ ∗

Ref. 248 Proteins, the molecules that make up many cell structures, are chains of aminoacids. It seems that very few proteins are knotted, and that most of these form trefoil knots. However, a figure-eight knotted protein has been discovered in 2000 by William Taylor.

∗ ∗

One of the most incredible discoveries of recent years is related to knots in DNA molecules. The *DNA molecules* inside cell nuclei can be hundreds of millions of base pairs long; they regularly need to be packed and unpacked. When this is done, often the same happens as when a long piece of rope or a long cable is taken out of a closet.

It is well known that you can roll up a rope and put it into a closet in such a way that it looks orderly stored, but when it is pulled out at one end, a large number of knots Ref. 250 is suddenly found. In 2007, this effects was finally explored in detail. Strings of a few metres in length were put into square boxes and shaken, in order to speed up the effect. The result, shown partly in Figure 162, was astonishing: almost every imaginable knot – up to a certain complexity that depends on the length and flexibility of the string – was formed in this way.

To make a long story short, this also happens to nature when it unpacks DNA in cell nuclei. Life requires that DNA molecules move inside the cell nucleus without hindrance. So what does nature do? Nature takes a simpler approach: when there are unwanted crossings, it cuts the DNA, moves it over and puts the ends together again. In cell nuclei, there are special enzymes, the so-called topoisomerases, which perform this process. The details of this fascinating process are still object of modern research.

∗ ∗

The great mathematician Carl-Friedrich Gauß – often written as 'Gauss' in English – was the first person to ask what happens when an electrical current I flows along a wire A Ref. 251 that is *linked* with a wire B. He discovered a beautiful result by calculating the effect of the magnetic field of one wire onto the other:

$$\frac{1}{4\pi I} \int_A d\boldsymbol{x}_A \cdot \boldsymbol{B}_B = \frac{1}{4\pi} \int_A d\boldsymbol{x}_A \cdot \int_B d\boldsymbol{x}_B \times \frac{(\boldsymbol{x}_A - \boldsymbol{x}_B)}{|\boldsymbol{x}_A - \boldsymbol{x}_B|^3} = n \,, \tag{128}$$

where the integrals are performed along the wires. Gauss found that the number n does not depend on the precise shape of the wires, but only on the way they are linked. Deforming the wires does not change the resulting number n. Mathematicians call such a number a *topological invariant*. In short, Gauss discovered a physical method to calculate a mathematical invariant for links; the research race to do the same for other invariants,

FIGURE 163 A large raindrop falling downwards.

FIGURE 164 Is this possible?

and in particular for knots and braids, is still going on today.

In the 1980s, Edward Witten was able to generalize this approach to include the nuclear interactions, and to define more elaborate knot invariants, a discovery that brought him the Fields medal.

* *

If we move along a knot and count the crossings where we stay above and subtract the number of crossings where we pass below, we get a number called the *writhe* of the knot. It is not an invariant, but usually a tool in building them. Indeed, the writhe is not necessarily invariant under one of the three Reidemeister moves. Can you see which one, using Figure 157? However, the writhe is invariant under flypes.

Challenge 157 ny

* *

Modern knot research is still a topic with many open questions. A recent discovery is the *quasi-quantization of three-dimensional writhe* in tight knots. Many discoveries are still expected in the domain of geometric knot theory.

Ref. 252

* *

There are two ways to tie your shoes. Can you find them?

Challenge 158 e

* *

What is the shape of raindrops? Try to picture it. However, use your reason, not your prejudice! By the way, it turns out that there is a maximum size for raindrops, with a value of about 4 mm. The shape of such a large raindrop is shown in Figure 163. Can you imagine where the limit comes from?

Challenge 159 s

Ref. 253

For comparison, the drops in clouds, fog or mist are in the range of 1 to 100 μm, with a peak at 10 to 15 μm. In situations where all droplets are of similar size and where light is scattered only once by the droplets, one can observe coronae, glories or fogbows.

Vol. III, page 143

Vol. III, page 117

* *

What is the entity shown in Figure 164 – a knot, a braid or a link?

Challenge 160 s

* *

Can you find a way to classify tie knots?

Challenge 161 d

FIGURE 165 A flying snake, *Chrysopelea paradisii*, performing the feat that gave it its name (QuickTime film © Jake Socha).

* *

Challenge 162 s   Are you able to find a way to classify the way shoe laces can be threaded?

* *

A striking example of how wobbly entities can behave is given in Figure 165. There is indeed a family of snakes that like to jump off a tree and sail through the air to a neighbouring tree. Both the jump and the sailing technique have been studied in recent years. The website www.flyingsnake.org by Jake Socha provides additional films. His fascinat-
Ref. 254   ing publications tell more about these intriguing reptiles.

* *

Vol. I, page 283   When a plane moves at supersonic speed through humid air, sometimes a conical cloud forms and moves with the plane. How does this cloud differ from the ones studied above?

Challenge 163 e

* *

Challenge 164 ny   One of the toughest challenges about clouds: is it possible to make rain on demand? So far, there are almost no positive results. Inventing a method, possibly based on hygroscopic salt injection or with the help of lasers, will be a great help to mankind.

* *

Do knots have a relation to elementary particles? The question is about 150 years old. It was first investigated by William Thomson-Kelvin and Peter Tait in the late nineteenth century. So far, no proof of a relation has been found. Knots might be of importance at Planck scales, the smallest dimensions possible in nature. We will explore how knots and the structure of elementary particles might be related in the last volume of this adventure.

SUMMARY ON WOBBLY OBJECTS

We can sum up the possible motions of extended systems in a few key themes. In earlier chapters we studied waves, solitons and interpenetration. These observations are described by wave equations. In this chapter we explored the way to move through shape change, explored eversion, studied vortices, fluids, polymers, knots and their rearrangement, and explored the motion of dislocations in solids. We found that shape change is described by gauge theory, eversion is described by space-duality, vortices follow the Schrödinger equation, fluids and polymers resemble general relativity and black holes, knot shapes are hard to calculate and dislocations behave relativistically.

Vol. I, page 273

The motion of wobbly objects is often a neglected topic in textbooks on motion. Research is progressing at full speed; it is expected that many beautiful analogies with traditional physics will be discovered in the near future. For example, in this chapter we have not explored any possible analogy for the motion of light. Similarly, including quantum theory into the description of wobbly bodies' motion remains a fascinating issue for anybody aiming to publish in a new field.

Challenge 165 r

In summary, we found that wobbly entities can reproduce most fields of modern physics. Are there wobbly entities that reproduce all of modern physics? We will explore the question in the last volume.

CHAPTER 12

QUANTUM PHYSICS IN A NUTSHELL – AGAIN

COMPARED to classical physics, quantum theory is definitely more complex. The basic idea however, is simple: in nature there is a smallest change, or a smallest action, with the value $\hbar = 1.1 \cdot 10^{-34}$ Js. The smallest action value leads to all the strange observations made in the microscopic domain, such as wave behaviour of matter, indeterminacy relations, decoherence, randomness in measurements, indistinguishability, quantization of angular momentum, tunnelling, pair creation, decay, particle reactions and virtual particle exchange.

QUANTUM FIELD THEORY IN A FEW SENTENCES

> " Deorum offensae diis curae.
> Voltaire, *Traité sur la tolérance.*

All of quantum theory can be resumed in a few sentences.

> ▷ *In nature, actions smaller than $\hbar = 1.1 \cdot 10^{-34}$ Js are not observed.*

The existence of a smallest action in nature directly leads to the main lesson we learned about motion in the quantum part of our adventure:

> ▷ *If something moves, it is made of quantons, or quantum particles.*

> ▷ *There are* elementary *quantum particles.*

These statements apply to everything, thus to all objects and to all images, i.e., to matter and to radiation. Moving stuff is made of *quantons*. Stones, water waves, light, sound waves, earthquakes, tooth paste and everything else we can interact with is made of moving quantum particles. Experiments show:

> ▷ All intrinsic object properties observed in nature – such as electric charge, weak charge, colour charge, spin, parity, lepton number, etc., with the only exception of mass – appear as *integer* numbers of a smallest unit; in composed systems they either add or multiply.

▷ An *elementary quantum particle* or *elementary quanton* is a countable entity, smaller than its own Compton wavelength, described by energy–momentum, mass, spin, C, P and T parity, electric charge, colour, weak isospin, isospin, strangeness, charm, topness, beauty, lepton number and baryon number.

Page 239

All moving entities are made of elementary quantum particles. To see how deep this result is, you can apply it to all those moving entities for which it is usually forgotten, such as ghosts, spirits, angels, nymphs, daemons, devils, gods, goddesses and souls. You can check yourself what happens when their particle nature is taken into account.

Challenge 166 e

Quantum particles are never at rest, cannot be localized, move probabilistically, behave as particles or as waves, interfere, can be polarized, can tunnel, are indistinguishable, have antiparticles, interact locally, define length and time scales and they limit measurement precision.

Quantum particles come in two types:

▷ Matter is composed of *fermions:* quarks and leptons. There are 6 quarks that make up nuclei, and 6 leptons – 3 charged leptons, including the electron, and 3 uncharged neutrinos. Elementary fermions have spin 1/2 and obey the Pauli exclusion principle.

▷ Radiation is due to the three gauge interactions and is composed of *bosons:* photons, the weak vector bosons and the 8 gluons. These elementary bosons all have spin 1.

In our adventure, we wanted to know what matter and interactions are. Now we know: they are due to elementary quantum particles. The exploration of motion inside matter, including particle reactions and virtual particle exchange, showed us that matter is made of a *finite number* of elementary quantum particles. Experiments show:

▷ In flat space, elementary particles interact in one of three ways: there is the *electromagnetic interaction*, the *strong nuclear interaction* and the *weak nuclear interaction*.

▷ The three interactions are *exchanges of virtual bosons*.

▷ The three interactions are described by the gauge symmetries U(1), SU(3) and a broken, i.e., approximate SU(2) symmetry.

The three gauge symmetries fix the Lagrangian of every physical system in flat space-time. The most simple description of the Lagrangians is with the help of Feynman diagrams and the gauge groups.

▷ In all interactions, energy, momentum, angular momentum, electric charge, colour charge, CPT parity, lepton number and baryon number are conserved.

The list of conserved quantities implies:

▷ Quantum *field* theory is the part of quantum physics that includes the description of *particle transformations*.

The possibility of particle transformations – including particle reactions, particle emission and particle absorption – results from the existence of a minimum action and of a maximum speed in nature. Emission of light, radioactivity, the burning of the Sun and the history of the composite matter we are made of are due to particle transformations.

Vol. IV, page 29

Due to the possibility of particle transformations, quantum field theory introduces a limit for the localization of particles. In fact, any object of mass m can be localized only within intervals of the Compton wavelength

$$\lambda_C = \frac{h}{mc} = \frac{2\pi\hbar}{mc} \; , \tag{129}$$

where c is the speed of light. At the latest at this distance we have to abandon the classical description and use quantum field theory. If we approach the Compton wavelength, particle transformations become so important that classical physics and even simple quantum theory are not sufficient.

Quantum electrodynamics is the quantum field description of electromagnetism. It includes and explains all particle transformations that involve photons. The Lagrangian of QED is determined by the electromagnetic gauge group U(1), the requirements of space-time (Poincaré) symmetry, permutation symmetry and renormalizability. The latter requirement follows from the continuity of space-time. Through the effects of virtual particles, QED describes electromagnetic decay, lamps, lasers, pair creation, Unruh radiation for accelerating observers, vacuum energy and the Casimir effect, i.e., the attraction of neutral conducting bodies. Particle transformations due to quantum electrodynamics also introduce corrections to classical electrodynamics; among others, particle transformations produce small departures from the superposition principle for electromagnetic fields, including the possibility of photon-photon scattering.

The theory of *weak nuclear interaction* describes parity violation, quark mixings, neutrino mixings, massive vector bosons and the Higgs field for the breaking of the weak SU(2) gauge symmetry. The weak interaction explains a large part of radioactivity, including the heat production inside the Earth, and describes processes that make the Sun shine.

Quantum chromodynamics, the field theory of the strong nuclear interaction, describes all particle transformations that involve gluons. At fundamental scales, the strong interaction is mediated by eight elementary gluons. At larger, femtometre scales, the strong interaction effectively acts through the exchange of spin 0 pions, is strongly attractive, and leads to the formation of atomic nuclei. The strong interaction determines nuclear fusion and fission. Quantum chromodynamics, or QCD, explains the masses of

Page 183

mesons and baryons through their description as bound quark states.

By including particle transformations, quantum field theory provides a common basis of concepts and descriptions to materials science, nuclear physics, chemistry, biology, medicine and to most of astronomy. For example, the same concepts allow us to answer questions such as why water is liquid at room temperature, why copper is red, why the rainbow is coloured, why the Sun and the stars continue to shine, why there are about

110 elements, where a tree takes the material to make its wood and why we are able to move our right hand at our own will. Quantum theory explains the origin of material properties and the origin of the properties of life.

Quantum field theory describes all material properties, be they mechanical, optical, electric or magnetic. It describes all waves that occur in materials, such as sound and phonons, mangetic waves and magnons, light, plasmons, and all localized excitations. Quantum field theory also describes collective effects in matter, such as superconductivity, semiconductor effects and superfluidity. Finally, quantum field theory describes all interactions between matter and radiation, from colour to antimatter creation.

Quantum field theory also clarifies that the particle description of nature, including the conservation of particle number – defined as the difference between particles and antiparticles – follows from the possibility to describe interactions *perturbatively*. A perturbative description of nature is possible only at low energies. At extremely high energies, higher than those observed in experiments, the situation is expected to change and non-perturbative effects should come into play. These situations will be explored in the next volume.

ACHIEVEMENTS IN PRECISION

Classical physics is unable to predict *any* property of matter. Quantum field theory predicts all properties of matter, and to the full number of digits – sometimes thirteen – that can be measured today. The precision is usually *not* limited by the inaccuracy of theory, it is limited by the measurement accuracy. In other words, the agreement between quantum field theory and experiment is only limited by the amount of money one is willing to spend. Table 23 shows some predictions of classical physics and of quantum field theory. The predictions are deduced from the properties of nature collected in the millennium list, which is given in the next section.

TABLE 23 Selected comparisons between classical physics, quantum theory and experiment.

OBSERVABLE	CLASSICAL PREDICTION	PREDICTION OF QUANTUM THEORY[a]	MEASUREMENT	COST ESTIMATE[b]
Simple motion of bodies				
Indeterminacy	0	$\Delta x \Delta p \geqslant \hbar/2$	$(1 \pm 10^{-2})\,\hbar/2$	10 k€
Matter wavelength	none	$\lambda p = 2\pi\hbar$	$(1 \pm 10^{-2})\,\hbar$	10 k€
Tunnelling rate in α decay	0	$1/\tau$ is finite	$(1 \pm 10^{-2})\,\tau$	5 k€
Compton wavelength	none	$\lambda_c = h/m_e c$	$(1 \pm 10^{-3})\,\lambda$	20 k€
Pair creation rate	0	σE	agrees	100 k€
Radiative decay time in hydrogen	none	$\tau \sim 1/n^3$	(1 ± 10^{-2})	5 k€
Smallest angular momentum	0	$\hbar/2$	$(1 \pm \pm 10^{-6})\,\hbar/2$	10 k€
Casimir effect/pressure	0	$p = (\pi^2 \hbar c)/(240 r^4)$	(1 ± 10^{-3})	30 k€

TABLE 23 (Continued) Selected comparisons between classical physics, quantum theory and experiment.

OBSERVABLE	CLASSICAL PREDICTION	PREDICTION OF QUANTUM THEORY[a]	MEASUREMENT	COST ESTIMATE[b]
Colours of objects				
Spectrum of hot objects	diverges	$\lambda_{max} = hc/(4.956\,kT)$	$(1 \pm 10^{-4})\,\Delta\lambda$	10 k€
Lamb shift	none	$\Delta\lambda = 1057.86(1)$ MHz	$(1 \pm 10^{-6})\,\Delta\lambda$	50 k€
Rydberg constant	none	$R_\infty = m_e c \alpha^2/2h$	$(1 \pm 10^{-9})\,R_\infty$	50 k€
Stefan–Boltzmann constant	none	$\sigma = \pi^2 k^4/60\hbar^3 c^2$	$(1 \pm 3 \cdot 10^{-8})\,\sigma$	20 k€
Wien's displacement constant	none	$b = \lambda_{max} T$	$(1 \pm 10^{-5})\,b$	20 k€
Refractive index of water	none	1.34	a few %	1 k€
Photon-photon scattering	0	from QED: finite	agrees	50 M€
Particle and interaction properties				
Electron gyromagnetic ratio	1 or 2	2.002 319 304 3(1)	2.002 319 304 3737(82)	30 M€
Z boson mass	none	$m_Z^2 = m_W^2(1 + \sin^2\theta_W)$	$(1 \pm 10^{-3})\,m_Z$	100 M€
Proton mass	none	$(1 \pm 5\,\%)\,m_p$	$m_p = 1.67$ yg	1 M€
Proton lifetime	≈ 1 μs	∞	$> 10^{35}$ a	100 M€
Chemical reaction rate	0	from QED	correct within errors	2 k€
Composite matter properties				
Atom lifetime	≈ 1 μs	∞	$> 10^{20}$ a	1 €
Molecular size	none	from QED	within 10^{-3}	20 k€
Von Klitzing constant	∞	$h/e^2 = \mu_0 c/2\alpha$	$(1 \pm 10^{-7})\,h/e^2$	1 M€
AC Josephson constant	0	$2e/h$	$(1 \pm 10^{-6})\,2e/h$	5 M€
Heat capacity of metals at 0 K	25 J/K	0	$< 10^{-3}$ J/K	10 k€
Heat capacity of diatomic gas at 0 K	25 J/K	0	$< 10^{-3}$ J/K	10 k€
Water density	none	1000.00 kg/m^3 at 4°C	agrees	10 k€
Minimum electr. conductivity	0	$G = 2e^2/\hbar$	$G(1 \pm 10^{-3})$	3 k€
Ferromagnetism	none	exists	exists	2 €
Superfluidity	none	exists	exists	200 k€
Bose–Einsein condensation	none	exists	exists	2 M€
Superconductivity (metal)	none	exists	exists	100 k€

TABLE 23 (Continued) Selected comparisons between classical physics, quantum theory and experiment.

OBSERVABLE	CLASSI-CAL PREDIC-TION	PREDICTION OF QUANTUM THEORY[a]	MEASURE-MENT	COST ESTI-MATE[b]
Superconductivity (high T)	none	*none yet*	exists	100 k€

a. All these predictions are calculated from the fundamental quantities given in the millennium list.

b. Sometimes the cost for the calculation of the prediction is higher than that of the experimental observation. (Can you spot the examples?) The sum of the two is given. Challenge 167 s

We notice that the values predicted by quantum theory do not differ from the measured ones. In contrast, classical physics does not allow us to calculate any of the observed values. This shows the progress that quantum physics has brought in the description of nature.

In short, in the microscopic domain quantum theory is in *perfect* correspondence with nature; despite prospects of fame and riches, despite the largest number of researchers ever, no contradiction with observation has been found yet. But despite this impressive agreement, there still are *unexplained* observations; they form the so-called *millennium list*.

WHAT IS UNEXPLAINED BY QUANTUM THEORY AND GENERAL RELATIVITY?

The material gathered in this quantum part of our mountain ascent, together with the earlier summary of general relativity, allows us to describe *all* observed phenomena connected to motion. For the first time, there are no known differences between theory and practice. Vol. II, page 268

Despite the precision of the description of nature, some things are missing. Whenever we ask 'why?' about an observation and continue doing so after each answer, we arrive at one of the *unexplained* properties of nature listed in Table 24. The table lists all issues about fundamental motion that were unexplained in the year 2000, so that we can call it the *millennium list* of open problems.

TABLE 24 The millennium list: *everything* the standard model and general relativity *cannot* explain; thus, also the list of the *only* experimental data available to test the final, unified description of motion.

OBSERVABLE	PROPERTY UNEXPLAINED IN THE YEAR 2000

Local quantities unexplained by the standard model: particle properties

$\alpha = 1/137.036(1)$	the low energy value of the electromagnetic coupling constant
α_{w} or θ_{w}	the low energy value of the weak coupling constant or the value of the weak mixing angle
α_{s}	the value of the strong coupling constant at one specific energy value
m_{q}	the values of the 6 quark masses

TABLE 24 (Continued) *Everything* the standard model and general relativity *cannot* explain.

OBSERVABLE	PROPERTY UNEXPLAINED IN THE YEAR 2000
m_l	the values of 6 lepton masses
m_W	the value of the mass of the W vector boson
m_H	the value of the mass of the scalar Higgs boson
$\theta_{12}, \theta_{13}, \theta_{23}$	the value of the three quark mixing angles
δ	the value of the CP violating phase for quarks
$\theta^v_{12}, \theta^v_{13}, \theta^v_{23}$	the value of the three neutrino mixing angles
$\delta^v, \alpha_1, \alpha_2$	the value of the three CP violating phases for neutrinos
$3 \cdot 4$	the number of fermion generations and of particles in each generation
J, P, C, etc.	the origin of all quantum numbers of each fermion and each boson

Local mathematical structures unexplained by the standard model

$c, \hbar, k$	the origin of the invariant Planck units of quantum field theory
$3 + 1$	the number of dimensions of physical space and time
SO(3,1)	the origin of Poincaré symmetry, i.e., of spin, position, energy, momentum
$S(n)$	the origin of particle identity, i.e., of permutation symmetry
Gauge symmetry	the origin of the gauge groups, in particular:
U(1)	the origin of the electromagnetic gauge group, i.e., of the quantization of electric charge, as well as the vanishing of magnetic charge
SU(2)	the origin of weak interaction gauge group, its breaking and P violation
SU(3)	the origin of strong interaction gauge group and its CP conservation
Ren. group	the origin of renormalization properties
$\delta W = 0$	the origin of wave functions and the least action principle in quantum theory
$W = \int L_{SM}\, dt$	the origin of the Lagrangian of the standard model of particle physics

Global quantities unexplained by general relativity and cosmology

0	the observed flatness, i.e., vanishing curvature, of the universe
$1.2(1) \cdot 10^{26}$ m	the distance of the horizon, i.e., the 'size' of the universe (if it makes sense)
$\rho_{de} = \Lambda c^4/(8\pi G)$ $\approx 0.5\, \text{nJ/m}^3$	the value and nature of the observed vacuum energy density, dark energy or cosmological constant
$(5 \pm 4) \cdot 10^{79}$	the number of baryons in the universe (if it makes sense), i.e., the average visible matter density in the universe
$f_0(1, ..., c.\, 10^{90})$	the initial conditions for $c.\, 10^{90}$ particle fields in the universe (if or as long as they make sense), including the homogeneity and isotropy of matter distribution, and the density fluctuations at the origin of galaxies
ρ_{dm}	the density and nature of dark matter

Global mathematical structures unexplained by general relativity and cosmology

c, G	the origin of the invariant Planck units of general relativity
$\delta \int L_{GR} dt = 0$	the origin of the least action principle and the Lagrangian of general relativity
$R \times S^3$	the observed topology of the universe

The millennium list has several notable aspects. First of all, neither quantum mechanics

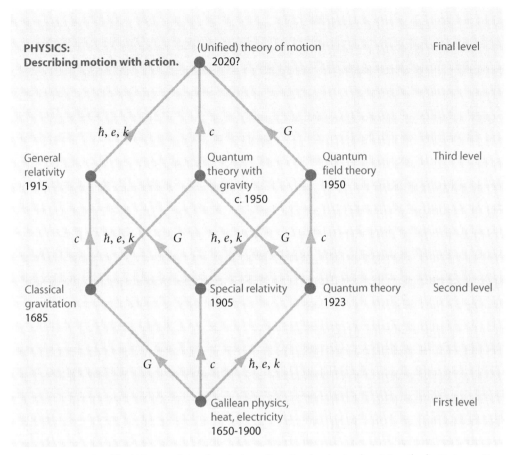

FIGURE 166 A simplified history of the description of motion in physics, by giving the limits to motion included in each description. The arrows show which constant of nature needs to be added and taken into account to reach the next level of description. (The electric charge *e* is taken to represent all three gauge coupling constants.)

nor general relativity explain any property unexplained in the other field. The two theories do not help each other; the unexplained parts of both fields simply add up. Secondly, both in quantum theory and in general relativity, motion still remains the change of position with time. In short, so far, we did not achieve our goal: we still do not understand motion! We are able to describe motion with full precision, but we still do not know what it is. Our basic questions remain: What are time and space? What is mass? What is charge and what are the other properties of objects? What are fields? Why are all the electrons the same?

We also note that the millennium list of open questions, Table 24, contains extremely *different* concepts. This means that at this point of our walk there is *a lot* we do not understand. Finding the answers will require effort. Page 291

On the other hand, the millennium list of unexplained properties of nature is also *short*. The description of nature that our adventure has produced so far is concise and precise. No discrepancies from experiments are known. In other words, we have a good description of motion *in practice*. Going further is unnecessary if we only want to improve measurement precision. Simplifying the above list is mainly important from the *conceptual* point of view. For this reason, the study of physics at university often stops at this point. However, as the millennium list shows, even though we have *no* known discrepancies with experiments, we are *not* at the top of Motion Mountain.

THE PHYSICS CUBE

Page 8 Another review of the progress and of the open issues of physics, already given in the introduction, is shown in Figure 166: the *physics cube*. From the lowest corner of the cube, representing Galilean physics and related topics from everyday life, three edges – labelled c, G and $\hbar$, e, k – lead to classical gravity, special relativity and quantum theory. Each constant implies a limit to motion; in the corresponding theory, this *one* limit is taken into account. From these second level theories, similar edges lead upwards to general relativity, quantum field theory and quantum theory with gravity; each of these third level theories takes into account *two* of the limits.* The present volume completes the third level. We stress that each theory in the second and third level is *exact*, though only *in its domain*. No differences between experiment and theory are known in their domains.

From the third level theories, the edges lead to the last missing corner: the (unified) theory of motion that takes into account *all limits* of nature. Only this theory is a complete and unified description of nature. Since we already know all limits to motion, in order to arrive at the last level, we do not new experiments. We do not need new knowledge. We only have to advance, in the right direction, with careful thinking. And we can start from three different points. Reaching the final theory of motion is the topic of the last volume of our adventure.

THE INTENSE EMOTIONS DUE QUANTUM FIELD THEORY AND GENERAL RELATIVITY

It is sometimes deemed chic to pretend that the adventure is over at the stage we have just reached,** the third level of Figure 166. The reasoning given is as follows. If we change the Page 291 values of the unexplained constants in the millennium list of Table 24 only ever so slightly, Ref. 257 nature would look completely different from what it does. Indeed, these consequences have been studied in great detail; an overview of the connections is given in the following table.

---

* Of course, Figure 166 gives a simplified view of the history of physics. A more precise diagram would use three different arrows for $\hbar$, e and k, making the figure a five-dimensional cube. However, not all of its Challenge 168 e corners would have dedicated theories (can you confirm this?). The diagram would be much less appealing. And most of all, the conclusions mentioned in the text would not change.

** Actually this attitude is not new. Only the arguments have changed. Maybe the greatest physicist ever, Ref. 256 James Clerk Maxwell, already fought against this attitude over a hundred years ago: 'The opinion seems to have got abroad that, in a few years, all great physical constants will have been approximately estimated, and that the only occupation which will be left to men of science will be to carry these measurements to another place of decimals. [...] The history of science shows that even during that phase of her progress in which she devotes herself to improving the accuracy of the numerical measurement of quantities with which she has long been familiar, she is preparing the materials for the subjugation of new regions, which would have remained unknown if she had been contented with the rough methods of her early pioneers.'

TABLE 25 A selection of the consequences of changing the properties of nature.

OBSERVABLE	CHANGE	RESULT
Local quantities, from quantum theory		
α_{em}	smaller:	Only short lived, smaller and hotter stars; no Sun.
	larger:	Darker Sun, animals die of electromagnetic radiation, too much proton decay, no planets, no stellar explosions, no star formation, no galaxy formation.
	+60%:	Quarks decay into leptons.
	+200%:	Proton-proton repulsion makes nuclei impossible.
α_w	−50%:	Carbon nucleus unstable.
	very weak:	No hydrogen, no p-p cycle in stars, no C-N-O cycle.
	+2%:	No protons from quarks.
	$G_F m_e^2 \neq \sqrt{G m_e^2}$:	Either no or only helium in the universe.
	much larger:	No stellar explosions, faster stellar burning.
α_s	−9%:	No deuteron, stars much less bright.
	−1%:	No C resonance, no life.
	+3.4%:	Diproton stable, faster star burning.
	much larger:	Carbon unstable, heavy nuclei unstable, widespread leukaemia.
n-p mass difference	larger:	Neutron decays in proton inside nuclei; no elements.
	smaller:	Free neutron not unstable, all protons into neutrons during big bang; no elements.
	smaller than m_e:	Protons would capture electrons, no hydrogen atoms, star life much shorter.
m_l changes:		
e-p mass ratio	much different:	No molecules.
	much smaller:	No solids.
3 generations	6-8:	Only helium in nature.
	>8:	No asymptotic freedom and confinement.
Global quantities, from general relativity		
horizon size	much smaller:	No people.
baryon number	very different:	No smoothness .
	much higher:	No solar system.
Initial condition changes:		
Moon mass	smaller:	Small Earth magnetic field; too much cosmic radiation; widespread child skin cancer.
Moon mass	larger:	Large Earth magnetic field; too little cosmic radiation; no evolution into humans.
Sun's mass	smaller:	Too cold for the evolution of life.

TABLE 25 (Continued) A selection of the consequences of changing the properties of nature.

OBSERVABLE	CHANGE	RESULT
Sun's mass	larger:	Sun too short lived for the evolution of life.
Jupiter mass	smaller:	Too many comet impacts on Earth; extinction of animal life.
Jupiter mass	larger:	Too little comet impacts on Earth; no Moon; no dinosaur extinction.
Oort cloud object number	smaller:	No comets; no irregular asteroids; no Moon; still dinosaurs.
galaxy centre distance	smaller:	Irregular planet motion; supernova dangers.
initial cosmic speed	+0.1%:	1000 times faster universe expansion.
	−0.0001%:	Universe recollapses after 10 000 years.
vacuum energy density	change by 10^{-55}:	No flatness.
3 + 1 dimensions	different:	No atoms, no planetary systems.

Local structures, from quantum theory

permutation symmetry	none:	No matter.
Lorentz symmetry	none:	No communication possible.
U(1)	different:	No Huygens principle, no way to *see* anything.
SU(2)	different:	No radioactivity, no Sun, no life.
SU(3)	different:	No stable quarks and nuclei.

Global structures, from general relativity

topology	other:	Unknown; possibly correlated γ ray bursts or star images at the antipodes.

Some researchers speculate that the whole of Table 25 can be condensed into a single sentence: if any parameter in nature is changed, the universe would either have too many or too few black holes. However, the proof of this condensed summary is not complete yet. But it is a beautiful hypothesis.

The effects of changing nature that are listed in Table 25 lead us to a profound experience: even the tiniest changes in the properties of nature are incompatible with our existence. What does this experience mean? Answering this question too rapidly is dangerous. Many have fallen into one of several traps:

— The first trap is to deduce, incorrectly, that the unexplained numbers and other properties from the millennium list do not need to be explained, i.e., deduced from more general principles.
— The second trap is to deduce, incorrectly, that the universe has been *created* or *designed*.
— The third trap is to deduce, incorrectly, that the universe is *designed for people*.

Ref. 258
Challenge 169 r

— The fourth trap is to deduce, incorrectly, that the universe is *one of many*.

All these traps are irrational and incorrect beliefs. All these beliefs have in common that they have no factual basis, that they discourage further search and that they sell many books.

The first trap is due to a combination of pessimism and envy; it is a type of wishful thinking. But wishful thinking has no place in the study of motion. The second trap works because many physicists incorrectly speak of *fine tuning* in nature. Many researchers succumb to the belief in 'creation' and are unable to steer clear from the logical errors contained in it. We discussed them earlier on. The third trap, is often, again in- Vol. III, page 276 correctly, called the *anthropic principle*. The name is a mistake, because we saw that the anthropic principle is indistinguishable both from the simian principle and from the sim- Vol. III, page 282 ple request that statements be based on observations. Around 2000, the third trap has even become fashionable among frustrated particle theorists. The fourth trap, the belief in *multiple universes*, is a minority view, but sells many books. Most people that hold this view are found in institutions, and that is indeed where they belong.

Stopping our mountain ascent with an incorrect belief at the present stage is not different from doing so directly at the beginning. Such a choice has been taken in various societies that lacked the passion for rational investigation, and still is taken in circles that discourage the use of reason among their members. Looking for beliefs instead of looking for answers means to give up the ascent of Motion Mountain while pretending to have reached the top. Every such case is a tragedy, sometimes a small one, sometimes a larger one.

In fact, Table 25 purveys only one message: all evidence implies that we are only a *tiny part* of the universe, but that we are *linked* with all other aspects of it. Due to our small size and due to all the connections with our environment, any imagined tiny change would make us disappear, like a water droplet is swept away by large wave. Our walk has repeatedly reminded us of this smallness and dependence, and overwhelmingly does so again at this point.

In our adventure, accepting the powerful message of Table 25 is one of the most awe-inspiring, touching and motivating moments. It shows clearly how vast the universe is. It also shows how much we are dependent on many different and distant aspects of nature. Having faced this powerful experience, everybody has to make up his own mind on whether to proceed with the adventure or not. Of course, there is no obligation to do Challenge 170 s so.

What awaits us?

Assuming that you have decided to continue the adventure, it is natural to ask what awaits you. The shortness of the millennium list of unexplained aspects of nature, given in Table 24, means that *no additional experimental data* are available as check of the final Page 291 description of nature. Everything we need to arrive at the final description of motion will be deduced from the experimental data given in the millennium list, and from nothing else. In other words, future experiments will *not* help us – except if they change something in the millennium list. Accelerator experiments might do this with the particle list or astronomical experiments with the topology issue. Fantasy provides no limits; fortunately, nature does.

The lack of new experimental data means that to continue the walk is a *conceptual* adventure only. Nevertheless, storms rage near the top of Motion Mountain. We have to walk keeping our eyes open, without any other guidance except our reason. This is not an adventure of action, but an adventure of the mind. And it is a fascinating one, as we shall soon find out. To provide an impression of what awaits us, we rephrase the remaining issues in five simple challenges.

1 – What determines *colours*? In other words, what relations of nature fix the famous fine structure constant? Like the hero of Douglas Adams' books, physicists know the answer to the greatest of questions: it is 137.036. But they do not know the question.

2 – What fixes the contents of a *teapot*? It is given by its size to the third power. But why are there only three dimensions? Why is the tea content limited in this way?

3 – Was Democritus *right*? Our adventure has confirmed his statement up to this point: nature is indeed well described by the concepts of *particle* and of *vacuum*. At large scales, relativity has added a horizon, and at small scales, quantum field theory added vacuum energy and pair creation. Nevertheless, both theories *assume* the existence of particles and the existence of space-time, and neither *predicts* them. Even worse, both theories completely fail to predict the existence of *any* of the properties either of space-time – such as its dimensionality – or of particles – such as their masses and other quantum numbers. A lot is missing.

4 – Was Democritus *wrong*? It is often said that the standard model has only about twenty unknown parameters; this common mistake negates about 10^{93} initial conditions! To get an idea of the problem, we simply estimate the number N of possible states of all particles in the universe by

$$N = n \, v \, d \, p \, f \tag{130}$$

where n is the number of particles, v is the number of variables (position, momentum, spin), d is the number of different values each of them can take (limited by the maximum of 61 decimal digits), p is the number of visible space-time points (about 10^{183}) and f is a factor expressing how many of all these initial conditions are actually independent of each other. We thus get the following number of possible states of all particles in the universe:

$$N = 10^{92} \cdot 8 \cdot 10^{61} \cdot 10^{183} \cdot f = 10^{336} \cdot f \tag{131}$$

from which the 10^{93} initial conditions have to be explained. But no explanation is known. Worse, there is also the additional problem that we know nothing whatsoever about f. Its value could be 0, if all data were interdependent, or 1, if none were. Even worse, above Vol. IV, page 157 we noted that initial conditions cannot be defined for the universe at all; thus f should be undefined and not be a number at all! Whatever the case, we need to understand how all the visible particles acquire their present 10^{93} states.

Vol. I, page 375 5 – Were our efforts up to this point *in vain*? Quite at the beginning of our walk we noted that in classical physics, space and time are defined using matter, whereas matter is defined using space-time. Hundred years of general relativity and of quantum theory, including dozens of geniuses, have not solved this oldest paradox of all. The issue is still Challenge 171 e open at this point of our walk, as you might want to check by yourself.

The answers to these five challenges define the top of Motion Mountain. Answering them means to know *everything* about motion. In summary, our quest for the unravelling of the essence of motion gets really interesting from this point onwards!

> " That is why Leucippus and Democritus, who say that the atoms move always in the void and the unlimited, must say what movement is, and in what their natural motion consists. "
>
> Aristotle, *Treaty of the Heaven*   Ref. 259

UNITS, MEASUREMENTS AND CONSTANTS

M EASUREMENTS are comparisons with standards. Standards are based on *units*. any different systems of units have been used throughout the world. ost of these standards confer power to the organization in charge of them. Such power can be misused; this is the case today, for example in the computer industry, and was so in the distant past. The solution is the same in both cases: organize an independent and global standard. For measurement units, this happened in the eighteenth century: in order to avoid misuse by authoritarian institutions, to eliminate problems with differing, changing and irreproducible standards, and – this is not a joke – to simplify tax collection and to make it more just, a group of scientists, politicians and economists agreed on a set of units. It is called the *Système International d'Unités*, abbreviated *SI*, and is defined by an international treaty, the 'Convention du Mètre'. The units are maintained by an international organization, the 'Conférence Générale des Poids et Mesures', and its daughter organizations, the 'Commission Internationale des Poids et Mesures' and the 'Bureau International des Poids et Mesures' (BIPM). All originated in the times just before the French revolution.

Ref. 260

SI units

All SI units are built from seven *base units*, whose official definitions, translated from French into English, are given below, together with the dates of their formulation:

▪ 'The *second* is the duration of 9 192 631 770 periods of the radiation corresponding to the transition between the two hyperfine levels of the ground state of the caesium 133 atom.' (1967)*

▪ 'The *metre* is the length of the path travelled by light in vacuum during a time interval of 1/299 792 458 of a second.' (1983)*

▪ 'The *kilogram* is the unit of mass; it is equal to the mass of the international prototype of the kilogram.' (1901)*

▪ 'The *ampere* is that constant current which, if maintained in two straight parallel conductors of infinite length, of negligible circular cross-section, and placed 1 metre apart in vacuum, would produce between these conductors a force equal to $2 \cdot 10^{-7}$ newton per metre of length.' (1948)*

▪ 'The *kelvin*, unit of thermodynamic temperature, is the fraction 1/273.16 of the thermodynamic temperature of the triple point of water.' (1967)*

▪ 'The *mole* is the amount of substance of a system which contains as many elementary entities as there are atoms in 0.012 kilogram of carbon 12.' (1971)*

▪ 'The *candela* is the luminous intensity, in a given direction, of a source that emits monochromatic radiation of frequency $540 \cdot 10^{12}$ hertz and has a radiant intensity in that direction of (1/683) watt per steradian.' (1979)[*]

We note that both time and length units are defined as certain properties of a standard example of motion, namely light. In other words, also the Conférence Générale des Poids et Mesures makes the point that the observation of motion is a *prerequisite* for the definition and construction of time and space. *Motion is the fundament of every observation and measurement.* By the way, the use of light in the definitions had been proposed already in 1827 by Jacques Babinet.[**]

From these basic units, all other units are defined by multiplication and division. Thus, all SI units have the following properties:

▪ SI units form a system with *state-of-the-art precision*: all units are defined with a precision that is higher than the precision of commonly used measurements. Moreover, the precision of the definitions is regularly being improved. The present relative uncertainty of the definition of the second is around 10^{-14}, for the metre about 10^{-10}, for the kilogram about 10^{-9}, for the ampere 10^{-7}, for the mole less than 10^{-6}, for the kelvin 10^{-6} and for the candela 10^{-3}.

▪ SI units form an *absolute* system: all units are defined in such a way that they can be reproduced in every suitably equipped laboratory, independently, and with high precision. This avoids as much as possible any misuse by the standard-setting organization. (The kilogram, still defined with the help of an artefact, is the last exception to this requirement; extensive research is under way to eliminate this artefact from the definition – an international race that will take a few more years. There are two approaches: counting particles, or fixing $\hbar$. The former can be achieved in crystals, e.g., crystals made of pure silicon, the latter using any formula where $\hbar$ appears, such as the formula for the de Broglie wavelength or that of the Josephson effect.)

▪ SI units form a *practical* system: the base units are quantities of everyday magnitude. Frequently used units have standard names and abbreviations. The complete list includes the seven base units just given, the supplementary units, the derived units and the admitted units.

The *supplementary* SI units are two: the unit for (plane) angle, defined as the ratio of arc length to radius, is the *radian* (rad). For solid angle, defined as the ratio of the subtended area to the square of the radius, the unit is the *steradian* (sr).

The *derived* units with special names, in their official English spelling, i.e., without capital letters and accents, are:

---

[*] The respective symbols are s, m, kg, A, K, mol and cd. The international prototype of the kilogram is a platinum–iridium cylinder kept at the BIPM in Sèvres, in France. For more details on the levels of the caesium atom, consult a book on atomic physics. The Celsius scale of temperature θ is defined as: $\theta/°C = T/K - 273.15$; note the small difference with the number appearing in the definition of the kelvin. SI also states: 'When the mole is used, the elementary entities must be specified and may be atoms, molecules, ions, electrons, other particles, or specified groups of such particles.' In the definition of the mole, it is understood that the carbon 12 atoms are unbound, at rest and in their ground state. In the definition of the candela, the frequency of the light corresponds to 555.5 nm, i.e., green colour, around the wavelength to which the eye is most sensitive.

Vol. I, page 92
Ref. 261

[**] Jacques Babinet (b. 1794 Lusignan, d. 1874 Paris), French physicist who published important work in optics.

NAME	ABBREVIATION	NAME	ABBREVIATION
hertz	$Hz = 1/s$	newton	$N = kg\,m/s^2$
pascal	$Pa = N/m^2 = kg/m\,s^2$	joule	$J = Nm = kg\,m^2/s^2$
watt	$W = kg\,m^2/s^3$	coulomb	$C = As$
volt	$V = kg\,m^2/As^3$	farad	$F = As/V = A^2s^4/kg\,m^2$
ohm	$\Omega = V/A = kg\,m^2/A^2s^3$	siemens	$S = 1/\Omega$
weber	$Wb = Vs = kg\,m^2/As^2$	tesla	$T = Wb/m^2 = kg/As^2 = kg/Cs$
henry	$H = Vs/A = kg\,m^2/A^2s^2$	degree Celsius	$°C$ (see definition of kelvin)
lumen	$lm = cd\,sr$	lux	$lx = lm/m^2 = cd\,sr/m^2$
becquerel	$Bq = 1/s$	gray	$Gy = J/kg = m^2/s^2$
sievert	$Sv = J/kg = m^2/s^2$	katal	$kat = mol/s$

The *admitted* non-SI units are *minute, hour, day* (for time), *degree* $1° = \pi/180$ rad, *minute* $1' = \pi/10\,800$ rad, *second* $1'' = \pi/648\,000$ rad (for angles), *litre* and *tonne*. All other units are to be avoided.

All SI units are made more practical by the introduction of standard names and abbreviations for the powers of ten, the so-called *prefixes*:*

POWER	NAME		POWER	NAME		POWER	NAME		POWER	NAME	
10^1	deca	da	10^{-1}	deci	d	10^{18}	Exa	E	10^{-18}	atto	a
10^2	hecto	h	10^{-2}	centi	c	10^{21}	Zetta	Z	10^{-21}	zepto	z
10^3	kilo	k	10^{-3}	milli	m	10^{24}	Yotta	Y	10^{-24}	yocto	y
10^6	Mega	M	10^{-6}	micro	μ	unofficial:			Ref. 262		
10^9	Giga	G	10^{-9}	nano	n	10^{27}	Xenta	X	10^{-27}	xenno	x
10^{12}	Tera	T	10^{-12}	pico	p	10^{30}	Wekta	W	10^{-30}	weko	w
10^{15}	Peta	P	10^{-15}	femto	f	10^{33}	Vendekta	V	10^{-33}	vendeko	v
						10^{36}	Udekta	U	10^{-36}	udeko	u

▪ SI units form a *complete* system: they cover in a systematic way the full set of observables of physics. Moreover, they fix the units of measurement for all other sciences as well.

▪ SI units form a *universal* system: they can be used in trade, in industry, in commerce,

---

* Some of these names are invented (yocto to sound similar to Latin *octo* 'eight', zepto to sound similar to Latin *septem*, yotta and zetta to resemble them, exa and peta to sound like the Greek words ἑξάκις and πεντάκις for 'six times' and 'five times', the unofficial ones to sound similar to the Greek words for nine, ten, eleven and twelve); some are from Danish/Norwegian (atto from *atten* 'eighteen', femto from *femten* 'fifteen'); some are from Latin (from *mille* 'thousand', from *centum* 'hundred', from *decem* 'ten', from *nanus* 'dwarf'); some are from Italian (from *piccolo* 'small'); some are Greek (micro is from μικρός 'small', deca/deka from δέκα 'ten', hecto from ἑκατόν 'hundred', kilo from χίλιοι 'thousand', mega from μέγας 'large', giga from γίγας 'giant', tera from τέρας 'monster').

Translate: I was caught in such a traffic jam that I needed a microcentury for a picoparsec and that my car's fuel consumption was two tenths of a square millimetre.

Challenge 172 e

at home, in education and in research. They could even be used by extraterrestrial civilizations, if they existed.

- SI units form a *coherent* system: the product or quotient of two SI units is also an SI unit. This means that in principle, the same abbreviation, e.g. 'SI', could be used for every unit.

The SI units are not the only possible set that could fulfil all these requirements, but they are the only existing system that does so.* In the near future, the BIPM plans to use the cube of physical constants, shown in Figure 1, to define SI units. This implies fixing the values of e and k in addition to the already fixed value for c. The only exception will remain the fixing of a basic time unit with the help of an atomic transition, not with the constant G, because this constant cannot be measured with high precision.

THE MEANING OF MEASUREMENT

Every measurement is a comparison with a standard. Therefore, any measurement requires *matter* to realize the standard (even for a speed standard), and *radiation* to achieve the comparison. The concept of measurement thus assumes that matter and radiation exist and can be clearly separated from each other.

Challenge 173 e

Every measurement is a comparison. Measuring thus implies that space and time exist, and that they differ from each other.

Every measurement produces a measurement result. Therefore, every measurement implies the *storage* of the result. The process of measurement thus implies that the situation before and after the measurement can be distinguished. In other terms, every measurement is an *irreversible* process.

Every measurement is a process. Thus every measurement takes a certain amount of time and a certain amount of space.

All these properties of measurements are simple but important. Beware of anybody who denies them.

PLANCK'S NATURAL UNITS

Since the exact form of many equations depends on the system of units used, theoretical physicists often use unit systems optimized for producing simple equations. The chosen units and the values of the constants of nature are related. In microscopic physics, the system of *Planck's natural units* is frequently used. They are defined by setting $c = 1$, $\hbar = 1$, $G = 1$, $k = 1$, $\varepsilon_0 = 1/4\pi$ and $\mu_0 = 4\pi$. Planck units are thus defined from combinations of fundamental constants; those corresponding to the fundamental SI units are given in

---

* Apart from international units, there are also *provincial* units. Most provincial units still in use are of Roman origin. The mile comes from *milia passum*, which used to be one thousand (double) strides of about 1480 mm each; today a nautical mile, once defined as minute of arc on the Earth's surface, is exactly 1852 m). The inch comes from *uncia/onzia* (a twelfth – now of a foot). The pound (from *pondere* 'to weigh') is used as a translation of *libra* – balance – which is the origin of its abbreviation lb. Even the habit of counting in dozens instead of tens is Roman in origin. These and all other similarly funny units – like the system in which all units start with 'f', and which uses furlong/fortnight as its unit of velocity – are now officially defined as multiples of SI units.

Table 27.* The table is also useful for converting equations written in natural units back
Challenge 174 e to SI units: just substitute every quantity X by X/X_{Pl}.

TABLE 27 Planck's (uncorrected) natural units.

NAME	DEFINITION			VALUE
Basic units				
the Planck length	l_{Pl}	$=$	$\sqrt{\hbar G/c^3}$	$=$ $1.616\,0(12) \cdot 10^{-35}$ m
the Planck time	t_{Pl}	$=$	$\sqrt{\hbar G/c^5}$	$=$ $5.390\,6(40) \cdot 10^{-44}$ s
the Planck mass	m_{Pl}	$=$	$\sqrt{\hbar c/G}$	$=$ $21.767(16)\,\mu$g
the Planck current	I_{Pl}	$=$	$\sqrt{4\pi\varepsilon_0 c^6/G}$	$=$ $3.479\,3(22) \cdot 10^{25}$ A
the Planck temperature	T_{Pl}	$=$	$\sqrt{\hbar c^5/Gk^2}$	$=$ $1.417\,1(91) \cdot 10^{32}$ K
Trivial units				
the Planck velocity	v_{Pl}	$=$	c	$=$ 0.3 Gm/s
the Planck angular momentum	L_{Pl}	$=$	$\hbar$	$=$ $1.1 \cdot 10^{-34}$ Js
the Planck action	S_{aPl}	$=$	$\hbar$	$=$ $1.1 \cdot 10^{-34}$ Js
the Planck entropy	S_{ePl}	$=$	k	$=$ 13.8 yJ/K
Composed units				
the Planck mass density	ρ_{Pl}	$=$	$c^5/G^2\hbar$	$=$ $5.2 \cdot 10^{96}$ kg/m^3
the Planck energy	E_{Pl}	$=$	$\sqrt{\hbar c^5/G}$	$=$ 2.0 GJ $= 1.2 \cdot 10^{28}$ eV
the Planck momentum	p_{Pl}	$=$	$\sqrt{\hbar c^3/G}$	$=$ 6.5 Ns
the Planck power	P_{Pl}	$=$	c^5/G	$=$ $3.6 \cdot 10^{52}$ W
the Planck force	F_{Pl}	$=$	c^4/G	$=$ $1.2 \cdot 10^{44}$ N
the Planck pressure	p_{Pl}	$=$	$c^7/G\hbar$	$=$ $4.6 \cdot 10^{113}$ Pa
the Planck acceleration	a_{Pl}	$=$	$\sqrt{c^7/\hbar G}$	$=$ $5.6 \cdot 10^{51}$ m/s^2
the Planck frequency	f_{Pl}	$=$	$\sqrt{c^5/\hbar G}$	$=$ $1.9 \cdot 10^{43}$ Hz
the Planck electric charge	q_{Pl}	$=$	$\sqrt{4\pi\varepsilon_0 c\hbar}$	$=$ 1.9 aC $= 11.7$ e
the Planck voltage	U_{Pl}	$=$	$\sqrt{c^4/4\pi\varepsilon_0 G}$	$=$ $1.0 \cdot 10^{27}$ V
the Planck resistance	R_{Pl}	$=$	$1/4\pi\varepsilon_0 c$	$=$ $30.0\,\Omega$
the Planck capacitance	C_{Pl}	$=$	$4\pi\varepsilon_0\sqrt{\hbar G/c^3}$	$=$ $1.8 \cdot 10^{-45}$ F
the Planck inductance	L_{Pl}	$=$	$(1/4\pi\varepsilon_0)\sqrt{\hbar G/c^7}$	$=$ $1.6 \cdot 10^{-42}$ H
the Planck electric field	E_{Pl}	$=$	$\sqrt{c^7/4\pi\varepsilon_0\hbar G^2}$	$=$ $6.5 \cdot 10^{61}$ V/m
the Planck magnetic flux density	B_{Pl}	$=$	$\sqrt{c^5/4\pi\varepsilon_0\hbar G^2}$	$=$ $2.2 \cdot 10^{53}$ T

---

* The natural units x_{Pl} given here are those commonly used today, i.e., those defined using the constant $\hbar$, and not, as Planck originally did, by using the constant $h = 2\pi\hbar$. The electromagnetic units can also be defined with other factors than $4\pi\varepsilon_0$ in the expressions: for example, using $4\pi\varepsilon_0\alpha$, with the *fine-structure* Vol. IV, page 182 *constant* α, gives $q_{Pl} = e$. For the explanation of the numbers between brackets, see below.

The natural units are important for another reason: whenever a quantity is sloppily called 'infinitely small (or large)', the correct expression is 'as small (or as large) as the corresponding corrected Planck unit'. As explained throughout the text, and especially in the final part, this substitution is possible because almost all Planck units provide, within a correction factor of order 1, the extremal value for the corresponding observable – some an upper and some a lower limit. Unfortunately, these correction factors are not yet widely known. The exact extremal value for each observable in nature is obtained when G is substituted by $4G$ and $4\pi\varepsilon_0$ by $4\pi\varepsilon_0\alpha$ in all Planck quantities. These extremal values, or *corrected Planck units*, are the *true natural units*. To exceed the extremal values is possible only for some extensive quantities. (Can you find out which ones?)

Vol. VI, page 33

Challenge 175 s

OTHER UNIT SYSTEMS

A central aim of research in high-energy physics is the calculation of the strengths of all interactions; therefore it is not practical to set the gravitational constant G to unity, as in the Planck system of units. For this reason, high-energy physicists often only set $c = \hbar = k = 1$ and $\mu_0 = 1/\varepsilon_0 = 4\pi$,* leaving only the gravitational constant G in the equations.

In this system, only one fundamental unit exists, but its choice is free. Often a standard length is chosen as the fundamental unit, length being the archetype of a measured quantity. The most important physical observables are then related by

$$
\begin{aligned}
1/[l^2] &= [E]^2 = [F] = [B] = [E_{\text{electric}}] \,, \\
1/[l] &= [E] = [m] = [p] = [a] = [f] = [I] = [U] = [T] \,, \\
1 &= [v] = [q] = [e] = [R] = [S_{\text{action}}] = [S_{\text{entropy}}] = \hbar = c = k = [\alpha] \,, \quad (132) \\
[l] &= 1/[E] = [t] = [C] = [L] \quad \text{and} \\
[l]^2 &= 1/[E]^2 = [G] = [P]
\end{aligned}
$$

where we write $[x]$ for the unit of quantity x. Using the same unit for time, capacitance and inductance is not to everybody's taste, however, and therefore electricians do not use this system.**

Often, in order to get an impression of the energies needed to observe an effect under study, a standard energy is chosen as fundamental unit. In particle physics the most common energy unit is the *electronvolt* (eV), defined as the kinetic energy acquired by an electron when accelerated by an electrical potential difference of 1 volt ('protonvolt'

---

* Other definitions for the proportionality constants in electrodynamics lead to the Gaussian unit system often used in theoretical calculations, the Heaviside–Lorentz unit system, the electrostatic unit system, and the electromagnetic unit system, among others.

Ref. 263

** In the list, l is length, E energy, F force, E_{electric} the electric and B the magnetic field, m mass, p momentum, a acceleration, f frequency, I electric current, U voltage, T temperature, v speed, q charge, R resistance, P power, G the gravitational constant.

The web page www.chemie.fu-berlin.de/chemistry/general/units_en.html provides a tool to convert various units into each other.

Researchers in general relativity often use another system, in which the *Schwarzschild radius* $r_{\text{s}} = 2Gm/c^2$ is used to measure masses, by setting $c = G = 1$. In this case, mass and length have the *same* dimension, and $\hbar$ has the dimension of an area.

would be a better name). Therefore one has $1\,\text{eV} = 1.6 \cdot 10^{-19}\,\text{J}$, or roughly

$$1\,\text{eV} \approx \tfrac{1}{6}\,\text{aJ} \qquad\qquad (133)$$

which is easily remembered. The simplification $c = \hbar = 1$ yields $G = 6.9 \cdot 10^{-57}\,\text{eV}^{-2}$ and allows one to use the unit eV also for mass, momentum, temperature, frequency, time and length, with the respective correspondences $1\,\text{eV} \equiv 1.8 \cdot 10^{-36}\,\text{kg} \equiv 5.4 \cdot 10^{-28}\,\text{Ns}$ $\equiv 242\,\text{THz} \equiv 11.6\,\text{kK}$ and $1\,\text{eV}^{-1} \equiv 4.1\,\text{fs} \equiv 1.2\,\mu\text{m}$.

Challenge 176 e

To get some feeling for the unit eV, the following relations are useful. Room temperature, usually taken as $20°\text{C}$ or $293\,\text{K}$, corresponds to a kinetic energy per particle of $0.025\,\text{eV}$ or $4.0\,\text{zJ}$. The highest particle energy measured so far belongs to a cosmic ray with an energy of $3 \cdot 10^{20}\,\text{eV}$ or $48\,\text{J}$. Down here on the Earth, an accelerator able to produce an energy of about $105\,\text{GeV}$ or $17\,\text{nJ}$ for electrons and antielectrons has been built, and one able to produce an energy of $14\,\text{TeV}$ or $2.2\,\mu\text{J}$ for protons will be finished soon. Both are owned by CERN in Geneva and have a circumference of $27\,\text{km}$.

Ref. 264

The lowest temperature measured up to now is $280\,\text{pK}$, in a system of rhodium nuclei held inside a special cooling system. The interior of that cryostat may even be the coolest point in the whole universe. The kinetic energy per particle corresponding to that temperature is also the smallest ever measured: it corresponds to $24\,\text{feV}$ or $3.8\,\text{vJ} = 3.8 \cdot 10^{-33}\,\text{J}$. For isolated particles, the record seems to be for neutrons: kinetic energies as low as $10^{-7}\,\text{eV}$ have been achieved, corresponding to de Broglie wavelengths of $60\,\text{nm}$.

Ref. 265

CURIOSITIES AND FUN CHALLENGES ABOUT UNITS

The Planck length is roughly the de Broglie wavelength $\lambda_\text{B} = h/mv$ of a man walking comfortably ($m = 80\,\text{kg}$, $v = 0.5\,\text{m/s}$); this motion is therefore aptly called the 'Planck stroll.'

Ref. 266

∗ ∗

The Planck mass is equal to the mass of about 10^{19} protons. This is roughly the mass of a human embryo at about ten days of age.

∗ ∗

The most precisely measured quantities in nature are the frequencies of certain millisecond pulsars, the frequency of certain narrow atomic transitions, and the Rydberg constant of *atomic* hydrogen, which can all be measured as precisely as the second is defined. The caesium transition that defines the second has a finite line width that limits the achievable precision: the limit is about 14 digits.

Ref. 267

∗ ∗

The most precise clock ever built, using microwaves, had a stability of 10^{-16} during a running time of $500\,\text{s}$. For longer time periods, the record in 1997 was about 10^{-15}; but values around 10^{-17} seem within technological reach. The precision of clocks is limited for short measuring times by noise, and for long measuring times by drifts, i.e., by systematic effects. The region of highest stability depends on the clock type; it usually lies between $1\,\text{ms}$ for optical clocks and $5000\,\text{s}$ for masers. Pulsars are the only type of clock

Ref. 268
Ref. 269

for which this region is not known yet; it certainly lies at more than 20 years, the time elapsed at the time of writing since their discovery.

∗ ∗

The shortest times measured are the lifetimes of certain 'elementary' particles. In particular, the lifetime of certain D mesons have been measured at less than 10^{-23} s. Such times are measured using a bubble chamber, where the track is photographed. Can you estimate how long the track is? (This is a trick question – if your length cannot be observed with an optical microscope, you have made a mistake in your calculation.)

Ref. 270

Challenge 177 s

∗ ∗

The longest times encountered in nature are the lifetimes of certain radioisotopes, over 10^{15} years, and the lower limit of certain proton decays, over 10^{32} years. These times are thus much larger than the age of the universe, estimated to be fourteen thousand million years.

Ref. 271

PRECISION AND ACCURACY OF MEASUREMENTS

Measurements are the basis of physics. Every measurement has an *error*. Errors are due to lack of precision or to lack of accuracy. *Precision* means how well a result is reproduced when the measurement is repeated; *accuracy* is the degree to which a measurement corresponds to the actual value.

Lack of precision is due to accidental or *random errors*; they are best measured by the *standard deviation*, usually abbreviated σ; it is defined through

$$\sigma^2 = \frac{1}{n-1} \sum_{i=1}^{n} (x_i - \bar{x})^2 \, ,$$ (134)

where $\bar{x}$ is the average of the measurements x_i. (Can you imagine why $n-1$ is used in the formula instead of n?)

Challenge 178 s

For most experiments, the distribution of measurement values tends towards a normal distribution, also called *Gaussian distribution*, whenever the number of measurements is increased. The distribution, shown in Figure 167, is described by the expression

$$N(x) \approx e^{-\frac{(x-\bar{x})^2}{2\sigma^2}} \, .$$ (135)

The square σ^2 of the standard deviation is also called the *variance*. For a Gaussian distribution of measurement values, 2.35σ is the full width at half maximum.

Challenge 179 e

Lack of accuracy is due to *systematic errors*; usually these can only be estimated. This estimate is often added to the random errors to produce a *total experimental error*, sometimes also called *total uncertainty*. The *relative* error or uncertainty is the ratio between the error and the measured value.

Ref. 272

For example, a professional measurement will give a result such as 0.312(6) m. The number between the parentheses is the standard deviation σ, in units of the last digits. As above, a Gaussian distribution for the measurement results is assumed. Therefore, a

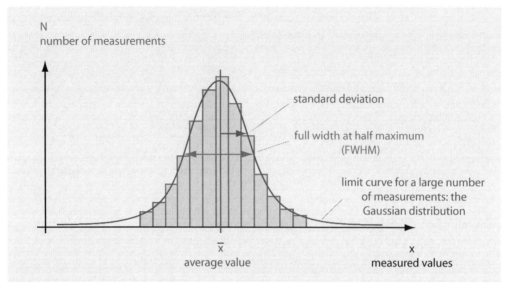

FIGURE 167 A precision experiment and its measurement distribution. The precision is high if the width of the distribution is narrow; the accuracy is high if the centre of the distribution agrees with the actual value.

Challenge 180 e   value of 0.312(6) m implies that the actual value is expected to lie

— within 1σ with 68.3% probability, thus in this example within 0.312 ± 0.006 m;
— within 2σ with 95.4% probability, thus in this example within 0.312 ± 0.012 m;
— within 3σ with 99.73% probability, thus in this example within 0.312 ± 0.018 m;
— within 4σ with 99.9937% probability, thus in this example within 0.312 ± 0.024 m;
— within 5σ with 99.999 943% probability, thus in this example within 0.312 ± 0.030 m;
— within 6σ with 99.999 999 80% probability, thus within 0.312 ± 0.036 m;
— within 7σ with 99.999 999 999 74% probability, thus within 0.312 ± 0.041 m.

Challenge 181 s   (Do the latter numbers make sense?)

Note that standard deviations have one digit; you must be a world expert to use two, and a fool to use more. If no standard deviation is given, a (1) is assumed. As a result, among professionals, 1 km and 1000 m are *not* the same length!

What happens to the errors when two measured values A and B are added or subtracted? If the all measurements are independent – or uncorrelated – the standard deviation of the sum *and* that of difference is given by $\sigma = \sqrt{\sigma_A^2 + \sigma_B^2}$. For both the product or ratio of two measured and uncorrelated values C and D, the result is $\rho = \sqrt{\rho_C^2 + \rho_D^2}$, where the ρ terms are the *relative* standard deviations.

Challenge 182 s   Assume you measure that an object moves 1.0 m in 3.0 s: what is the measured speed value?

LIMITS TO PRECISION

What are the limits to accuracy and precision? There is no way, even in principle, to measure a length x to a *precision* higher than about 61 digits, because in nature, the ratio between the largest and the smallest measurable length is $\Delta x/x > l_{\text{Pl}}/d_{\text{horizon}} = 10^{-61}$.

(Is this ratio valid also for force or for volume?) In the final volume of our text, studies Challenge 183 e of clocks and metre bars strengthen this theoretical limit. Vol. VI, page 87

But it is not difficult to deduce more stringent practical limits. No imaginable machine can measure quantities with a higher precision than measuring the diameter of the Earth within the smallest length ever measured, about 10^{-19} m; that is about 26 digits of precision. Using a more realistic limit of a 1000 m sized machine implies a limit of 22 digits. If, as predicted above, time measurements really achieve 17 digits of precision, then they are nearing the practical limit, because apart from size, there is an additional practical restriction: cost. Indeed, an additional digit in measurement precision often means an additional digit in equipment cost.

PHYSICAL CONSTANTS

In physics, general observations are deduced from more fundamental ones. As a consequence, many measurements can be deduced from more fundamental ones. The most fundamental measurements are those of the physical constants.

The following tables give the world's best values of the most important physical constants and particle properties – in SI units and in a few other common units – as published in the standard references. The values are the world averages of the best measurements made up to the present. As usual, experimental errors, including both random and estimated systematic errors, are expressed by giving the standard deviation in the last digits. In fact, behind each of the numbers in the following tables there is a long story which is worth telling, but for which there is not enough room here. Ref. 273

In principle, all quantitative properties of matter can be calculated with quantum theory and the values of certain physical constants. For example, colour, density and elastic properties can be predicted using the equations of the standard model of particle physics and the values of the following basic constants. Ref. 274
Ref. 273
Page 239

TABLE 28 Basic physical constants.

QUANTITY	SYMBOL	VALUE IN SI UNITS	UNCERT.[a]
Constants that define the SI measurement units			
Vacuum speed of light[c]	c	299 792 458 m/s	0
Vacuum permeability[c]	μ_0	$4\pi \cdot 10^{-7}$ H/m	0
		$= 1.256\,637\,061\,435\ldots$ µH/m0	
Vacuum permittivity[c]	$\varepsilon_0 = 1/\mu_0 c^2$	$8.854\,187\,817\,620\ldots$ pF/m	0
Original Planck constant	h	$6.626\,069\,57(52) \cdot 10^{-34}$ Js	$4.4 \cdot 10^{-8}$
Reduced Planck constant, quantum of action	$\hbar$	$1.054\,571\,726(47) \cdot 10^{-34}$ Js	$4.4 \cdot 10^{-8}$
Positron charge	e	$0.160\,217\,656\,5(35)$ aC	$2.2 \cdot 10^{-8}$
Boltzmann constant	k	$1.380\,6488(13) \cdot 10^{-23}$ J/K	$9.1 \cdot 10^{-7}$
Gravitational constant	G	$6.673\,84(80) \cdot 10^{-11}$ Nm²/kg²	$1.2 \cdot 10^{-4}$
Gravitational coupling constant	$\kappa = 8\pi G/c^4$	$2.076\,50(25) \cdot 10^{-43}$ s²/kg m	$1.2 \cdot 10^{-4}$
Fundamental constants (of unknown origin)			
Number of space-time dimensions		$3+1$	0^{b}

TABLE 28 (Continued) Basic physical constants.

QUANTITY	SYMBOL	VALUE IN SI UNITS	UNCERT.[a]
Fine-structure constant[d] or	$\alpha = \frac{e^2}{4\pi\varepsilon_0 \hbar c}$	$1/137.035\,999\,074(44)$	$3.2 \cdot 10^{-10}$
e.m. coupling constant	$= g_{em}(m_e^2 c^2)$	$= 0.007\,297\,352\,5698(24)$	$3.2 \cdot 10^{-10}$
Fermi coupling constant[d] or	$G_F/(\hbar c)^3$	$1.166\,364(5) \cdot 10^{-5}\,\text{GeV}^{-2}$	$4.3 \cdot 10^{-6}$
weak coupling constant	$\alpha_w(M_Z) = g_w^2/4\pi$	$1/30.1(3)$	$1 \cdot 10^{-2}$
Weak mixing angle	$\sin^2\theta_W(\overline{MS})$	$0.231\,24(24)$	$1.0 \cdot 10^{-3}$
	$\sin^2\theta_W$ (on shell)	$0.2224(19)$	$8.7 \cdot 10^{-3}$
	$= 1 - (m_W/m_Z)^2$		
Strong coupling constant[d]	$\alpha_s(M_Z) = g_s^2/4\pi$	$0.118(3)$	$25 \cdot 10^{-3}$
CKM quark mixing matrix	$|V|$	$\begin{pmatrix} 0.97428(15) & 0.2253(7) & 0.00347(16) \\ 0.2252(7) & 0.97345(16) & 0.0410(11) \\ 0.00862(26) & 0.0403(11) & 0.999152(45) \end{pmatrix}$	
Jarlskog invariant	J	$2.96(20) \cdot 10^{-5}$	
PMNS neutrino mixing m.	P	$\begin{pmatrix} 0.82 & 0.55 & -0.15 + 0.038i \\ -0.36 + 0.020i & 0.70 + 0.013i & 0.61 \\ 0.44 + 0.026i & -0.45 + 0.017i & 0.77 \end{pmatrix}$	

Elementary particle masses (of unknown origin)

Electron mass	m_e	$9.109\,382\,91(40) \cdot 10^{-31}\,\text{kg}$	$4.4 \cdot 10^{-8}$
		$5.485\,799\,0946(22) \cdot 10^{-4}\,\text{u}$	$4.0 \cdot 10^{-10}$
		$0.510\,998\,928(11)\,\text{MeV}$	$2.2 \cdot 10^{-8}$
Muon mass	m_μ	$1.883\,531\,475(96) \cdot 10^{-28}\,\text{kg}$	$5.1 \cdot 10^{-8}$
		$0.113\,428\,9267(29)\,\text{u}$	$2.5 \cdot 10^{-8}$
		$105.658\,3715(35)\,\text{MeV}$	$3.4 \cdot 10^{-8}$
Tau mass	m_τ	$1.776\,82(16)\,\text{GeV}/c^2$	
El. neutrino mass	m_{ν_e}	$< 2\,\text{eV}/c^2$	
Muon neutrino mass	m_{ν_e}	$< 2\,\text{eV}/c^2$	
Tau neutrino mass	m_{ν_e}	$< 2\,\text{eV}/c^2$	
Up quark mass	u	1.8 to $3.0\,\text{MeV}/c^2$	
Down quark mass	d	4.5 to $5.5\,\text{MeV}/c^2$	
Strange quark mass	s	$95(5)\,\text{MeV}/c^2$	
Charm quark mass	c	$1.275(25)\,\text{GeV}/c^2$	
Bottom quark mass	b	$4.18(17)\,\text{GeV}/c^2$	
Top quark mass	t	$173.5(1.4)\,\text{GeV}/c^2$	
Photon mass	γ	$< 2 \cdot 10^{-54}\,\text{kg}$	
W boson mass	$W^\pm$	$80.385(15)\,\text{GeV}/c^2$	
Z boson mass	Z^0	$91.1876(21)\,\text{GeV}/c^2$	
Higgs mass	H	$126(1)\,\text{GeV}/c^2$	
Gluon mass	$g_{1...8}$	$c.\,0\,\text{MeV}/c^2$	

Composite particle masses

TABLE 28 (Continued) Basic physical constants.

QUANTITY	SYMBOL	VALUE IN SI UNITS	UNCERT.[a]
Proton mass	m_p	$1.672\,621\,777(74) \cdot 10^{-27}$ kg	$4.4 \cdot 10^{-8}$
		$1.007\,276\,466\,812(90)$ u	$8.9 \cdot 10^{-11}$
		$938.272\,046(21)$ MeV	$2.2 \cdot 10^{-8}$
Neutron mass	m_n	$1.674\,927\,351(74) \cdot 10^{-27}$ kg	$4.4 \cdot 10^{-8}$
		$1.008\,664\,916\,00(43)$ u	$4.2 \cdot 10^{-10}$
		$939.565\,379(21)$ MeV	$2.2 \cdot 10^{-8}$
Atomic mass unit	$m_u = m_{12_C}/12 = 1$ u	$1.660\,538\,921(73)$ yg	$4.4 \cdot 10^{-8}$

a. Uncertainty: standard deviation of measurement errors.

b. Only measured from to 10^{-19} m to 10^{26} m.

c. Defining constant.

d. All coupling constants depend on the 4-momentum transfer, as explained in the section on renormalization. *Fine-structure constant* is the traditional name for the electromagnetic coupling constant g_{em} in the case of a 4-momentum transfer of $Q^2 = m_e^2 c^2$, which is the smallest one possible. At higher momentum transfers it has larger values, e.g., $g_{em}(Q^2 = M_W^2 c^2) \approx 1/128$. In contrast, the strong coupling constant has lover values at higher momentum transfers; e.g., $\alpha_s(34\,\text{GeV}) = 0.14(2)$.

Page 116

Why do all these constants have the values they have? For any constant *with a dimension*, such as the quantum of action $\hbar$, the numerical value has only historical meaning. It is $1.054 \cdot 10^{-34}$ Js because of the SI definition of the joule and the second. The question why the value of a dimensional constant is not larger or smaller therefore always requires one to understand the origin of some dimensionless number giving the ratio between the constant and the corresponding *natural unit* that is defined with c, G, $\hbar$ and α. More details and the values of the natural units were given earlier on. Understanding the Page 303 sizes of atoms, people, trees and stars, the duration of molecular and atomic processes, or the mass of nuclei and mountains, implies understanding the ratios between these values and the corresponding natural units. The key to understanding nature is thus the understanding of all ratios, and thus of all dimensionless constants. The quest of understanding all ratios, including the fine structure constant α itself, is completed only in the final volume of our adventure.

The basic constants yield the following useful high-precision observations.

TABLE 29 Derived physical constants.

QUANTITY	SYMBOL	VALUE IN SI UNITS	UNCERT.
Vacuum wave resistance	$Z_0 = \sqrt{\mu_0/\varepsilon_0}$	$376.730\,313\,461\,77...$ Ω	0
Avogadro's number	N_A	$6.022\,141\,29(27) \cdot 10^{23}$	$4.4 \cdot 10^{-8}$
Loschmidt's number at 273.15 K and 101 325 Pa	N_L	$2.686\,7805(24) \cdot 10^{23}$	$9.1 \cdot 10^{-7}$
Faraday's constant	$F = N_A e$	$96\,485.3365(21)$ C/mol	$2.2 \cdot 10^{-8}$
Universal gas constant	$R = N_A k$	$8.314\,4621(75)$ J/mol K	$9.1 \cdot 10^{-7}$
Molar volume of an ideal gas	$V = RT/p$	$22.413\,968(20)$ l/mol	$9.1 \cdot 10^{-7}$

TABLE 29 (Continued) Derived physical constants.

QUANTITY	SYMBOL	VALUE IN SI UNITS	UNCERT.
at 273.15 K and 101 325 Pa			
Rydberg constant [a]	$R_\infty = m_e c \alpha^2 / 2h$	$10\,973\,731.568\,539(55)$ m^{-1}	$5 \cdot 10^{-12}$
Conductance quantum	$G_0 = 2e^2/h$	$77.480\,917\,346(25)$ µS	$3.2 \cdot 10^{-10}$
Magnetic flux quantum	$\varphi_0 = h/2e$	$2.067\,833\,758(46)$ pWb	$2.2 \cdot 10^{-8}$
Josephson frequency ratio	$2e/h$	$483.597\,870(11)$ THz/V	$2.2 \cdot 10^{-8}$
Von Klitzing constant	$h/e^2 = \mu_0 c/2\alpha$	$25\,812.807\,4434(84)$ Ω	$3.2 \cdot 10^{-10}$
Bohr magneton	$\mu_B = e\hbar/2m_e$	$9.274\,009\,68(20)$ yJ/T	$2.2 \cdot 10^{-8}$
Classical electron radius	$r_e = e^2/4\pi\varepsilon_0 m_e c^2$	$2.817\,940\,3267(27)$ fm	$9.7 \cdot 10^{-10}$
Compton wavelength	$\lambda_C = h/m_e c$	$2.426\,310\,2389(16)$ pm	$6.5 \cdot 10^{-10}$
of the electron	$\lambda_c = \hbar/m_e c = r_e/\alpha$	$0.386\,159\,268\,00(25)$ pm	$6.5 \cdot 10^{-10}$
Bohr radius [a]	$a_\infty = r_e/\alpha^2$	$52.917\,721\,092(17)$ pm	$3.2 \cdot 10^{-10}$
Quantum of circulation	$h/2m_e$	$3.636\,947\,5520(24) \cdot 10^{-4}$ m^2/s	$6.5 \cdot 10^{-10}$
Specific positron charge	e/m_e	$1.758\,820\,088(39) \cdot 10^{11}$ C/kg	$2.2 \cdot 10^{-8}$
Cyclotron frequency	$f_c/B = e/2\pi m_e$	$27.992\,491\,10(62)$ GHz/T	$2.2 \cdot 10^{-8}$
of the electron			
Electron magnetic moment	μ_e	$-9.284\,764\,30(21) \cdot 10^{-24}$ J/T	$2.2 \cdot 10^{-8}$
	μ_e/μ_B	$-1.001\,159\,652\,180\,76(27)$	$2.6 \cdot 10^{-13}$
	μ_e/μ_N	$-1.838\,281\,970\,90(75) \cdot 10^3$	$4.1 \cdot 10^{-10}$
Electron g-factor	g_e	$-2.002\,319\,304\,361\,53(53)$	$2.6 \cdot 10^{-13}$
Muon–electron mass ratio	m_μ/m_e	$206.768\,2843(52)$	$2.5 \cdot 10^{-8}$
Muon magnetic moment	μ_μ	$-4.490\,448\,07(15) \cdot 10^{-26}$ J/T	$3.4 \cdot 10^{-8}$
muon g-factor	g_μ	$-2.002\,331\,8418(13)$	$6.3 \cdot 10^{-10}$
Proton–electron mass ratio	m_p/m_e	$1\,836.152\,672\,45(75)$	$4.1 \cdot 10^{-10}$
Specific proton charge	e/m_p	$9.578\,833\,58(21) \cdot 10^7$ C/kg	$2.2 \cdot 10^{-8}$
Proton Compton wavelength	$\lambda_{C,p} = h/m_p c$	$1.321\,409\,856\,23(94)$ fm	$7.1 \cdot 10^{-10}$
Nuclear magneton	$\mu_N = e\hbar/2m_p$	$5.050\,783\,53(11) \cdot 10^{-27}$ J/T	$2.2 \cdot 10^{-8}$
Proton magnetic moment	μ_p	$1.410\,606\,743(33) \cdot 10^{-26}$ J/T	$2.4 \cdot 10^{-8}$
	μ_p/μ_B	$1.521\,032\,210(12) \cdot 10^{-3}$	$8.1 \cdot 10^{-9}$
	μ_p/μ_N	$2.792\,847\,356(23)$	$8.2 \cdot 10^{-9}$
Proton gyromagnetic ratio	$\gamma_p = 2\mu_p/\hbar$	$2.675\,222\,005(63) \cdot 10^8$ Hz/T	$2.4 \cdot 10^{-8}$
Proton g factor	g_p	$5.585\,694\,713(46)$	$8.2 \cdot 10^{-9}$
Neutron–electron mass ratio	m_n/m_e	$1\,838.683\,6605(11)$	$5.8 \cdot 10^{-10}$
Neutron–proton mass ratio	m_n/m_p	$1.001\,378\,419\,17(45)$	$4.5 \cdot 10^{-10}$
Neutron Compton wavelength	$\lambda_{C,n} = h/m_n c$	$1.319\,590\,9068(11)$ fm	$8.2 \cdot 10^{-10}$
Neutron magnetic moment	μ_n	$-0.966\,236\,47(23) \cdot 10^{-26}$ J/T	$2.4 \cdot 10^{-7}$
	μ_n/μ_B	$-1.041\,875\,63(25) \cdot 10^{-3}$	$2.4 \cdot 10^{-7}$
	μ_n/μ_N	$-1.913\,042\,72(45)$	$2.4 \cdot 10^{-7}$
Stefan–Boltzmann constant	$\sigma = \pi^2 k^4/60\hbar^3 c^2$	$56.703\,73(21)$ nW/m^{2}K^4	$3.6 \cdot 10^{-6}$
Wien's displacement constant	$b = \lambda_{max} T$	$2.897\,7721(26)$ mmK	$9.1 \cdot 10^{-7}$

TABLE 29 (Continued) Derived physical constants.

QUANTITY	SYMBOL	VALUE IN SI UNITS	UNCERT.
		58.789 254(53) GHz/K	$9.1 \cdot 10^{-7}$
Electron volt	eV	$1.602\,176\,565(35) \cdot 10^{-19}$ J	$2.2 \cdot 10^{-8}$
Bits to entropy conversion const.	$k \ln 2$	10^{23} bit = 0.956 994 5(9) J/K	$9.1 \cdot 10^{-7}$
TNT energy content		3.7 to 4.0 MJ/kg	$4 \cdot 10^{-2}$

a. For infinite mass of the nucleus.

Some useful properties of our local environment are given in the following table.

TABLE 30 Astronomical constants.

QUANTITY	SYMBOL	VALUE
Tropical year 1900 [a]	a	31 556 925.974 7 s
Tropical year 1994	a	31 556 925.2 s
Mean sidereal day	d	$23^h 56' 4.090\,53''$
Average distance Earth–Sun [b]		149 597 870.691(30) km
Astronomical unit [b]	AU	149 597 870 691 m
Light year, based on Julian year [b]	al	9.460 730 472 5808 Pm
Parsec	pc	30.856 775 806 Pm = 3.261 634 al
Earth's mass	$M_\oplus$	$5.973(1) \cdot 10^{24}$ kg
Geocentric gravitational constant	GM	$3.986\,004\,418(8) \cdot 10^{14}$ m^3/s^2
Earth's gravitational length	$l_\oplus = 2GM/c^2$	8.870 056 078(16) mm
Earth's equatorial radius [c]	$R_{\oplus\text{eq}}$	6378.1366(1) km
Earth's polar radius [c]	$R_{\oplus\text{p}}$	6356.752(1) km
Equator–pole distance [c]		10 001.966 km (average)
Earth's flattening [c]	$e_\oplus$	1/298.25642(1)
Earth's av. density	$\rho_\oplus$	5.5 Mg/m^3
Earth's age	$T_\oplus$	4.54(5) Ga = 143(2) Ps
Earth's normal gravity	g	9.806 65 m/s^2
Earth's standard atmospher. pressure	p_0	101 325 Pa
Moon's radius	$R_{\leftmoon\,\text{v}}$	1738 km in direction of Earth
Moon's radius	$R_{\leftmoon\,\text{h}}$	1737.4 km in other two directions
Moon's mass	$M_{\leftmoon}$	$7.35 \cdot 10^{22}$ kg
Moon's mean distance [d]	$d_{\leftmoon}$	384 401 km
Moon's distance at perigee [d]		typically 363 Mm, historical minimum 359 861 km
Moon's distance at apogee [d]		typically 404 Mm, historical maximum 406 720 km
Moon's angular size [e]		average $0.5181° = 31.08'$, minimum $0.49°$, maximum - shortens line $0.55°$
Moon's average density	$\rho_{\leftmoon}$	3.3 Mg/m^3

TABLE 30 (Continued) Astronomical constants.

QUANTITY	SYMBOL	VALUE
Moon's surface gravity	$g_{\mathbb{C}}$	$1.62\,\text{m/s}^2$
Moons's atmospheric pressure	$p_{\mathbb{C}}$	from $10^{-10}\,\text{Pa}$ (night) to $10^{-7}\,\text{Pa}$ (day)
Jupiter's mass	$M_{\mathcal{L}}$	$1.90 \cdot 10^{27}\,\text{kg}$
Jupiter's radius, equatorial	$R_{\mathcal{L}}$	$71.398\,\text{Mm}$
Jupiter's radius, polar	$R_{\mathcal{L}}$	$67.1(1)\,\text{Mm}$
Jupiter's average distance from Sun	$D_{\mathcal{L}}$	$778\,412\,020\,\text{km}$
Jupiter's surface gravity	$g_{\mathcal{L}}$	$24.9\,\text{m/s}^2$
Jupiter's atmospheric pressure	$p_{\mathcal{L}}$	from $20\,\text{kPa}$ to $200\,\text{kPa}$
Sun's mass	$M_{\odot}$	$1.988\,43(3) \cdot 10^{30}\,\text{kg}$
Sun's gravitational length	$2GM_{\odot}/c^2$	$2.953\,250\,08(5)\,\text{km}$
Sun's luminosity	$L_{\odot}$	$384.6\,\text{YW}$
Solar equatorial radius	$R_{\odot}$	$695.98(7)\,\text{Mm}$
Sun's angular size		$0.53°$ average; minimum on fourth of July (aphelion) $1888''$, maximum on fourth of January (perihelion) $1952''$
Sun's average density	$\rho_{\odot}$	$1.4\,\text{Mg/m}^3$
Sun's average distance	AU	$149\,597\,870.691(30)\,\text{km}$
Sun's age	$T_{\odot}$	$4.6\,\text{Ga}$
Solar velocity around centre of galaxy	$v_{\odot g}$	$220(20)\,\text{km/s}$
Solar velocity against cosmic background	$v_{\odot b}$	$370.6(5)\,\text{km/s}$
Sun's surface gravity	$g_{\odot}$	$274\,\text{m/s}^2$
Sun's lower photospheric pressure	$p_{\odot}$	$15\,\text{kPa}$
Distance to Milky Way's centre		$8.0(5)\,\text{kpc} = 26.1(1.6)\,\text{kal}$
Milky Way's age		$13.6\,\text{Ga}$
Milky Way's size		$c.\ 10^{21}\,\text{m}$ or $100\,\text{kal}$
Milky Way's mass		10^{12} solar masses, $c.\ 2 \cdot 10^{42}\,\text{kg}$
Most distant galaxy cluster known	SXDF-XCLJ 0218-0510	$9.6 \cdot 10^9\,\text{al}$

a. Defining constant, from vernal equinox to vernal equinox; it was once used to define the second. (Remember: π seconds is about a nanocentury.) The value for 1990 is about 0.7 s less, corresponding to a slowdown of roughly 0.2 ms/a. (Watch out: why?) There is even an empirical formula for the change of the length of the year over time.

b. The truly amazing precision in the average distance Earth–Sun of only 30 m results from time averages of signals sent from Viking orbiters and Mars landers taken over a period of over twenty years. Note that the International Astronomical Union distinguishes the average distance Earth–Sun from the *astronomical unit* itself; the latter is defined as a fixed and exact length. Also the *light year* is a unit defined as an exact number by the IAU. For more details, see www.iau.org/public/measuring.

Challenge 184 s
Ref. 275

c. The shape of the Earth is described most precisely with the World Geodetic System. The last edition dates from 1984. For an extensive presentation of its background and its details, see the www.wgs84.com website. The International Geodesic Union refined the data in 2000. The radii and the flattening given here are those for the 'mean tide system'. They differ from those of the 'zero tide system' and other systems by about 0.7 m. The details constitute a science in itself.

d. Measured centre to centre. To find the precise position of the Moon at a given date, see the www.fourmilab.ch/earthview/moon_ap_per.html page. For the planets, see the page www.fourmilab.ch/solar/solar.html and the other pages on the same site.

e. Angles are defined as follows: 1 degree = $1° = \pi/180$ rad, 1 (first) minute = $1' = 1°/60$, 1 second (minute) = $1'' = 1'/60$. The ancient units 'third minute' and 'fourth minute', each 1/60th of the preceding, are not in use any more. ('Minute' originally means 'very small', as it still does in modern English.)

Some properties of nature at large are listed in the following table. (If you want a challenge, can you determine whether any property of the universe itself is listed?)

Challenge 185 s

TABLE 31 Cosmological constants.

QUANTITY	SYMBOL	VALUE
Cosmological constant	Λ	$c. \ 1 \cdot 10^{-52}$ m^{-2}
Age of the universe [a]	t_0	$4.333(53) \cdot 10^{17}$ s $= 13.73(0.17) \cdot 10^9$ a
(determined from space-time, via expansion, using general relativity)		
Age of the universe [a]	t_0	over $3.5(4) \cdot 10^{17}$ s $= 11.5(1.5) \cdot 10^9$ a
(determined from matter, via galaxies and stars, using quantum theory)		
Hubble parameter [a]	H_0	$2.3(2) \cdot 10^{-18}$ s$^{-1} = 0.73(4) \cdot 10^{-10}$ a^{-1}
		$= h_0 \cdot 100$ km/s Mpc $= h_0 \cdot 1.0227 \cdot 10^{-10}$ a^{-1}
Reduced Hubble parameter [a]	h_0	$0.71(4)$
Deceleration parameter [a]	$q_0 = -(\ddot{a}/a)_0/H_0^2$	$-0.66(10)$
Universe's horizon distance [a]	$d_0 = 3ct_0$	$40.0(6) \cdot 10^{26}$ m $= 13.0(2)$ Gpc
Universe's topology		trivial up to 10^{26} m
Number of space dimensions		3, for distances up to 10^{26} m
Critical density	$\rho_c = 3H_0^2/8\pi G$	$h_0^2 \cdot 1.878\,82(24) \cdot 10^{-26}$ kg/m^3
of the universe		$= 0.95(12) \cdot 10^{-26}$ kg/m^3
(Total) density parameter [a]	$\Omega_0 = \rho_0/\rho_c$	$1.02(2)$
Baryon density parameter [a]	$\Omega_{B0} = \rho_{B0}/\rho_c$	$0.044(4)$
Cold dark matter density parameter [a]	$\Omega_{CDM0} = \rho_{CDM0}/\rho_c$	$0.23(4)$
Neutrino density parameter [a]	$\Omega_{\nu 0} = \rho_{\nu 0}/\rho_c$	0.001 to 0.05
Dark energy density parameter [a]	$\Omega_{X0} = \rho_{X0}/\rho_c$	$0.73(4)$
Dark energy state parameter	$w = p_X/\rho_X$	$-1.0(2)$
Baryon mass	m_b	$1.67 \cdot 10^{-27}$ kg
Baryon number density		$0.25(1)$ /m^3
Luminous matter density		$3.8(2) \cdot 10^{-28}$ kg/m^3
Stars in the universe	n_s	$10^{22\pm1}$
Baryons in the universe	n_b	$10^{81\pm1}$
Microwave background temperature [b]	T_0	$2.725(1)$ K

TABLE 31 (Continued) Cosmological constants.

QUANTITY	SYMBOL	VALUE
Photons in the universe	n_γ	10^{89}
Photon energy density	$\rho_\gamma = \pi^2 k^4/15T_0^4$	$4.6 \cdot 10^{-31}$ kg/m^3
Photon number density		410.89 /cm^3 or 400 /cm$^3(T_0/2.7\,\mathrm{K})^3$
Density perturbation amplitude	$\sqrt{S}$	$5.6(1.5) \cdot 10^{-6}$
Gravity wave amplitude	$\sqrt{T}$	$< 0.71\sqrt{S}$
Mass fluctuations on 8 Mpc	σ_8	$0.84(4)$
Scalar index	n	$0.93(3)$
Running of scalar index	$\mathrm{d}n/\mathrm{d}\ln k$	$-0.03(2)$
Planck length	$l_{\mathrm{Pl}} = \sqrt{\hbar G/c^3}$	$1.62 \cdot 10^{-35}$ m
Planck time	$t_{\mathrm{Pl}} = \sqrt{\hbar G/c^5}$	$5.39 \cdot 10^{-44}$ s
Planck mass	$m_{\mathrm{Pl}} = \sqrt{\hbar c/G}$	$21.8\,\mu$g
Instants in history [a]	t_0/t_{Pl}	$8.7(2.8) \cdot 10^{60}$
Space-time points inside the horizon [a]	$N_0 = (R_0/l_{\mathrm{Pl}})^3 \cdot$ (t_0/t_{Pl})	$10^{244\pm1}$
Mass inside horizon	M	$10^{54\pm1}$ kg

a. The index 0 indicates present-day values.

b. The radiation originated when the universe was 380 000 years old and had a temperature of about 3000 K; the fluctuations ΔT_0 which led to galaxy formation are today about $16 \pm 4\,\mu$K = $6(2) \cdot 10^{-6}\,T_0$.

Vol. II, page 213

USEFUL NUMBERS

Ref. 276

π	3.14159 26535 89793 23846 26433 83279 50288 41971 69399 37510$_5$
e	2.71828 18284 59045 23536 02874 71352 66249 77572 47093 69995$_9$
γ	0.57721 56649 01532 86060 65120 90082 40243 10421 59335 93992$_3$
$\ln 2$	0.69314 71805 59945 30941 72321 21458 17656 80755 00134 36025$_5$
$\ln 10$	2.30258 50929 94045 68401 79914 54684 36420 76011 01488 62877$_2$
$\sqrt{10}$	3.16227 76601 68379 33199 88935 44432 71853 37195 55139 32521$_6$

COMPOSITE PARTICLE PROPERTIES

$\mathbf{T}$HE following table lists the most important composite particles. The list has not changed much recently, mainly because of the vast progress that was achieved already in the middle of the twentieth century. In principle, using the standard model of particle physics, *all* properties of composite matter and radiation can be deduced. In particular, all properties of objects encountered in everyday life follow. (Can you explain how the size of an apple follows from the standard model?) The most important examples of composites are grouped in the following table.

Page 239

Challenge 186 s

TABLE 32 Properties of selected composites.

COMPOSITE	MASS m, QUANTUM NUMBERS[a]	LIFETIME τ, MAIN DECAY MODES	SIZE (DIAM.)
Mesons (hadrons, bosons) (selected from over 130 known types)			
Pion $\pi^0 (u\bar{u} - d\bar{d})/\sqrt{2}$	$134.9764(6)\,\text{MeV}/c^2$ $I^G(J^{PC}) = 1^-(0^{-+})$, $S = C = B = 0$	$84(6)$ as, $2\gamma\ 98.798(32)\%$	$\sim 1\,\text{fm}$
Pion $\pi^+ (u\bar{d})$	$139.56995(35)\,\text{MeV}/c^2$ $I^G(J^P) = 1^-(0^-)$, $S = C = B = 0$	$26.030(5)\,\text{ns}$, $\mu^+ \nu_\mu\ 99.9877(4)\%$	$\sim 1\,\text{fm}$
Kaon K_S^0	$m_{K_S^0}$	$89.27(9)\,\text{ps}$	$\sim 1\,\text{fm}$
Kaon K_L^0	$m_{K_S^0} + 3.491(9)\,\mu\text{eV}/c^2$	$51.7(4)\,\text{ns}$	$\sim 1\,\text{fm}$
Kaon $K^\pm (u\bar{s}, \bar{u}s)$	$493.677(16)\,\text{MeV}/c^2$	$12.386(24)\,\text{ns}$, $\mu^+ \nu_\mu\ 63.51(18)\%$ $\pi^+ \pi^0\ 21.16(14)\%$	$\sim 1\,\text{fm}$
Kaon $K^0 (d\bar{s})$ (50 % K_S, 50 % K_L)	$497.672(31)\,\text{MeV}/c^2$	n.a.	$\sim 1\,\text{fm}$
All kaons $K^\pm, K^0, K_S^0, K_L^0$:	$I(J^P) = \frac{1}{2}(0^-)$, $S = \pm 1$, $B = C = 0$		
Baryons (hadrons, fermions) (selected from over 100 known types)			
Proton p or $N^+ (uud)$	$1.67262158(13)\,\text{yg}$ $= 1.00727646688(13)\,\text{u}$ $= 938.271998(38)\,\text{MeV}/c^2$ $I(J^P) = \frac{1}{2}(\frac{1}{2}^+)$, $S = 0$ gyromagnetic ratio $\mu_p/\mu_N = 2.792847337(29)$ electric dipole moment $d = (-4 \pm 6) \cdot 10^{-26} e\,\text{m}$	$\tau_{\text{total}} > 1.6 \cdot 10^{25}\,\text{a}$, $\tau(p \to e^+ \pi^0) > 5.5 \cdot 10^{32}\,\text{a}$ Ref. 277	$0.89(1)\,\text{fm}$

TABLE 32 (Continued) Properties of selected composites.

COMPOSITE	MASS m, QUANTUM NUMBERS[a]	LIFETIME τ, MAIN DECAY MODES	SIZE (DIAM.)
	electric polarizability $\alpha_e = 12.1(0.9) \cdot 10^{-4}\,\text{fm}^3$		
	magnetic polarizability $\alpha_m = 2.1(0.9) \cdot 10^{-4}\,\text{fm}^3$		
Neutron[b] n or N^0 (udd)	$1.674\,927\,16(13)\,\text{yg}$	$887.0(2.0)\,\text{s}$, $pe^-\bar{\nu}_e$ 100 %	~ 1 fm
	$= 1.008\,664\,915\,78(55)\,\text{u} = 939.565\,330(38)\,\text{MeV}/c^2$		
	$I(J^P) = \frac{1}{2}(\frac{1}{2}^+)$, $S = 0$		
	gyromagnetic ratio $\mu_n/\mu_N = -1.913\,042\,72(45)$		
	electric dipole moment $d_n = (-3.3 \pm 4.3) \cdot 10^{-28}\,e\,\text{m}$		
	electric polarizability $\alpha = 0.98(23) \cdot 10^{-3}\,\text{fm}^3$		
Omega Ω^- (sss)	$1672.43(32)\,\text{MeV}/c^2$	$82.2(1.2)\,\text{ps}$,	~ 1 fm
		ΛK^- 67.8(7)%,	
		$\Xi^0\pi^-$ 23.6(7)%	
	gyromagnetic ratio $\mu_\Omega/\mu_N = -1.94(22)$		

composite radiation: glueballs

glueball candidate $f_0(1500)$, status unclear	1503(11) MeV	full width 120(19) MeV	~ 1 fm
	$I^G(J^{PC}) = 0^+(0^{++})$		

Atoms (selected from 114 known elements with over 2000 known nuclides)   Ref. 278

Hydrogen (^{1}H) [lightest]	$1.007\,825\,032(1)\,\text{u} = 1.6735\,\text{yg}$		$2 \cdot 53\,\text{pm}$
Antihydrogen[c]	$1.007\,\text{u} = 1.67\,\text{yg}$		$2 \cdot 53\,\text{pm}$
Helium (^{4}He) [smallest]	$4.002\,603250(1)\,\text{u} = 6.6465\,\text{yg}$		$2 \cdot 31\,\text{pm}$
Carbon (^{12}C)	$12\,\text{u} = 19.926\,482(12)\,\text{yg}$		$2 \cdot 77\,\text{pm}$
Bismuth (^{209}Bi*) [shortest living and rarest]	209 u	0.1 ps Ref. 279	
Tantalum (^{180m}Ta) [second longest living radioactive]	180 u	$> 10^{15}\,\text{a}$ Ref. 280	
Bismuth (^{209}Bi) [longest living radioactive]	209 u	$1.9(2)10^{19}\,\text{a}$ Ref. 279	
Francium (^{223}Fr) [largest]	223 u	22 min	$2 \cdot 0.28\,\text{nm}$
Atom 118 (^{289}Uuo) [heaviest]	294 u	0.9 ms	

Molecules[d] (selected from over 10^7 known types)

Hydrogen (H_2)	~ 2 u	$> 10^{25}\,\text{a}$	
Water (H_2O)	~ 18 u	$> 10^{25}\,\text{a}$	
ATP (adenosinetriphosphate)	507 u	$> 10^{10}\,\text{a}$	$c.\ 3\,\text{nm}$
Human Y chromosome	$70 \cdot 10^6$ base pairs	$> 10^6\,\text{a}$	$c.\ 50\,\text{mm}$ (uncoiled)

TABLE 32 (Continued) Properties of selected composites.

COMPOSITE	MASS m, QUANTUM NUMBERS[a]	LIFETIME τ, MAIN DECAY MODES	SIZE (DIAM.)
Other composites			
Blue whale nerve cell	~ 1 kg	~ 50 a	20 m
Cell (red blood)	0.1 ng	7 plus 120 days	~ 10 µm
Cell (sperm)	10 pg	not fecundated: ~ 5 d	length 60 µm, head 3 µm × 5 µm
Cell (ovule)	1 µg	fecundated: over 4000 million years	~ 120 µm
Cell (*E. coli*)	1 pg	4000 million years	body: 2 µm
Apple	0.1 kg	4 weeks	0.1 m
Adult human	35 kg < m < 350 kg	$\tau \approx 2.5 \cdot 10^9$ s Ref. 281 $\approx$ 600 million breaths $\approx$ 2 500 million heartbeats < 122 a, 60 % H_2O and 40 % dust	~ 1.7 m
Heaviest living thing: colony of aspen trees	$6.6 \cdot 10^6$ kg	> 130 a	> 4 km
Larger composites	See the table on page 226 in volume I.		

Notes (see also the notes of Table 9):

Page 241

a. The charge parity C is defined only for certain neutral particles, namely those that are different from their antiparticles. For neutral mesons, the charge parity is given by $C = (-1)^{L+S}$, where L is the orbital angular momentum.

P is the parity under space inversion $r \rightarrow -r$. For mesons, it is related to the orbital angular momentum L through $P = (-1)^{L+1}$.

The electric polarizability, defined on page 65 in volume III, is predicted to vanish for all elementary particles.

 G-parity is defined only for mesons and given by $G = (-1)^{L+S+I} = (-1)^I C$.

 b. Neutrons bound in nuclei have a lifetime of at least 10^{20} years.

 c. The first *anti-atoms*, made of antielectrons and antiprotons, were made in January 1996 at CERN in Geneva. All properties of antimatter checked so far are consistent with theoretical predictions. Ref. 283

 d. The number of existing molecules is several orders of magnitude larger than the number of molecules that have been analysed and named.

The most important matter composites are the *atoms*. Their size, structure and interactions determine the properties and colour of everyday objects. Atom types, also called *elements* in chemistry, are most usefully set out in the so-called *periodic table*, which groups together atoms with similar properties in rows and columns. It is given in Table 33 and results from the various ways in which protons, neutrons and electrons can combine to form aggregates.

Comparable to the periodic table of the atoms, there are tables for the mesons (made of two quarks) and the baryons (made of three quarks). Neither the meson nor the baryon table is included here; they can both be found in the *Review of Particle Physics* at pdg.web.cern.ch. In fact, the baryon table still has a number of vacant spots. The missing baryons are extremely heavy and short-lived (which means expensive to make and detect), and their discovery is not expected to yield deep new insights.

TABLE 33 The periodic table of the elements known in 2006, with their atomic numbers.

Group

	1	2	3	4	5	6	7	8	9	10	11	12	13	14	15	16	17	18
	I	II	IIIa	IVa	Va	VIa	VIIa		VIIIa		Ia	IIa	III	IV	V	VI	VII	VIII

Period

Period	1	2		3	4	5	6	7	8	9	10	11	12	13	14	15	16	17	18
1	1 H																		2 He
2	3 Li	4 Be												5 B	6 C	7 N	8 O	9 F	10 Ne
3	11 Na	12 Mg												13 Al	14 Si	15 P	16 S	17 Cl	18 Ar
4	19 K	20 Ca	21 Sc	22 Ti	23 V	24 Cr	25 Mn	26 Fe	27 Co	28 Ni	29 Cu	30 Zn	31 Ga	32 Ge	33 As	34 Se	35 Br	36 Kr
5	37 Rb	38 Sr	39 Y	40 Zr	41 Nb	42 Mo	43 Tc	44 Ru	45 Rh	46 Pd	47 Ag	48 Cd	49 In	50 Sn	51 Sb	52 Te	53 I	54 Xe
6	55 Cs	56 Ba	*	72 Hf	73 Ta	74 W	75 Re	76 Os	77 Ir	78 Pt	79 Au	80 Hg	81 Tl	82 Pb	83 Bi	84 Po	85 At	86 Rn
7	87 Fr	88 Ra	**	104 Rf	105 Db	106 Sg	107 Bh	108 Hs	109 Mt	110 Ds	111 Rg	112 Cn	113 Uut	114 Uuq	115 Uup	116 Uuh	117 Uus	118 Uuo

Lanthanoids *	57 La	58 Ce	59 Pr	60 Nd	61 Pm	62 Sm	63 Eu	64 Gd	65 Tb	66 Dy	67 Ho	68 Er	69 Tm	70 Yb	71 Lu
Actinoids **	89 Ac	90 Th	91 Pa	92 U	93 Np	94 Pu	95 Am	96 Cm	97 Bk	98 Cf	99 Es	100 Fm	101 Md	102 No	103 Lr

More elaborate periodic tables can be found on the chemlab.pc.maricopa.edu/periodic website. The most beautiful of them all can be found on page 56. The *atomic number* gives the number of protons (and electrons) found in an atom of a given element. This number

B COMPOSITE PARTICLE PROPERTIES

determines the chemical behaviour of an element. Most – but not all – elements up to 92 are found on Earth; the others can be produced in laboratories. The highest element discovered is element 118. In a famous case of research fraud, a scientist in the 1990s tricked two whole research groups into claiming to have made and observed elements 116 and 118. Both elements were independently made and observed later on.

Nowadays, extensive physical and chemical data are available for every element. Photographs of the pure elements are shown in Figure 18. Elements in the same *group* behave similarly in chemical reactions. The *periods* define the repetition of these similarities.

Ref. 284
Page 57

The elements of group 1 are the alkali metals (though hydrogen is a gas), those of group 2 are the Earth-alkali metals. Actinoids, lanthanoids and groups 3 to 13 are metals; in particular, groups 3 to 12 are transition or heavy metals. The elements of group 16 are called *chalkogens*, i.e., ore-formers; group 17 are the *halogens*, i.e., the salt-formers, and group 18 are the inert *noble gases*, which form (almost) no chemical compounds. The groups 13, 14 and 15 contain metals, semimetals, a liquid and gases; they have no special name. Groups 1 and 13 to 17 are central for the chemistry of life; in fact, 96 % of living matter is made of C, O, N, H;* almost 4 % of P, S, Ca, K, Na, Cl; trace elements such as Mg, V, Cr, Mn, Fe, Co, Ni, Cu, Zn, Cd, Pb, Sn, Li, Mo, Se, Si, I, F, As, B form the rest. Over 30 elements are known to be essential for animal life. The full list is not yet known; candidate elements to extend this list are Al, Br, Ge and W.

Many elements exist in versions with different numbers of neutrons in their nucleus, and thus with different mass; these various *isotopes* – so called because they are found at the *same place* in the periodic table – behave identically in chemical reactions. There are over 2000 of them.

Ref. 278, Ref. 285

TABLE 34 The elements, with their atomic number, average mass, atomic radius and main properties.

NAME	SYM-BOL	AT. N.	AVER. MASS[a] IN U (ERROR), LONGEST LIFETIME	ATO-MIC[e] RA-DIUS IN PM	MAIN PROPERTIES, (NAMING)[h] DIS-COVERY DATE AND USE
Actinium[b]	Ac	89	(227.0277(1)) 21.77(2) a	(188)	Highly radioactive metallic rare Earth (Greek *aktis* ray) 1899, used as alpha-emitting source.
Aluminium	Al	13	26.981 538 (8) stable	118c, 143m	Light metal (Latin *alumen* alum) 1827, used in machine construction and living beings.
Americium[b]	Am	95	(243.0614(1)) 7.37(2) ka	(184)	Radioactive metal (Italian *America* from Amerigo) 1945, used in smoke detectors.
Antimony	Sb	51	121.760(1)[f] stable	137c, 159m, 205v	Toxic semimetal (via Arabic from Latin *stibium*, itself from Greek, Egyptian for one of its minerals) antiquity, colours rubber, used in medicines, constituent of enzymes.

* The 'average formula' of life is approximately $C_5H_{40}O_{18}N$.

TABLE 34 (Continued) The elements, with their atomic number, average mass, atomic radius and main properties.

NAME	SYM-BOL	AT. N.	AVER. MASS[a] IN U (ERROR), LONGEST LIFETIME	ATO-MIC[e] RA-DIUS IN PM	MAIN PROPERTIES, (NAMING)[h] DISCOVERY DATE AND USE
Argon	Ar	18	39.948(1)[f] stable	(71n)	Noble gas (Greek *argos* inactive, from *anergos* without energy) 1894, third component of air, used for welding and in lasers.
Arsenic	As	33	74.921 60(2) stable	120c, 185v	Poisonous semimetal (Greek *arsenikon* tamer of males) antiquity, for poisoning pigeons and doping semiconductors.
Astatine[b]	At	85	(209.9871(1)) 8.1(4) h	(140)	Radioactive halogen (Greek *astatos* unstable) 1940, no use.
Barium	Ba	56	137.327(7) stable	224m	Earth-alkali metal (Greek *bary* heavy) 1808, used in vacuum tubes, paint, oil industry, pyrotechnics and X-ray diagnosis.
Berkelium[b]	Bk	97	(247.0703(1)) 1.4(3) ka	n.a.	Made in lab, probably metallic (Berkeley, US town) 1949, no use because rare.
Beryllium	Be	4	9.012 182(3) stable	106c, 113m	Toxic Earth-alkali metal (Greek *beryllos*, a mineral) 1797, used in light alloys, in nuclear industry as moderator.
Bismuth	Bi	83	208.980 40(1) stable	170m, 215v	Diamagnetic metal (Latin via German *weisse Masse* white mass) 1753, used in magnets, alloys, fire safety, cosmetics, as catalyst, nuclear industry.
Bohrium[b]	Bh	107	(264.12(1)) 0.44 s[g]	n.a.	Made in lab, probably metallic (after Niels Bohr) 1981, found in nuclear reactions, no use.
Boron	B	5	10.811(7)[f] stable	83c	Semimetal, semiconductor (Latin *borax*, from Arabic and Persian for brilliant) 1808, used in glass, bleach, pyrotechnics, rocket fuel, medicine.
Bromine	Br	35	79.904(1) stable	120c, 185v	Red-brown liquid (Greek *bromos* strong odour) 1826, fumigants, photography, water purification, dyes, medicines.
Cadmium	Cd	48	112.411(8)[f] stable	157m	Heavy metal, cuttable and screaming (Greek *kadmeia*, a zinc carbonate mineral where it was discovered) 1817, electroplating, solder, batteries, TV phosphors, dyes.
Caesium	Cs	55	132.905 4519(2) stable	273m	Alkali metal (Latin *caesius* sky blue) 1860, getter in vacuum tubes, photoelectric cells, ion propulsion, atomic clocks.

TABLE 34 (Continued) The elements, with their atomic number, average mass, atomic radius and main properties.

Name	Symbol	At. N.	Aver. mass[a] in u (error), longest lifetime	Atomic[e] radius in pm	Main properties, (naming)[h] discovery date and use
Calcium	Ca	20	40.078(4)[f] stable	197m	Earth-alkali metal (Latin *calcis* chalk) antiquity, pure in 1880, found in stones and bones, reducing agent, alloying.
Californium[b]	Cf	98	(251.0796(1)) 0.90(5) ka	n.a.	Made in lab, probably metallic, strong neutron emitter (Latin *calor* heat and *fornicare* have sex, the land of hot sex :-) 1950, used as neutron source, for well logging.
Carbon	C	6	12.0107(8)[f] stable	77c	Makes up coal and diamond (Latin *carbo* coal) antiquity, used to build most life forms.
Cerium	Ce	58	140.116(1)[f] stable	183m	Rare Earth metal (after asteroid Ceres, Roman goddess) 1803, cigarette lighters, incandescent gas mantles, glass manufacturing, self-cleaning ovens, carbon-arc lighting in the motion picture industry, catalyst, metallurgy.
Chlorine	Cl	17	35.453(2)[f] stable	102c, 175v	Green gas (Greek *chloros* yellow-green) 1774, drinking water, polymers, paper, dyes, textiles, medicines, insecticides, solvents, paints, rubber.
Chromium	Cr	24	51.9961(6) stable	128m	Transition metal (Greek *chromos* colour) 1797, hardens steel, makes steel stainless, alloys, electroplating, green glass dye, catalyst.
Cobalt	Co	27	58.933195(5) stable	125m	Ferromagnetic transition metal (German *Kobold* goblin) 1694, part of vitamin B_{12}, magnetic alloys, heavy-duty alloys, enamel dyes, ink, animal nutrition.
Copernicium[b]	Cn	112	(285) 34 s[g]	n.a.	Made in lab, 1996, no use.
Copper	Cu	29	63.546(3)[f] stable	128m	Red metal (Latin *cuprum* from Cyprus island) antiquity, part of many enzymes, electrical conductors, bronze, brass and other alloys, algicides, etc.
Curium[b]	Cm	96	(247.0704(1)) 15.6(5) Ma	n.a.	Highly radioactive, silver-coloured (after Pierre and Marie Curie) 1944, used as radioactivity source.
Darmstadtium[b]	Ds	110	(271) 1.6 min[g]	n.a.	Made in lab (after the German city) 1994, no use.

TABLE 34 (Continued) The elements, with their atomic number, average mass, atomic radius and main properties.

Name	Symbol	At. N.	Aver. mass[a] in u (error), longest lifetime	Atomic[e] radius in pm	Main properties, (naming)[h] discovery date and use
Dubnium[b]	Db	105	(262.1141(1)) 34(5) s	n.a.	Made in lab in small quantities, radioactive (Dubna, Russian city) 1967, no use (once known as hahnium).
Dysprosium	Dy	66	162.500(1)[f] stable	177m	Rare Earth metal (Greek *dysprositos* difficult to obtain) 1886, used in laser materials, as infrared source material, and in nuclear industry.
Einsteinium[b]	Es	99	(252.0830(1)) 472(2) d	n.a.	Made in lab, radioactive (after Albert Einstein) 1952, no use.
Erbium	Er	68	167.259(3)[f] stable	176m	Rare Earth metal (Ytterby, Swedish town) 1843, used in metallurgy and optical fibres.
Europium	Eu	63	151.964(1)[f] stable	204m	Rare Earth metal (named after the continent) 1901, used in red screen phosphor for TV tubes.
Fermium[b]	Fm	100	(257.0901(1)) 100.5(2) d	n.a.	Made in lab (after Enrico Fermi) 1952, no use.
Flerovium[b]	Uuq	114	(289)	2.7 s[g]	1999, no use.
Fluorine	F	9	18.998 4032(5) stable	62c, 147v	Gaseous halogen (from fluorine, a mineral, from Greek *fluo* flow) 1886, used in polymers and toothpaste.
Francium[b]	Fr	87	(223.0197(1)) 22.0(1) min	(278)	Radioactive metal (from France) 1939, no use.
Gadolinium	Gd	64	157.25(3)[f] stable	180m	Rare-earth metal (after Johan Gadolin) 1880, used in lasers and phosphors.
Gallium	Ga	31	69.723(1) stable	125c, 141m	Almost liquid metal (Latin for both the discoverer's name and his nation, France) 1875, used in optoelectronics.
Germanium	Ge	32	72.64(1) stable	122c, 195v	Semiconductor (from Germania, as opposed to gallium) 1886, used in electronics.
Gold	Au	79	196.966 569(4) stable	144m	Heavy noble metal (Sanskrit *jval* to shine, Latin aurum) antiquity, electronics, jewels.
Hafnium	Hf	72	178.49(2)[c] stable	158m	Metal (Latin for Copenhagen) 1923, alloys, incandescent wire.
Hassium[b]	Hs	108	(277) 16.5 min[g]	n.a.	Radioactive element (Latin form of German state Hessen) 1984, no use .

TABLE 34 (Continued) The elements, with their atomic number, average mass, atomic radius and main properties.

Name	Sym- bol	At. N.	Aver. mass[a] in u (error), longest lifetime	Ato- mic[e] ra- dius in pm	Main properties, (naming)[h] dis- covery date and use
Helium	He	2	$4.002\,602(2)^f$ stable	(31n)	Noble gas (Greek *helios* Sun) where it was discovered 1895, used in balloons, stars, diver's gas and cryogenics.
Holmium	Ho	67	$164.930\,32(2)$ stable	177m	Metal (Stockholm, Swedish capital) 1878, alloys.
Hydrogen	H	1	$1.007\,94(7)^f$ stable	30c	Reactive gas (Greek for water-former) 1766, used in building stars and universe.
Indium	In	49	$114.818(3)$ stable	141c, 166m	Soft metal (Greek *indikon* indigo) 1863, used in solders and photocells.
Iodine	I	53	$126.904\,47(3)$ stable	140c, 198v	Blue-black solid (Greek *iodes* violet) 1811, used in photography.
Iridium	Ir	77	$192.217(3)$ stable	136m	Precious metal (Greek *iris* rainbow) 1804, electrical contact layers.
Iron	Fe	26	$55.845(2)$ stable	127m	Metal (Indo-European *ayos* metal, Latin ferrum) antiquity, used in metallurgy.
Krypton	Kr	36	$83.798(2)^f$ stable	(88n)	Noble gas (Greek *kryptos* hidden) 1898, used in lasers.
Lanthanum	La	57	$138.905\,47(7)^{c,f}$ stable	188m	Reactive rare Earth metal (Greek *lanthanein* to be hidden) 1839, used in lamps and in special glasses.
Lawrencium[b]	Lr	103	$(262.110\,97(1))$ 3.6(3) h	n.a.	Appears in reactions (after Ernest Lawrence) 1961, no use.
Lead	Pb	82	$207.2(1)^{c,f}$ stable	175m	Poisonous, malleable heavy metal (Latin *plumbum*) antiquity, used in car batteries, radioactivity shields, paints.
Lithium	Li	3	$6.941(2)^f$ stable	156m	Light alkali metal with high specific heat (Greek *lithos* stone) 1817, used in batteries, anti-depressants, alloys, nuclear fusion and many chemicals.
Livermorium[b]	Uuh	116	(293)	61 ms[g]	False claim from 1999, claim from 2000, no use.
Lutetium	Lu	71	$174.967(1)^f$ stable	173m	Rare-earth metal (Latin *Lutetia* for Paris) 1907, used as catalyst.
Magnesium	Mg	12	$24.3050(6)$ stable	160m	Light common alkaline Earth metal (from Magnesia, a Greek district in Thessalia) 1755, used in alloys, pyrotechnics, chemical synthesis and medicine, found in chlorophyll.

TABLE 34 (Continued) The elements, with their atomic number, average mass, atomic radius and main properties.

Name	Sym-bol	At. N.	Aver. mass[a] in u (error), longest lifetime	Ato-mic[e] radius in pm	Main properties, (naming)[h] discovery date and use
Manganese	Mn	25	54.938 045(5) stable	126m	Brittle metal (Italian *manganese*, a mineral) 1774, used in alloys, colours amethyst and permanganate.
Meitnerium[b]	Mt	109	(268.1388(1)) 0.070 s[g]	n.a.	Appears in nuclear reactions (after Lise Meitner) 1982, no use.
Mendelevium[b]	Md	101	(258.0984(1)) 51.5(3) d	n.a.	Appears in nuclear reactions (after Дмитрии Иванович Менделеев Dmitriy Ivanovich Mendeleyev) 1955, no use.
Mercury	Hg	80	200.59(2) stable	157m	Liquid heavy metal (Latin god Mercurius, Greek *hydrargyrum* liquid silver) antiquity, used in switches, batteries, lamps, amalgam alloys.
Molybdenum	Mo	42	95.94(2)[f] stable	140m	Metal (Greek *molybdos* lead) 1788, used in alloys, as catalyst, in enzymes and lubricants.
Neodymium	Nd	60	144.242(3)[c,f] stable	182m	(Greek *neos* and *didymos* new twin) 1885.
Neon	Ne	10	20.1797(6)[f] stable	(36n)	Noble gas (Greek *neos* new) 1898, used in lamps, lasers and cryogenics.
Neptunium[b]	Np	93	(237.0482(1)) 2.14(1) Ma	n.a.	Radioactive metal (planet Neptune, after Uranus in the solar system) 1940, appears in nuclear reactors, used in neutron detection and by the military.
Nickel	Ni	28	58.6934(2) stable	125m	Metal (German *Nickel* goblin) 1751, used in coins, stainless steels, batteries, as catalyst.
Niobium	Nb	41	92.906 38(2) stable	147m	Ductile metal (Greek Niobe, mythical daughter of Tantalos) 1801, used in arc welding, alloys, jewellery, superconductors.
Nitrogen	N	7	14.0067(2)[f] stable	70c, 155v	Diatomic gas (Greek for nitre-former) 1772, found in air, in living organisms, Viagra, fertilizers, explosives.
Nobelium[b]	No	102	(259.1010(1)) 58(5) min	n.a.	(after Alfred Nobel) 1958, no use.
Osmium	Os	76	190.23(3)[f] stable	135m	Heavy metal (from Greek *osme* odour) 1804, used for fingerprint detection and in very hard alloys.

TABLE 34 (Continued) The elements, with their atomic number, average mass, atomic radius and main properties.

Name	Sym-bol	At. N.	Aver. mass[a] in u (error), longest lifetime	Ato-mic[e] ra-dius in pm	Main properties, (naming)[h] dis-covery date and use
Oxygen	O	8	15.9994(3)[f] stable	66c, 152v	Transparent, diatomic gas (formed from Greek to mean 'acid former') 1774, used for combustion, blood regeneration, to make most rocks and stones, in countless compounds, colours auroras red.
Palladium	Pd	46	106.42(1)[f] stable	138m	Heavy metal (from asteroid Pallas, after the Greek goddess) 1802, used in alloys, white gold, catalysts, for hydride storage.
Phosphorus	P	15	30.973 762(2) stable	109c, 180v	Poisonous, waxy, white solid (Greek *phosphoros* light bearer) 1669, fertilizers, glasses, porcelain, steels and alloys, living organisms, bones.
Platinum	Pt	78	195.084(9) stable	139m	Silvery-white, ductile, noble heavy metal (Spanish *platina* little silver) pre-Columbian, again in 1735, used in corrosion-resistant alloys, magnets, furnaces, catalysts, fuel cells, cathodic protection systems for large ships and pipelines; being a catalyst, a fine platinum wire glows red hot when placed in vapour of methyl alcohol, an effect used in hand warmers.
Plutonium	Pu	94	(244.0642(1)) 80.0(9) Ma	n.a.	Extremely toxic alpha-emitting metal (after the planet) synthesized 1940, found in nature 1971, used as nuclear explosive, and to power space equipment, such as satellites and the measurement equipment brought to the Moon by the Apollo missions.
Polonium	Po	84	(208.9824(1)) 102(5) a	(140)	Alpha-emitting, volatile metal (from Poland) 1898, used as thermoelectric power source in space satellites, as neutron source when mixed with beryllium; used in the past to eliminate static charges in factories, and on brushes for removing dust from photographic films.

TABLE 34 (Continued) The elements, with their atomic number, average mass, atomic radius and main properties.

NAME	SYM-BOL	AT. N.	AVER. MASS[a] IN U (ERROR), LONGEST LIFETIME	ATO-MIC[e] RA-DIUS IN PM	MAIN PROPERTIES, (NAMING)[h] DIS-COVERY DATE AND USE
Potassium	K	19	39.0983(1) stable	238m	Reactive, cuttable light metal (German *Pottasche*, Latin *kalium* from Arabic *quilyi*, a plant used to produce potash) 1807, part of many salts and rocks, essential for life, used in fertilizers, essential to chemical industry.
Praeseodymium	Pr	59	140.907 65(2) stable	183m	White, malleable rare Earth metal (Greek *praesos didymos* green twin) 1885, used in cigarette lighters, material for carbon arcs used by the motion picture industry for studio lighting and projection, glass and enamel dye, darkens welder's goggles.
Promethium[b]	Pm	61	(144.9127(1)) 17.7(4) a	181m	Radioactive rare Earth metal (from the Greek mythical figure of Prometheus) 1945, used as β source and to excite phosphors.
Protactinium	Pa	91	(231.035 88(2)) 32.5(1) ka	n.a.	Radioactive metal (Greek *protos* first, as it decays into actinium) 1917, found in nature, no use.
Radium	Ra	88	(226.0254(1)) 1599(4) a	(223)	Highly radioactive metal (Latin *radius* ray) 1898, no use any more; once used in luminous paints and as radioactive source and in medicine.
Radon	Rn	86	(222.0176(1)) 3.823(4) d	(130n)	Radioactive noble gas (from its old name 'radium emanation') 1900, no use (any more), found in soil, produces lung cancer .
Rhenium	Re	75	186.207(1)[c] stable	138m	TRansition metal (Latin *rhenus* for Rhine river) 1925, used in filaments for mass spectrographs and ion gauges, superconductors, thermocouples, flash lamps, and as catalyst.
Rhodium	Rh	45	102.905 50(2) stable	135m	White metal (Greek *rhodon* rose) 1803, used to harden platinum and palladium alloys, for electroplating, and as catalyst.
Roentgenium[b]	Rg	111	(272.1535(1)) 1.5 ms[g]	n.a.	Made in lab (after Conrad Roentgen) 1994, no use.
Rubidium	Rb	37	85.4678(3)[f] stable	255m	Silvery-white, reactive alkali metal (Latin *rubidus* red) 1861, used in photocells, optical glasses, solid electrolytes.

TABLE 34 (Continued) The elements, with their atomic number, average mass, atomic radius and main properties.

NAME	SYM-BOL	AT. N.	AVER. MASS[a] IN U (ERROR), LONGEST LIFETIME	ATO-MIC[e] RA-DIUS IN PM	MAIN PROPERTIES, (NAMING)[h] DIS-COVERY DATE AND USE
Ruthenium	Ru	44	101.107(2)[f] stable	134m	White metal (Latin *Rhuthenia* for Russia) 1844, used in platinum and palladium alloys, superconductors, as catalyst; the tetroxide is toxic and explosive.
Rutherfordium[b]	Rf	104	(261.1088(1)) 1.3 min[g]	n.a.	Radioactive transactinide (after Ernest Rutherford) 1964, no use.
Samarium	Sm	62	150.36(2)[c,f] stable	180m	Silver-white rare Earth metal (from the mineral samarskite, after Wassily Samarski) 1879, used in magnets, optical glasses, as laser dopant, in phosphors, in high-power light sources.
Scandium	Sc	21	44.955 912(6) stable	164m	Silver-white metal (from Latin *Scansia* Sweden) 1879, the oxide is used in high-intensity mercury vapour lamps, a radioactive isotope is used as tracer.
Seaborgium[b]	Sg	106	266.1219(1) 21 s[g]	n.a.	Radioactive transurane (after Glenn Seaborg) 1974, no use.
Selenium	Se	34	78.96(3)[f] stable	120c, 190v	Red or black or grey semiconductor (Greek *selene* Moon) 1818, used in xerography, glass production, photographic toners, as enamel dye.
Silicon	Si	14	28.0855(3)[f] stable	105c, 210v	Grey, shiny semiconductor (Latin *silex* pebble) 1823, Earth's crust, electronics, sand, concrete, bricks, glass, polymers, solar cells, essential for life.
Silver	Ag	47	107.8682(2)[f] stable	145m	White metal with highest thermal and electrical conductivity (Latin *argentum*, Greek *argyros*) antiquity, used in photography, alloys, to make rain.
Sodium	Na	11	22.989 769 28(2) stable	191m	Light, reactive metal (Arabic *souwad* soda, Egyptian and Arabic *natrium*) component of many salts, soap, paper, soda, salpeter, borax, and essential for life.
Strontium	Sr	38	87.62(1)[f] stable	215m	Silvery, spontaneously igniting light metal (Strontian, Scottish town) 1790, used in TV tube glass, in magnets, and in optical materials.

TABLE 34 (Continued) The elements, with their atomic number, average mass, atomic radius and main properties.

Name	Symbol	At. N.	Aver. mass[a] in u (error), longest lifetime	Atomic[e] radius in pm	Main properties, (naming)[h] discovery date and use
Sulphur	S	16	32.065(5)[f] stable	105c, 180v	Yellow solid (Latin) antiquity, used in gunpowder, in sulphuric acid, rubber vulcanization, as fungicide in wine production, and is essential for life; some bacteria use sulphur instead of oxygen in their chemistry.
Tantalum	Ta	73	180.947 88(2) stable	147m	Heavy metal (Greek Tantalos, a mythical figure) 1802, used for alloys, surgical instruments, capacitors, vacuum furnaces, glasses.
Technetium[b]	Tc	43	(97.9072(1)) 6.6(10) Ma	136m	Radioactive (Greek technetos artificial) 1939, used as radioactive tracer and in nuclear technology.
Tellurium	Te	52	127.60(3)[f] stable	139c, 206v	Brittle, garlic-smelling semiconductor (Latin tellus Earth) 1783, used in alloys and as glass component.
Terbium	Tb	65	158.925 35(2) stable	178m	Malleable rare Earth metal (Ytterby, Swedish town) 1843, used as dopant in optical material.
Thallium	Tl	81	204.3833(2) stable	172m	Soft, poisonous heavy metal (Greek thallos branch) 1861, used as poison and for infrared detection.
Thorium	Th	90	232.038 06(2)[d,f] 14.0(1) Ga	180m	Radioactive (Nordic god Thor, as in 'Thursday') 1828, found in nature, heats Earth, used as oxide in gas mantles for campers, in alloys, as coating, and in nuclear energy.
Thulium	Tm	69	168.934 21(2) stable	175m	Rare Earth metal (Thule, mythical name for Scandinavia) 1879, found in monazite, used in lasers and radiation detectors.
Tin	Sn	50	118.710(7)[f] stable	139c, 210v, 162m	Grey metal that, when bent, allows one to hear the 'tin cry' (Latin stannum) antiquity, used in paint, bronze and superconductors.
Titanium	Ti	22	47.867(1) stable	146m	Metal (Greek hero Titanos) 1791, alloys, fake diamonds.
Tungsten	W	74	183.84(1) stable	141m	Heavy, highest-melting metal (Swedish tung sten heavy stone, German name Wolfram) 1783, lightbulbs.

TABLE 34 (Continued) The elements, with their atomic number, average mass, atomic radius and main properties.

Name	Sym-bol	At. N.	Aver. mass[a] in u (error), longest lifetime	Atomic[e] radius in pm	Main properties, (naming)[h] discovery date and use
Ununtrium[b]	Uut	113	(284)	0.48 s[g]	2003, no use.
Ununpentium[b]	Uup	115	(288)	8/ ms[g]	2004, no use.
Ununseptium[b]	Uus	117	(294)	78 ms[g]	Claim from 2010, no use.
Ununoctium[b]	Uuo	118	(294)	0.9 ms[g]	False claim in 1999, new claim from 2006, no use.
Uranium	U	92	238.028 91(3)[d,f] $4.468(3) \cdot 10^9$ a	156m	Radioactive and of high density (planet Uranus, after the Greek sky god) 1789, found in pechblende and other minerals, used for nuclear energy.
Vanadium	V	23	50.9415(1) stable	135m	Metal (Vanadis, scandinavian goddess of beauty) 1830, used in steel.
Xenon	Xe	54	131.293(6)[f] stable	(103n) 200v	Noble gas (Greek *xenos* foreign) 1898, used in lamps and lasers.
Ytterbium	Yb	70	173.04(3)[f] stable	174m	Malleable heavy metal (Ytterby, Swedish town) 1878, used in superconductors.
Yttrium	Y	39	88.905 85(2) stable	180m	Malleable light metal (Ytterby, Swedish town) 1794, used in lasers.
Zinc	Zn	30	65.409(4) stable	139m	Heavy metal (German *Zinke* protuberance) antiquity, iron rust protection.
Zirconium	Zr	40	91.224(2)[f] stable	160m	Heavy metal (from the mineral zircon, after Arabic *zargum* golden colour) 1789, chemical and surgical instruments, nuclear industry.

a. The atomic mass unit is defined as $1\,u = \frac{1}{12}m(^{12}C)$, making $1\,u = 1.660\,5402(10)$ yg. For elements found on Earth, the *average* atomic mass for the naturally occurring isotope mixture is given, with the error in the last digit in brackets. For elements not found on Earth, the mass of the *longest living* isotope is given; as it is not an average, it is written in brackets, as is customary in this domain.   Ref. 285

b. The element is not found on Earth because of its short lifetime.

c. The element has at least one radioactive isotope.

d. The element has no stable isotopes.

e. Strictly speaking, the *atomic radius* does not exist. Because atoms are clouds, they have no boundary. Several approximate definitions of the 'size' of atoms are possible. Usually, the radius is defined in such a way as to be useful for the estimation of distances between atoms. This distance is different for different bond types. In the table, radii for metallic bonds are labelled m, radii   Ref. 286
for (single) covalent bonds with carbon c, and Van der Waals radii v. Noble gas radii are labelled   Ref. 286
n. Note that values found in the literature vary by about 10 %; values in brackets lack literature references.

The covalent radius can be up to 0.1 nm smaller than the metallic radius for elements on the (lower) left of the periodic table; on the (whole) right side it is essentially equal to the metallic radius. In between, the difference between the two decreases towards the right. Can you explain

Challenge 187 s why? By the way, ionic radii differ considerably from atomic ones, and depend both on the ionic charge and the element itself.

All these values are for atoms in their ground state. Excited atoms can be hundreds of times larger than atoms in the ground state; however, excited atoms do not form solids or chemical compounds.

f. The isotopic composition, and thus the average atomic mass, of the element varies depending on the place where it was mined or on subsequent human treatment, and can lie outside the values given. For example, the atomic mass of commercial lithium ranges between 6.939 and

Ref. 278 6.996 u. The masses of isotopes are known in atomic mass units to nine or more significant digits, and usually with one or two fewer digits in kilograms. The errors in the atomic mass are thus

Ref. 285 mainly due to the variations in isotopic composition.

g. The lifetime errors are asymmetric or not well known.

h. Extensive details on element names can be found on elements.vanderkrogt.net.

SPACES, ALGEBRAS AND SHAPES

ATHEMATICIANS are fond of generalizing concepts. One of the most generalized concepts of all is the concept of *space*. Understanding athematical definitions and generalizations means learning to think with precision. The following pages provide a simple introduction to the types of spaces that are of importance in physics.

VECTOR SPACES

Vector spaces, also called linear spaces, are mathematical generalizations of certain aspects of the intuitive three-dimensional space. A set of elements any two of which can be added together and any one of which can be multiplied by a number is called a vector space, if the result is again in the set and the usual rules of calculation hold.

More precisely, a *vector space* over a number field K is a set of elements, called *vectors*, for which a vector addition and a *scalar multiplication* is defined, such that for all vectors a, b, c and for all numbers s and r from K one has

$$
\begin{aligned}
(a + b) + c = a + (b + c) = a + b + c &\quad \textit{associativity of vector addition} \\
n + a = a &\quad \textit{existence of null vector} \\
(-a) + a = n &\quad \textit{existence of negative vector} \\
1a = a &\quad \textit{regularity of scalar multiplication} \\
(s + r)(a + b) = sa + sb + ra + rb &\quad \textit{complete distributivity of scalar multiplication}
\end{aligned}
\tag{136}
$$

If the field K, whose elements are called *scalars* in this context, is taken to be the real (or complex, or quaternionic) numbers, one speaks of a real (or complex, or quaternionic) vector space. Vector spaces are also called *linear vector spaces* or simply *linear spaces*.

The complex numbers, the set of all real functions defined on the real line, the set of all polynomials, the set of matrices with a given number of rows and columns, all form vector spaces. In mathematics, a vector is thus a more general concept than in physics. (What is the simplest possible mathematical vector space?)

Challenge 188 ny

In physics, the term 'vector' is reserved for elements of a more specialized type of vector space, namely normed inner product spaces. To define these, we first need the concept of a metric space.

A *metric space* is a set with a metric, i.e., a way to define distances between elements.

A real function $d(a, b)$ between elements is called a metric if

$$
\begin{aligned}
d(a, b) &\geqslant 0 \quad \textit{positivity of metric} \\
d(a, b) + d(b, c) &\geqslant d(a, c) \quad \textit{triangle inequality} \\
d(a, b) = 0 \quad \text{if and only if} \quad a &= b \quad \textit{regularity of metric}
\end{aligned}
\tag{137}
$$

A non-trivial example is the following. We define a special distance d between cities. If the two cities lie on a line going through Paris, we use the usual distance. In all other cases, we define the distance d by the shortest distance from one to the other travelling via Paris. This strange method defines a metric between all cities in France.

Challenge 189 s

A *normed* vector space is a linear space with a norm, or 'length', associated to each a vector. A *norm* is a non-negative number $\|a\|$ defined for each vector a with the properties

$$
\begin{aligned}
\|ra\| &= |r|\,\|a\| \quad \textit{linearity of norm} \\
\|a + b\| &\leqslant \|a\| + \|b\| \quad \textit{triangle inequality} \\
\|a\| = 0 \quad \text{only if} \quad a &= 0 \quad \textit{regularity}
\end{aligned}
\tag{138}
$$

Challenge 190 ny Usually there are many ways to define a norm for a given space. Note that a norm can always be used to define a metric by setting

$$
d(a, b) = \|a - b\|
\tag{139}
$$

so that all normed spaces are also metric spaces. This is the natural distance definition (in contrast to unnatural ones like that between French cities).

The norm is often defined with the help of an inner product. Indeed, the most special class of linear spaces are the *inner product spaces*. These are vector spaces with an *inner product*, also called *scalar product* $\cdot$ (not to be confused with the scalar multiplication!) which associates a number to each pair of vectors. An inner product space over $\mathbb{R}$ satisfies

$$
\begin{aligned}
a \cdot b &= b \cdot a \quad \textit{commutativity of scalar product} \\
(ra) \cdot (sb) &= rs(a \cdot b) \quad \textit{bilinearity of scalar product} \\
(a + b) \cdot c &= a \cdot c + b \cdot c \quad \textit{left distributivity of scalar product} \\
a \cdot (b + c) &= a \cdot b + a \cdot c \quad \textit{right distributivity of scalar product} \\
a \cdot a &\geqslant 0 \quad \textit{positivity of scalar product} \\
a \cdot a = 0 \quad \text{if and only if} \quad a &= 0 \quad \textit{regularity of scalar product}
\end{aligned}
\tag{140}
$$

for all vectors a, b, c and all scalars r, s. A *real* inner product space of finite dimension is also called a *Euclidean* vector space. The set of all velocities, the set of all positions, or the set of all possible momenta form such spaces.

An inner product space over $\mathbb{C}$ satisfies*

$$
\begin{aligned}
a \cdot b = \overline{b \cdot a} = \overline{b} \cdot \overline{a} \quad & \textit{Hermitean property} \\
(ra) \cdot (sb) = r\overline{s}(a \cdot b) \quad & \textit{sesquilinearity of scalar product} \\
(a + b) \cdot c = a \cdot c + b \cdot d \quad & \textit{left distributivity of scalar product} \\
a \cdot (b + c) = a \cdot b + a \cdot c \quad & \textit{right distributivity of scalar product} \qquad (141) \\
a \cdot a \geqslant 0 \quad & \textit{positivity of scalar product} \\
a \cdot a = 0 \quad \text{if and only if} \quad a = 0 \quad & \textit{regularity of scalar product}
\end{aligned}
$$

for all vectors a, b, c and all scalars r, s. A *complex* inner product space (of finite dimension) is also called a *unitary* or *Hermitean* vector space. If the inner product space is *complete*, it is called, especially in the infinite-dimensional complex case, a *Hilbert space*. Vol. IV, page 209 The space of all possible states of a quantum system forms a Hilbert space.

All inner product spaces are also metric spaces, and thus normed spaces, if the metric is defined by

$$
d(a, b) = \sqrt{(a - b) \cdot (a - b)} \, . \qquad (142)
$$

Only in the context of an inner product spaces we can speak about angles (or phase differences) between vectors, as we are used to in physics. Of course, like in normed spaces, inner product spaces also allows us to speak about the length of vectors and to define a *basis*, the mathematical concept necessary to define a coordinate system.

The *dimension* of a vector space is the number of linearly independent basis vectors. Can you define all these terms precisely? Challenge 191 ny

A *Hilbert space* is a real or complex inner product space that is also a *complete* metric space. In other terms, in a Hilbert space, distances vary continuously and behave as naively expected. Hilbert spaces can have an infinite number of dimensions.

Which vector spaces are of importance in physics? Challenge 192 ny

ALGEBRAS

The term *algebra* is used in mathematics with three different, but loosely related, meanings. First, it denotes a part of mathematics, as in 'I hated algebra at school'. Secondly, it denotes a set of formal rules that are obeyed by abstract objects, as in the expression 'tensor algebra'. Finally – and this is the only meaning used here – an algebra denotes a *specific* type of mathematical structure.

Intuitively, an algebra is a set of vectors with a vector multiplication defined on it. More precisely, a (unital, associative) algebra is a vector space (over a field K) that is also a (unital) ring. (The concept is due to Benjamin Peirce (b. 1809 Salem, d. 1880 Cambridge), father of Charles Sanders Peirce.) A ring is a set for which an addition and a multiplication is defined – like the integers. Thus, in an algebra, there are (often) *three* types of Vol. IV, page 208 multiplications:

---

* Two inequivalent forms of the sesquilinearity axiom exist. The other is $(ra) \cdot (sb) = \overline{r}s(a \cdot b)$. The term *sesquilinear* is derived from Latin and means for 'one-and-a-half-linear'.

— the (main) algebraic multiplication: the product of two vectors x and y is another vector $z = xy$;

— the scalar multiplication: the c-fold multiple of a vector x is another vector $y = cx$;

— if the vector space is a inner product space, the scalar product: the scalar product of two algebra elements (vectors) x and y is a scalar $c = x \cdot y$;

A precise definition of an algebra thus only needs to define properties of the (main) multiplication and to specify the number field K. An *algebra* is defined by the following axioms

$$x(y + z) = xy + xz \quad , \quad (x + y)z = xz + yz \quad \textit{distributivity of multiplication}$$
$$c(xy) = (cx)y = x(cy) \quad \textit{bilinearity} \tag{143}$$

for all vectors x, y, z and all scalars $c \in$ K. To stress their properties, algebras are also called *linear* algebras.

For example, the set of all linear transformations of an n-dimensional linear space (such as the translations on a plane, in space or in time) is a linear algebra, if the composition is taken as multiplication. So is the set of observables of a quantum mechanical system.[*]

An *associative algebra* is an algebra whose multiplication has the additional property that

$$x(yz) = (xy)z \quad \textit{associativity} . \tag{145}$$

Most algebras that arise in physics are associative[**] and unital. Therefore, in mathematical physics, a linear unital associative algebra is often simply called an algebra.

The set of multiples of the unit 1 of the algebra is called the *field of scalars* scal(A) of the algebra A. The field of scalars is also a subalgebra of A. The field of scalars and the scalars themselves behave in the same way.

We explore a few examples. The set of all polynomials in one variable (or in several variables) forms an algebra. It is commutative and infinite-dimensional. The constant

Challenge 194 e

Challenge 193 s

[*] *Linear transformations* are mappings from the vector space to itself, with the property that sums and scalar multiples of vectors are transformed into the corresponding sums and scalar multiples of the transformed vectors. Can you specify the set of all linear transformations of the plane? And of three-dimensional space? And of Minkowski space?

All linear transformations transform some special vectors, called *eigenvectors* (from the German word *eigen* meaning 'self') into multiples of themselves. In other words, if T is a transformation, e a vector, and

$$T(e) = \lambda e \tag{144}$$

Vol. IV, page 80

Vol. IV, page 146

where λ is a scalar, then the vector e is called an *eigenvector* of T, and λ is associated *eigenvalue*. The set of all eigenvalues of a transformation T is called the *spectrum* of T. Physicists did not pay much attention to these mathematical concepts until they discovered quantum theory. Quantum theory showed that observables are transformations in Hilbert space, because any measurement interacts with a system and thus transforms it. Quantum-mechanical experiments also showed that a measurement result for an observable must be an eigenvalue of the corresponding transformation. The state of the system after the measurement is given by the eigenvector corresponding to the measured eigenvalue. Therefore every expert on motion must know what an eigenvalue is.

[**] Note that a non-associative algebra does not possess a matrix representation.

polynomials form the field of scalars.

The set of $n \times n$ matrices, with the usual operations, also forms an algebra. It is n^2-dimensional. Those diagonal matrices (matrices with all off-diagonal elements equal to zero) whose diagonal elements all have the same value form the field of scalars. How is the scalar product of two matrices defined?

Challenge 195 ny

The set of all real-valued functions over a set also forms an algebra. Can you specify the multiplication? The constant functions form the field of scalars.

Challenge 196 s

A *star algebra*, also written *∗-algebra*, is an algebra over the *complex* numbers for which there is a mapping $* : A \rightarrow A$, $x \mapsto x^*$, called an *involution*, with the properties

$$(x^*)^* = x$$
$$(x + y)^* = x^* + y^*$$
$$(cx)^* = \bar{c}x^* \quad \text{for all} \quad c \in \mathbb{C}$$
$$(xy)^* = y^* x^* \tag{146}$$

valid for all elements x, y of the algebra A. The element x^* is called the *adjoint* of x. Star algebras are the main type of algebra used in quantum mechanics, since quantum-mechanical observables form a ∗-algebra.

A *C∗-algebra* is a Banach algebra over the complex numbers with an involution ∗ (a function that is its own inverse) such that the norm $\|x\|$ of an element x satisfies

$$\|x\|^2 = x^* \, x \, . \tag{147}$$

(A Banach algebra is a complete normed algebra; an algebra is *complete* if all Cauchy sequences converge.) In short, C∗-*algebra* is a nicely behaved algebra whose elements form a continuous set and a complex vector space. The name C comes from 'continuous functions'. Indeed, the *bounded* continuous functions form such an algebra, with a properly defined norm. Can you find it?

Challenge 197 s

Every C∗-algebra contains a space of Hermitean elements (which have a real spectrum), a set of normal elements, a multiplicative group of unitary elements and a set of positive elements (with non-negative spectrum).

We should mention one important type of algebra used in mathematics. A *division algebra* is an algebra for which the equations $ax = b$ and $ya = b$ are uniquely solvable in x or y for all b and all $a \neq 0$. Obviously, all type of continuous numbers must be division algebras. Division algebras are thus one way to generalize the concept of a number. One of the important results of modern mathematics states that (finite-dimensional) division algebras can *only* have dimension 1, like the reals, dimension 2, like the complex numbers, dimension 4, like the quaternions, or dimension 8, like the octonions. There is thus no way to generalize the concept of (continuous) 'number' to other dimensions.

And now for some fun. Imagine a ring A which contains a number field K as a subring (or 'field of scalars'). If the ring multiplication is defined in such a way that a general ring element multiplied with an element of K is the same as the scalar multiplication, then A is a vector space, and thus an algebra – *provided* that every element of K commutes with every element of A. (In other words, the subring K must be *central*.)

For example, the quaternions $\mathbb{H}$ are a four-dimensional real division algebra, but although $\mathbb{H}$ is a two-dimensional complex vector space, it is *not* a complex algebra, because i does not commute with j (one has $ij = -ji = k$). In fact, there are no finite-dimensional complex division algebras, and the only finite-dimensional real associative division algebras are $\mathbb{R}$, $\mathbb{C}$ and $\mathbb{H}$.

Now, if you are not afraid of getting a headache, think about this remark: every K-algebra is also an algebra over its field of scalars. For this reason, some mathematicians prefer to define an (associative) K-algebra simply as a ring which contains K as a central subfield.

In physics, it is the algebras related to symmetries which play the most important role. We study them next.

LIE ALGEBRAS

A Lie algebra is special type of algebra (and thus of vector space). Lie algebras are the most important type of non-associative algebras. A vector space L over the field $\mathbb{R}$ (or $\mathbb{C}$) with an additional binary operation $[\,,\,]$, called *Lie multiplication* or the *commutator*, is called a real (or complex) *Lie algebra* if this operation satisfies

$$[X, Y] = -[Y, X] \quad \textit{antisymmetry}$$
$$[aX + bY, Z] = a[X, Z] + b[Y, Z] \quad \textit{(left-)linearity}$$
$$[X, [Y,Z]] + [Y, [Z, X]] + [Z, [X, Y]] = 0 \quad \textit{Jacobi identity} \tag{148}$$

Challenge 198 e

for all elements $X, Y, Z \in L$ and for all $a, b \in \mathbb{R}$ (or $\mathbb{C}$). (Lie algebras are named after Sophus Lie.) The first two conditions together imply bilinearity. A Lie algebra is called *commutative* if $[X, Y] = 0$ for all elements X and Y. The *dimension* of the Lie algebra is the dimension of the vector space. A subspace N of a Lie algebra L is called an *ideal** if $[L, N] \subset N$; any ideal is also a *subalgebra*. A *maximal ideal* M which satisfies $[L, M] = 0$ is called the *centre* of L.

A Lie algebra is called a *linear* Lie algebra if its elements are linear transformations of another vector space V (intuitively, if they are 'matrices'). It turns out that every finite-dimensional Lie algebra is isomorphic to a linear Lie algebra. Therefore, there is no loss of generality in picturing the elements of finite-dimensional Lie algebras as matrices.

Page 348 The name 'Lie algebra' was chosen because the *generators*, i.e., the infinitesimal elements of every Lie group, form a Lie algebra. Since all important symmetries in nature form Lie groups, Lie algebras appear very frequently in physics. In mathematics, Lie algebras arise frequently because from any associative finite-dimensional algebra (in which the symbol $\cdot$ stands for its multiplication) a Lie algebra appears when we define the *commutator* by

$$[X, Y] = X \cdot Y - Y \cdot X . \tag{149}$$

(This fact gave the commutator its name.) Lie algebras are non-associative in general; but the above definition of the commutator shows how to build one from an associative

---

Challenge 199 ny * Can you explain the notation $[L, N]$? Can you define what a maximal ideal is and prove that there is only one?

algebra.

Since Lie algebras are vector spaces, the elements T_i of a *basis* of the Lie algebra always obey a relation of the form:

$$[T_i, T_j] = \sum_k c_{ij}^k T_k \; . \tag{150}$$

The numbers c_{ij}^k are called the *structure constants* of the Lie algebra. They depend on the choice of basis. The structure constants determine the Lie algebra completely. For example, the algebra of the Lie group SU(2), with the three generators defined by $T_a = \sigma^a/2i$, where the σ^a are the Pauli spin matrices, has the structure constants $C_{abc} = \varepsilon_{abc}$.[*] Vol. IV, page 215

CLASSIFICATION OF LIE ALGEBRAS

Finite-dimensional Lie algebras are classified as follows. Every finite-dimensional Lie algebra is the (semidirect) sum of a semisimple and a solvable Lie algebra.

A Lie algebra is called *solvable* if, well, if it is not semisimple. Solvable Lie algebras have not yet been classified completely. They are not important in physics.

A *semisimple* Lie algebra is a Lie algebra which has no non-zero solvable ideal. Other equivalent definitions are possible, depending on your taste:

— a semisimple Lie algebra does not contain non-zero Abelian ideals;
— its Killing form is non-singular, i.e., non-degenerate;
— it splits into the direct sum of non-Abelian simple ideals (this decomposition is unique);
— every finite-dimensional linear representation is completely reducible;
— the one-dimensional cohomology of g with values in an arbitrary finite-dimensional g-module is trivial.

Finite-dimensional semisimple Lie algebras have been completely classified. They decompose uniquely into a direct sum of *simple* Lie algebras. Simple Lie algebras can be complex

---

[*] Like groups, Lie algebras can be represented by matrices, i.e., by linear operators. Representations of Lie algebras are important in physics because many continuous symmetry groups are Lie groups.

The *adjoint representation* of a Lie algebra with basis $a_1...a_n$ is the set of matrices $\mathrm{ad}(a)$ defined for each element a by

$$[a, a_j] = \sum_c \mathrm{ad}(a)_{cj} a_c \; . \tag{151}$$

The definition implies that $\mathrm{ad}(a_i)_{jk} = c_{ij}^k$, where c_{ij}^k are the structure constants of the Lie algebra. For a real Lie algebra, all elements of $\mathrm{ad}(a)$ are real for all $a \in L$.

Note that for any Lie algebra, a scalar product can be defined by setting

$$X \cdot Y = \mathrm{Tr}(\, \mathrm{ad}X \cdot \mathrm{ad}Y \,) \; . \tag{152}$$

This scalar product is symmetric and bilinear. (Can you show that it is independent of the representation?) The corresponding bilinear form is also called the *Killing form*, after the German mathematician Wilhelm Killing (b. 1847 Burbach, d. 1923 Münster), the discoverer of the 'exceptional' Lie groups. The Killing form is invariant under the action of any automorphism of the Lie algebra L. In a given basis, one has

$$X \cdot Y = \mathrm{Tr}(\, (\mathrm{ad}X) \cdot (\mathrm{ad}Y)) = c_{lk}^i c_{si}^k x^l y^s = g_{ls} x^l y^s \tag{153}$$

where $g_{ls} = c_{lk}^i c_{si}^k$ is called the *Cartan metric tensor* of L.

or real.

The simple finite-dimensional *complex* Lie algebras all belong to four infinite classes and to five exceptional cases. The infinite classes are also called *classical*, and are: A_n for $n \geqslant 1$, corresponding to the Lie groups $SL(n + 1)$ and their compact 'cousins' $SU(n + 1)$; B_n for $n \geqslant 1$, corresponding to the Lie groups $SO(2n + 1)$; C_n for $n \geqslant 1$, corresponding to the Lie groups $Sp(2n)$; and D_n for $n \geqslant 4$, corresponding to the Lie groups $SO(2n)$. Thus A_n is the algebra of all skew-Hermitean matrices; B_n and D_n are the algebras of the symmetric matrices; and C_n is the algebra of the traceless matrices.

The exceptional Lie algebras are G_2, F_4, E_6, E_7, E_8. In all cases, the index gives the number of roots. The dimensions of these algebras are $A_n : n(n+2)$; B_n and $C_n : n(2n+1)$; $D_n : n(2n - 1)$; $G_2 : 14$; $F_4 : 32$; $E_6 : 78$; $E_7 : 133$; $E_8 : 248$.

The simple and finite-dimensional *real* Lie algebras are more numerous; their classification follows from that of the complex Lie algebras. Moreover, corresponding to each complex Lie group, there is always one compact real one. Real Lie algebras are not so important in fundamental physics.

Ref. 287

Of the large number of *infinite-dimensional* Lie algebras, only one is important in physics, the *Poincaré algebra*. A few other such algebras only appeared in failed attempts for unification.

TOPOLOGY – WHAT SHAPES EXIST?

> “ Topology is group theory. ”
> The Erlangen program

In a simplified view of topology that is sufficient for physicists, only one type of entity can possess shape: manifolds. *Manifolds* are generalized examples of pullovers: they are locally flat, can have holes and boundaries, and can often be turned inside out.

Pullovers are subtle entities. For example, can you turn your pullover inside out while your hands are tied together? (A friend may help you.) By the way, the same feat is also possible with your trousers, while your feet are tied together. Certain professors like to demonstrate this during topology lectures – of course with a carefully selected pair of underpants.

Challenge 200 s

Ref. 288
Challenge 201 s

Another good topological puzzle, the handcuff puzzle, is shown in Figure 168. Which of the two situations can be untied without cutting the ropes?

For a mathematician, pullovers and ropes are everyday examples of manifolds, and the operations that are performed on them are examples of deformations. Let us look at some more precise definitions. In order to define what a manifold is, we first need to define the concept of topological space.

Topological spaces

> “ En Australie, une mouche qui marche au
> plafond se trouve dans le même sens qu'une
> vache chez nous. ”
> Philippe Geluck, *La marque du chat.*

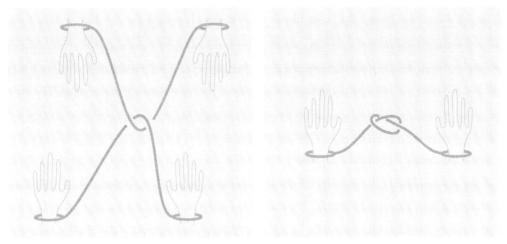

FIGURE 168 Which of the two situations can be untied without cutting?

Ref. 289

The study of shapes requires a good definition of a set made of 'points'. To be able to talk about shape, these sets must be structured in such a way as to admit a useful concept of 'neighbourhood' or 'closeness' between the elements of the set. The search for the most general type of set which allows a useful definition of neighbourhood has led to the concept of topological space. There are two ways to define a topology: one can define the concept of *open set* and then define the concept of *neighbourhood* with their help, or the other way round. We use the second option, which is somewhat more intuitive.

A *topological space* is a finite or infinite set X of elements, called *points*, together with the neighbourhoods for each point. A *neighbourhood* N of a point x is a collection of subsets Y_x of X with the properties that

— x is in every Y_x;
— if N and M are neighbourhoods of x, so is $N \cap M$;
— anything containing a neighbourhood of x is itself a neighbourhood of x.

The choice of the subsets Y_x is free. The subsets Y_x for all points x, chosen in a particular definition, contain a neighbourhood for each of their points; they are called *open sets*. (A neighbourhood and an open set usually differ, but all open sets are also neighbourhoods. Neighbourhoods of x can also be described as subsets of X that contain an open set that contains x.)

One also calls a topological space a 'set with a topology'. In effect, a *topology* specifies the systems of 'neighbourhoods' of every point of the set. 'Topology' is also the name of the branch of mathematics that studies topological spaces.

For example, the real numbers together with all open intervals form the *usual* topology of $\mathbb{R}$. Mathematicians have generalized this procedure. If one takes *all* subsets of $\mathbb{R}$ – or any other basis set – as open sets, one speaks of the *discrete* topology. If one takes *only* the full basis set and the empty set as open sets, one speaks of the *trivial* or *indiscrete* topology.

The concept of topological space allows us to define continuity. A mapping from one topological space X to another topological space Y is *continuous* if the inverse image

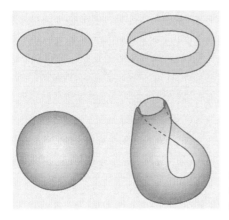

FIGURE 169 Examples of orientable and non-orientable manifolds of two dimensions: a disc, a Möbius strip, a sphere and a Klein bottle.

Challenge 202 e of every open set in Y is an open set in X. You may verify that this condition is not satisfied by a real function that makes a jump. You may also check that the term 'inverse' is necessary in the definition; otherwise a function with a jump would be continuous, as such a function may still map open sets to open sets.*

We thus need the concept of topological space, or of neighbourhood, if we want to express the idea that there are no jumps in nature. We also need the concept of topological space in order to be able to define limits.

Of the many special kinds of topological spaces that have been studied, one type is particularly important. A *Hausdorff space* is a topological space in which for any two points x and y there are disjoint open sets U and V such that x is in U and y is in V. A Hausdorff space is thus a space where, no matter how 'close' two points are, they can always be separated by open sets. This seems like a desirable property; indeed, non-Hausdorff spaces are rather tricky mathematical objects. (At Planck energy, it seems that vacuum appears to behave like a non-Hausdorff space; however, at Planck energy, vacuum is not really a space at all. So non-Hausdorff spaces play no role in physics.) A special case of Hausdorff space is well-known: the manifold.

MANIFOLDS

In physics, the most important topological spaces are differential manifolds. Loosely speaking, a *differential manifold* – physicists simply speak of a *manifold* – is a set of points that looks like $\mathbb{R}^n$ under the microscope – at small distances. For example, a sphere and a torus are both two-dimensional differential manifolds, since they look locally like a plane. Not all differential manifolds are that simple, as the examples of Figure 169 show.

A differential manifold is called *connected* if any two points can be joined by a path lying in the manifold. (The term has a more general meaning in topological spaces. But the notions of connectedness and pathwise connectedness coincide for differential manifolds.) We focus on connected manifolds in the following discussion. A manifold is called

---

* The Cauchy–Weierstass definition of continuity says that a real function $f(x)$ is continuous at a point a if (1) f is defined on an open interval containing a, (2) $f(x)$ tends to a limit as x tends to a, and (3) the limit is $f(a)$. In this definition, the continuity of f is defined using the intuitive idea that the real numbers form the Challenge 203 ny basic model of a set that has no gaps. Can you see the connection with the general definition given above?

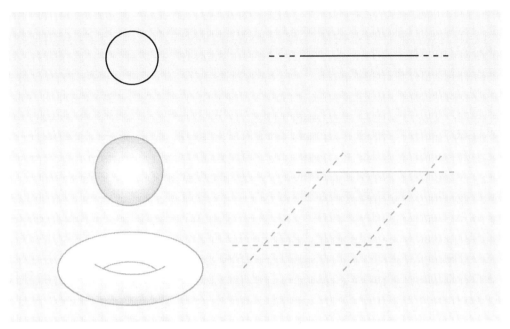

FIGURE 170 Compact (left) and non-compact (right) manifolds of various dimensions.

simply connected if every loop lying in the manifold can be contracted to a point. For example, a sphere is simply connected. A connected manifold which is not simply connected is called *multiply connected*. A torus is multiply connected.

Manifolds can be *non-orientable*, as the well-known Möbius strip illustrates. Non-orientable manifolds have only one surface: they do not admit a distinction between front and back. If you want to have fun, cut a paper Möbius strip into two along a centre line. You can also try this with paper strips with different twist values, and investigate the regularities.

Challenge 204 e

In two dimensions, closed manifolds (or surfaces), i.e., surfaces that are compact and without boundary, are always of one of three types:

— The simplest type are spheres with *n* attached handles; they are called *n-tori* or *surfaces of genus n*. They are orientable surfaces with Euler characteristic 2 − 2*n*.
— The *projective planes* with *n* handles attached are non-orientable surfaces with Euler characteristic 1 − 2*n*.
— The Klein bottles with *n* attached handles are non-orientable surfaces with Euler characteristic −2*n*.

Therefore Euler characteristic and orientability describe compact surfaces up to homeomorphism (and if surfaces are smooth, then up to diffeomorphism). Homeomorphisms are defined below.

The two-dimensional compact manifolds or surfaces with boundary are found by removing one or more discs from a surface in this list. A compact surface can be embedded in $\mathbb{R}^3$ if it is orientable or if it has non-empty boundary.

In physics, the most important manifolds are space-time and Lie groups of observables. We study Lie groups below. Strangely enough, the topology of space-time is not

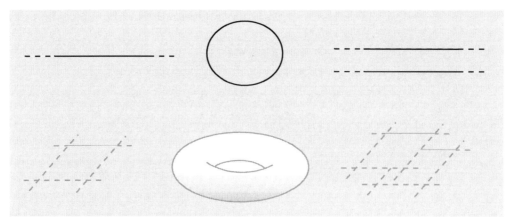

FIGURE 171 Simply connected (left), multiply connected (centre) and disconnected (right) manifolds of one (above) and two (below) dimensions.

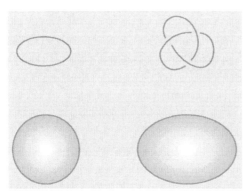

FIGURE 172 Examples of homeomorphic pairs of manifolds.

known. For example, it is unclear whether or not it is simply connected. Obviously, the reason is that it is difficult to observe what happens at large distances form the Earth. However, a similar difficulty appears near Planck scales.

If a manifold is imagined to consist of rubber, connectedness and similar global properties are not changed when the manifold is deformed. This fact is formalized by saying that two manifolds are *homeomorphic* (from the Greek words for 'same' and 'shape') if between them there is a continuous, one-to-one and onto mapping with a continuous inverse. The concept of homeomorphism is somewhat more general than that of rubber deformation, as can be seen from Figure 172.

HOLES, HOMOTOPY AND HOMOLOGY

Only 'well-behaved' manifolds play a role in physics: namely those which are orientable and connected. In addition, the manifolds associated with observables, are always compact. The main non-trivial characteristic of connected compact orientable manifolds is that they contain 'holes' (see Figure 173). It turns out that a proper description of the holes of manifolds allows us to distinguish between all different, i.e., non-homeomorphic, types of manifold.

FIGURE 173 The first four two-dimensional compact connected orientable manifolds: 0-, 1-, 2- and 3-tori.

There are three main tools to describe holes of manifolds and the relations among them: homotopy, homology and cohomology. These tools play an important role in the study of gauge groups, because any gauge group defines a manifold.

In other words, through homotopy and homology theory, mathematicians can *classify* manifolds. Given two manifolds, the properties of the holes in them thus determine whether they can be deformed into each other.

Physicists are now extending these results of standard topology. Deformation is a classical idea which assumes continuous space and time, as well as arbitrarily small action. In nature, however, quantum effects cannot be neglected. It is speculated that quantum effects can transform a physical manifold into one with a *different* topology: for example, a torus into a sphere. Can you find out how this can be achieved?

Challenge 205 d

Topological changes of physical manifolds happen via objects that are generalizations of manifolds. An *orbifold* is a space that is locally modelled by $\mathbb{R}^n$ modulo a finite group. Examples are the tear-drop or the half-plane. Orbifolds were introduced by Satake Ichiro in 1956; the name was coined by William Thurston. Orbifolds are heavily studied in string theory.

TYPES AND CLASSIFICATION OF GROUPS

We introduced mathematical *groups* early on because groups, especially symmetry groups, play an important role in many parts of physics, from the description of solids, molecules, atoms, nuclei, elementary particles and forces up to the study of shapes, cycles and patterns in growth processes.

Vol. I, page 238

Group theory is also one of the most important branches of modern mathematics, and is still an active area of research. One of the aims of group theory is the *classification* of all groups. This has been achieved only for a few special types. In general, one distinguishes between finite and infinite groups. Finite groups are better understood.

Every *finite* group is isomorphic to a subgroup of the symmetric group S_N, for some number N. Examples of finite groups are the crystalline groups, used to classify crystal structures, or the groups used to classify wallpaper patterns in terms of their symmetries. The symmetry groups of Platonic and many other regular solids are also finite groups.

Finite groups are a complex family. Roughly speaking, a general (finite) group can be seen as built from some fundamental bricks, which are groups themselves. These fundamental bricks are called *simple* (finite) groups. One of the high points of twentieth-century mathematics was the classification of the finite simple groups. It was a collaborative effort that took around 30 years, roughly from 1950 to 1980. The complete list of finite simple groups consists of

Ref. 290

1) the *cyclic groups* Z_p of prime group order;
2) the *alternating groups* A_n of degree n at least five;

3) the *classical linear groups*, $\mathrm{PSL}(n;q)$, $\mathrm{PSU}(n;q)$, $\mathrm{PSp}(2n;q)$ and $\mathrm{P}\Omega^{\varepsilon}(n;q)$;

4) the *exceptional* or *twisted groups* of Lie type ${}^{3}\mathrm{D}_4(q)$, $\mathrm{E}_6(q)$, ${}^{2}\mathrm{E}_6(q)$, $\mathrm{E}_7(q)$, $\mathrm{E}_8(q)$, $\mathrm{F}_4(q)$, ${}^{2}\mathrm{F}_4(2^n)$, $\mathrm{G}_2(q)$, ${}^{2}\mathrm{G}_2(3^n)$ and ${}^{2}\mathrm{B}_2(2^n)$;

5) the 26 *sporadic groups*, namely M_{11}, M_{12}, M_{22}, M_{23}, M_{24} (the Mathieu groups), J_1, J_2, J_3, J_4 (the Janko groups), Co_1, Co_2, Co_3 (the Conway groups), HS, Mc, Suz (the Co_1 'babies'), Fi_{22}, Fi_{23}, Fi'_{24} (the Fischer groups), $\mathrm{F}_1 = \mathrm{M}$ (the Monster), F_2, F_3, F_5, He ($= \mathrm{F}_7$) (the Monster 'babies'), Ru, Ly, and ON.

The classification was finished in the 1980s after over 10 000 pages of publications. The proof is so vast that a special series of books has been started to summarize and explain it. The first three families are infinite. The last family, that of the sporadic groups, is the most peculiar; it consists of those finite simple groups which do not fit into the other families. Some of these sporadic groups might have a role in particle physics: possibly even the largest of them all, the so-called *Monster* group. This is still a topic of research. (The Monster group has about $8.1 \cdot 10^{53}$ elements; more precisely, its order is 808 017 424 794 512 875 886 459 904 961 710 757 005 754 368 000 000 000 or $2^{46} \cdot 3^{20} \cdot 5^9 \cdot 7^6 \cdot 11^2 \cdot 13^3 \cdot 17 \cdot 19 \cdot 23 \cdot 29 \cdot 31 \cdot 41 \cdot 47 \cdot 59 \cdot 71$.)

Of the *infinite* groups, only those with some finiteness condition have been studied. It is only such groups that are of interest in the description of nature. Infinite groups are divided into *discrete* groups and *continuous* groups. Discrete groups are an active area of mathematical research, having connections with number theory and topology. Continuous groups are divided into *finitely generated* and *infinitely generated* groups. Finitely generated groups can be finite-dimensional or infinite-dimensional.

The most important class of finitely generated continuous groups are the Lie groups.

LIE GROUPS

In nature, the Lagrangians of the fundamental forces are invariant under gauge transformations and under continuous space-time transformations. These symmetry groups are examples of Lie groups, which are a special type of infinite continuous group. They are named after the great Norwegian mathematician Sophus Lie (b. 1842 Nordfjordeid, d. 1899 Kristiania). His name is pronounced like 'Lee'.

A (real) *Lie group* is an infinite symmetry group, i.e., a group with infinitely many elements, which is also an analytic manifold. Roughly speaking, this means that the elements of the group can be seen as points on a smooth (hyper-) surface whose shape can be described by an analytic function, i.e., by a function so smooth that it can be expressed as a power series in the neighbourhood of every point where it is defined. The points of the Lie group can be multiplied according to the group multiplication. Furthermore, the coordinates of the product have to be analytic functions of the coordinates of the factors, and the coordinates of the inverse of an element have to be analytic functions of the coordinates of the element. In fact, this definition is unnecessarily strict: it can be proved that a Lie group is just a topological group whose underlying space is a finite-dimensional, locally Euclidean manifold.

A *complex Lie group* is a group whose manifold is complex and whose group operations are holomorphic (instead of analytical) functions in the coordinates.

In short, a Lie group is a well-behaved manifold in which points can be multiplied (and technicalities). For example, the circle $\boldsymbol{T} = \{z \in \mathbb{C} : |z| = 1\}$, with the usual complex multiplication, is a real Lie group. It is Abelian. This group is also called S^1, as it is the one-dimensional sphere, or U(1), which means 'unitary group of one dimension'. The other one-dimensional Lie groups are the multiplicative group of non-zero real numbers and its subgroup, the multiplicative group of positive real numbers.

So far, in physics, only *linear* Lie groups have played a role – that is, Lie groups which act as linear transformations on some vector space. (The cover of SL(2,$\mathbb{R}$) or the complex compact torus are examples of non-linear Lie groups.) The important linear Lie groups for physics are the Lie subgroups of the general linear group GL(N,K), where K is a number field. This is defined as the set of all non-singular, i.e., invertible, N×N real, complex or quaternionic matrices. All the Lie groups discussed below are of this type.

Every *complex* invertible matrix A can be written in a unique way in terms of a unitary matrix U and a Hermitean matrix H:

$$A = U e^{H} . \tag{154}$$

(H is given by $H = \frac{1}{2} \ln A^{\dagger} A$, and U is given by $U = A e^{-H}$.)   Challenge 206 s

The simple Lie groups U(1) and SO(2,$\mathbb{R}$) and the Lie groups based on the real and complex numbers are Abelian (see Table 35); all others are non-Abelian.

Lie groups are manifolds. Therefore, in a Lie group one can define the distance between two points, the tangent plane (or tangent space) at a point, and the notions of integration and differentiations. Because Lie groups are manifolds, Lie groups have the same kind of structure as the objects of Figures 169, 170 and 171. Lie groups can have any number of dimensions. Like for any manifold, their global structure contains important information; let us explore it.

CONNECTEDNESS

It is not hard to see that the Lie groups SU(N) are simply connected for all N = 2, 3 ...; they have the topology of a 2N-dimensional sphere. The Lie group U(1), having the topology of the 1-dimensional sphere, or circle, is multiply connected.

The Lie groups SO(N) are *not* simply connected for any N = 2, 3 .... In general, SO(N,K) is connected, and GL(N,$\mathbb{C}$) is connected. All the Lie groups SL(N,K) are connected; and SL(N,$\mathbb{C}$) is simply connected. The Lie groups Sp(N,K) are connected; Sp(2N,$\mathbb{C}$) is simply connected. Generally, all semi-simple Lie groups are connected.

The Lie groups O(N,K), SO(N,M,K) and GL(N,$\mathbb{R}$) are not connected; they contain two connected components.

Note that the Lorentz group is not connected: it consists of four separate pieces. Like the Poincaré group, it is not compact, and neither is any of its four pieces. Broadly speaking, the non-compactness of the group of space-time symmetries is a consequence of the non-compactness of space-time.

COMPACTNESS

A Lie group is *compact* if it is closed and bounded when seen as a manifold. For a given parametrization of the group elements, the Lie group is compact if all parameter ranges are closed and finite intervals. Otherwise, the group is called *non-compact*. Both compact and non-compact groups play a role in physics. The distinction between the two cases is important, because representations of compact groups can be constructed in the same simple way as for finite groups, whereas for non-compact groups other methods have to be used. As a result, physical observables, which always belong to a representation of a symmetry group, have different properties in the two cases: if the symmetry group is compact, observables have *discrete* spectra; otherwise they do not.

All groups of internal gauge transformations, such as U(1) and SU(n), form compact groups. In fact, field theory *requires* compact Lie groups for gauge transformations. The only compact Lie groups are the torus groups T^n, O(n), U(n), SO(n) and SU(n), their double cover Spin(n) and the Sp(n). In contrast, SL(n,$\mathbb{R}$), GL(n,$\mathbb{R}$), GL(n,$\mathbb{C}$) and all others are not compact.

Besides being manifolds, Lie groups are obviously also groups. It turns out that most of their group properties are revealed by the behaviour of the elements which are very close (as points on the manifold) to the identity.

Every element of a compact and connected Lie group has the form exp(A) for some A. The elements A arising in this way form an algebra, called the *corresponding Lie algebra*. For any linear Lie group, every element of the connected subgroup can be expressed as a finite product of exponentials of elements of the corresponding Lie algebra. Mathematically, the vector space defined by the Lie algebra is tangent to the manifold defined by the Lie group, at the location of the unit element. In short, Lie algebras express the local properties of Lie groups near the identity. That is the reason for their importance in physics.

Page 338

TABLE 35 Properties of the most important real and complex Lie groups.

LIE GROUP	DESCRIP-TION	PROPERTIES[a]	LIE ALGEBRA	DESCRIPTION OF LIE ALGEBRA	DIMEN-SION
1. Real groups					real
$\mathbb{R}^n$	Euclidean space with addition	Abelian, simply connected, not compact; $\pi_0 = \pi_1 = 0$	$\mathbb{R}^n$	Abelian, thus Lie bracket is zero; not simple	n
$\mathbb{R}^\times$	non-zero real numbers with multiplication	Abelian, not connected, not compact; $\pi_0 = \mathbb{Z}_2$, no π_1	$\mathbb{R}$	Abelian, thus Lie bracket is zero	1
$\mathbb{R}^{>0}$	positive real numbers with multiplication	Abelian, simply connected, not compact; $\pi_0 = \pi_1 = 0$	$\mathbb{R}$	Abelian, thus Lie bracket is zero	1
$S^1 = \mathbb{R}/\mathbb{Z}$ $= U(1) =$ T $= SO(2)$ $= Spin(2)$	complex numbers of absolute value 1, with multiplication	Abelian, connected, not simply connected, compact; $\pi_0 = 0$, $\pi_1 = \mathbb{Z}$	$\mathbb{R}$	Abelian, thus Lie bracket is zero	1

TABLE 35 (Continued) Properties of the most important real and complex Lie groups.

LIE GROUP	DESCRIPTION	PROPERTIES[a]	LIE ALGEBRA	DESCRIPTION OF LIE ALGEBRA	DIMENSION
$\mathbb{H}^{\times}$	non-zero quaternions with multiplication	simply connected, not compact; $\pi_0 = \pi_1 = 0$	$\mathbb{H}$	quaternions, with Lie bracket the commutator	4
S^3	quaternions of absolute value 1, with multiplication, also known as Sp(1); topologically a 3-sphere	simply connected, compact; isomorphic to SU(2), Spin(3) and to double cover of SO(3); $\pi_0 = \pi_1 = 0$	$\text{Im}(\mathbb{H})$	quaternions with zero real part, with Lie bracket the commutator; simple and semi-simple; isomorphic to real 3-vectors, with Lie bracket the cross product; also isomorphic to su(2) and to so(3)	3
$GL(n, \mathbb{R})$	general linear group: invertible n-by-n real matrices	not connected, not compact; $\pi_0 = \mathbb{Z}_2$, no π_1	$M(n, \mathbb{R})$	n-by-n matrices, with Lie bracket the commutator	n^2
$GL^+(n, \mathbb{R})$	n-by-n real matrices with positive determinant	simply connected, not compact; $\pi_0 = 0$, for $n = 2$: $\pi_1 = \mathbb{Z}$, for $n \geq 2$: $\pi_1 = \mathbb{Z}_2$; $GL^+(1, \mathbb{R})$ isomorphic to $\mathbb{R}^{>0}$	$M(n, \mathbb{R})$	n-by-n matrices, with Lie bracket the commutator	n^2
$SL(n, \mathbb{R})$	special linear group: real matrices with determinant 1	simply connected, not compact if $n > 1$; $\pi_0 = 0$, for $n = 2$: $\pi_1 = \mathbb{Z}$, for $n \geq 2$: $\pi_1 = \mathbb{Z}_2$; $SL(1, \mathbb{R})$ is a single point, $SL(2, \mathbb{R})$ is isomorphic to SU(1, 1) and Sp(2, $\mathbb{R}$)	$sl(n, \mathbb{R})$ $= A_{n-1}$	n-by-n matrices with trace 0, with Lie bracket the commutator	$n^2 - 1$
$O(n, \mathbb{R})$ $= O(n)$	orthogonal group: real orthogonal matrices; symmetry of hypersphere	not connected, compact; $\pi_0 = \mathbb{Z}_2$, no π_1	$so(n, \mathbb{R})$	skew-symmetric n-by-n real matrices, with Lie bracket the commutator; so(3, $\mathbb{R}$) is isomorphic to su(2) and to $\mathbb{R}^3$ with the cross product	$n(n-1)/2$

TABLE 35 (Continued) Properties of the most important real and complex Lie groups.

LIE GROUP	DESCRIPTION	PROPERTIES[a]	LIE ALGEBRA	DESCRIPTION OF LIE ALGEBRA	DIMENSION
$SO(n, \mathbb{R})$ $= SO(n)$	special orthogonal group: real orthogonal matrices with determinant 1	connected, compact; for $n \geqslant 2$ not simply connected; $\pi_0 = 0$, for $n = 2$: $\pi_1 = \mathbb{Z}$, for $n \geq 2$: $\pi_1 = \mathbb{Z}_2$	$so(n, \mathbb{R})$ $= B_{\frac{n-1}{2}}$ or $D_{\frac{n}{2}}$	skew-symmetric n-by-n real matrices, with Lie bracket the commutator; for $n = 3$ and $n \geqslant 5$ simple and semisimple; $SO(4)$ is semisimple but not simple	$n(n-1)/2$
$Spin(n)$	spin group; double cover of $SO(n)$; $Spin(1)$ is isomorphic to $\mathbb{Q}_2$, $Spin(2)$ to S^1	simply connected for $n \geqslant 3$, compact; for $n = 3$ and $n \geqslant 5$ simple and semisimple; for $n > 1$: $\pi_0 = 0$, for $n > 2$: $\pi_1 = 0$	$so(n, \mathbb{R})$	skew-symmetric n-by-n real matrices, with Lie bracket the commutator	$n(n-1)/2$
$Sp(2n, \mathbb{R})$	symplectic group: real symplectic matrices	not compact; $\pi_0 = 0$, $\pi_1 = \mathbb{Z}$	$sp(2n, \mathbb{R})$ $= C_n$	real matrices A that satisfy $JA + A^T J = 0$ where J is the standard skew-symmetric matrix;[b] simple and semisimple	$n(2n+1)$
$Sp(n)$ for $n \geqslant 3$	compact symplectic group: quaternionic $n \times n$ unitary matrices	compact, simply connected; $\pi_0 = \pi_1 = 0$	$sp(n)$	n-by-n quaternionic matrices A satisfying $A = -A^*$, with Lie bracket the commutator; simple and semisimple	$n(2n+1)$
$U(n)$	unitary group: complex $n \times n$ unitary matrices	not simply connected, compact; it is *not* a complex Lie group/algebra; $\pi_0 = 0$, $\pi_1 = \mathbb{Z}$; isomorphic to S^1 for $n = 1$	$u(n)$	n-by-n complex matrices A satisfying $A = -A^*$, with Lie bracket the commutator	n^2
$SU(n)$	special unitary group: complex $n \times n$ unitary matrices with determinant 1	simply connected, compact; it is *not* a complex Lie group/algebra; $\pi_0 = \pi_1 = 0$	$su(n)$	n-by-n complex matrices A with trace 0 satisfying $A = -A^*$, with Lie bracket the commutator; for $n \geqslant 2$ simple and semisimple	$n^2 - 1$
2. Complex groups[c]					complex

TABLE 35 (Continued) Properties of the most important real and complex Lie groups.

Lie group	Description	Properties[a]	Lie algebra	Description of Lie algebra	Dimension
$\mathbb{C}^n$	group operation is addition	Abelian, simply connected, not compact; $\pi_0 = \pi_1 = 0$	$\mathbb{C}^n$	Abelian, thus Lie bracket is zero	n
$\mathbb{C}^\times$	non-zero complex numbers with multiplication	Abelian, not simply connected, not compact; $\pi_0 = 0$, $\pi_1 = \mathbb{Z}$	$\mathbb{C}$	Abelian, thus Lie bracket is zero	1
GL($n, \mathbb{C}$)	general linear group: invertible n-by-n complex matrices	simply connected, not compact; $\pi_0 = 0$, $\pi_1 = \mathbb{Z}$; for $n = 1$ isomorphic to $\mathbb{C}^\times$	M($n, \mathbb{C}$)	n-by-n matrices, with Lie bracket the commutator	n^2
SL($n, \mathbb{C}$)	special linear group: complex matrices with determinant 1	simply connected; for $n \geqslant 2$ not compact; $\pi_0 = \pi_1 = 0$; SL(2, $\mathbb{C}$) is isomorphic to Spin(3, $\mathbb{C}$) and Sp(2, $\mathbb{C}$)	sl($n, \mathbb{C}$)	n-by-n matrices with trace 0, with Lie bracket the commutator; simple, semisimple; sl(2, $\mathbb{C}$) is isomorphic to su(2, $\mathbb{C}$) $\otimes \mathbb{C}$	$n^2 - 1$
PSL(2, $\mathbb{C}$)	projective special linear group; isomorphic to the Möbius group, to the restricted Lorentz group SO$^+$(3, 1, $\mathbb{R}$) and to SO(3, $\mathbb{C}$)	not compact; $\pi_0 = 0$, $\pi_1 = \mathbb{Z}_2$	sl(2, $\mathbb{C}$)	2-by-2 matrices with trace 0, with Lie bracket the commutator; sl(2, $\mathbb{C}$) is isomorphic to su(2, $\mathbb{C}$) $\otimes \mathbb{C}$	3
O($n, \mathbb{C}$)	orthogonal group: complex orthogonal matrices	not connected; for $n \geqslant 2$ not compact; $\pi_0 = \mathbb{Z}_2$, no π_1	so($n, \mathbb{C}$)	skew-symmetric n-by-n complex matrices, with Lie bracket the commutator	$n(n-1)/2$

TABLE 35 (Continued) Properties of the most important real and complex Lie groups.

LIE GROUP	DESCRIPTION	PROPERTIES[a]	LIE ALGEBRA	DESCRIPTION OF LIE ALGEBRA	DIMENSION
$SO(n, \mathbb{C})$	special orthogonal group: complex orthogonal matrices with determinant 1	for $n \geqslant 2$ not compact; not simply connected; $\pi_0 = 0$, for $n = 2$: $\pi_1 = \mathbb{Z}$, for $n \geq 2$: $\pi_1 = \mathbb{Z}_2$; non-Abelian for $n > 2$, $SO(2, \mathbb{C})$ is Abelian and isomorphic to $\mathbb{C}^\times$	$so(n, \mathbb{C})$	skew-symmetric n-by-n complex matrices, with Lie bracket the commutator; for $n = 3$ and $n \geqslant 5$ simple and semisimple	$n(n-1)/2$
$Sp(2n, \mathbb{C})$	symplectic group: complex symplectic matrices	not compact; $\pi_0 = \pi_1 = 0$	$sp(2n, \mathbb{C})$	complex matrices that satisfy $JA + A^T J = 0$ where J is the standard skew-symmetric matrix;[b] simple and semi-simple	$n(2n+1)$

a. The *group of components* π_0 of a Lie group is given; the order of π_0 is the number of components of the Lie group. If the group is trivial (0), the Lie group is connected. The *fundamental group* π_1 of a connected Lie group is given. If the group π_1 is trivial (0), the Lie group is simply connected. This table is based on that in the Wikipedia, at en.wikipedia.org/wiki/Table_of_Lie_groups.
b. The standard skew-symmetric matrix J of rank $2n$ is $J_{kl} = \delta_{k,n+l} - \delta_{k+n,l}$.
c. Complex Lie groups and Lie algebras can be viewed as real Lie groups and real Lie algebras of twice the dimension.

MATHEMATICAL CURIOSITIES AND FUN CHALLENGES

Challenge 207 ny A theorem of topology says: you cannot comb a hairy football. Can you prove it?

$* *$

Topology is fun. If you want to laugh for half an hour, fix a modified pencil, as shown in Figure 174, to a button hole and let people figure out how to get it off again.

$* *$

There are at least six ways to earn a million dollars with mathematical research. The Clay Mathematics Institute at www.claymath.org offered such a prize for major advances in seven topics:

— proving the Birch and Swinnerton–Dyer conjecture about algebraic equations;
— proving the Poincaré conjecture about topological manifolds;
— solving the Navier–Stokes equations for fluids;

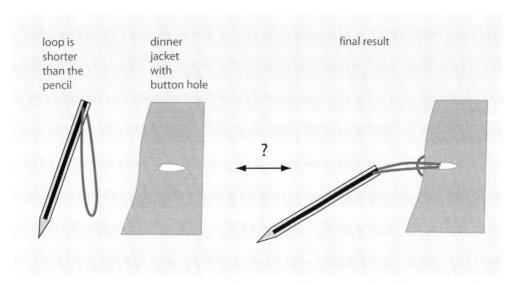

FIGURE 174 A well-known magic dexterity trick: make your friend go mad by adding a pencil to his dinner jacket.

— finding criteria distinguishing P and NP numerical problems;
— proving the Riemann hypothesis stating that the non-trivial zeros of the zeta function lie on a line;
— proving the Hodge conjectures;
— proving the connection between Yang–Mills theories and a mass gap in quantum field theory.

Vol. VI, page 300

The Poincaré conjecture was solved in 2002 by Grigori Perelman; on each of the other six topics, substantial progress can buy you a house.

CHALLENGE HINTS AND SOLUTIONS

Challenge 1, page 9: Do not hesitate to be demanding and strict. The next edition of the text will benefit from it.

Challenge 2, page 17: A *virus* is an example. It has no own metabolism. By the way, the ability of some viruses to form crystals is *not* a proof that they are not living beings, in contrast to what is often said.

Challenge 3, page 18: The navigation systems used by flies are an example.

Challenge 5, page 21: The thermal energy kT is about 4 zJ and a typical relaxation time is 0.1 ps.

Challenge 7, page 29: The argument is correct.

Challenge 8, page 29: This is not possible at present. If you know a way, publish it. It would help a sad single mother who has to live without financial help from the father, despite a lawsuit, as it was yet impossible to decide which of the two candidates is the right one.

Challenge 9, page 29: Also identical twins count as different persons and have different fates. Imprinting in the womb is different, so that their temperament will be different. The birth experience will be different; this is the most intense experience of every human, strongly determining his fears and thus his character. A person with an old father is also quite different from that with a young father. If the womb is not that of his biological mother, a further distinction of the earliest and most intensive experiences is given.

Challenge 10, page 29: Be sure to publish your results.

Challenge 11, page 30: Life's chemicals are synthesized inside the body; the asymmetry has been inherited along the generations. The common asymmetry thus shows that all life has a common origin.

Challenge 12, page 30: Well, men are more similar to chimpanzees than to women. More seriously, the above data, even though often quoted, are wrong. Newer measurements by Roy Britten in 2002 have shown that the difference in genome between humans and chimpanzees is about 5 % (See R. J. BRITTEN, *Divergence between samples of chimpanzee and human DNA sequences is 5 %, counting indels*, Proceedings of the National Academy of Sciences **99**, pp. 13633–13635, 15th of October, 2002.) In addition, though the difference between man and woman is smaller than one whole chromosome, the large size of the X chromosome, compared with the small size of the Y chromosome, implies that men have about 3 % less genetic material than women. However, all men have an X chromosome as well. That explains that still other measurements suggest that all humans share a pool of at least 99.9 % of common genes.

Challenge 15, page 33: Chemical processes, including diffusion and reaction rates, are strongly temperature dependent. They affect the speed of motion of the individual and thus its chance of survival. Keeping temperature in the correct range is thus important for evolved life forms.

Challenge 16, page 33: The first steps are not known at all.

Challenge 17, page 33: Since all the atoms we are made of originate from outer space, the answer is yes. But if one means that biological cells came to Earth from space, the answer is no, as most cells do not like vacuum. The same is true for DNA.

In fact, life and reproduction are properties of complex systems. In other words, asking whether life comes from outer space is like asking: 'Could car insurance have originated in outer space?'

Challenge 19, page 37: Haven't you tried yet? Physics is an experimental science.

Challenge 27, page 45: Exponential decays occur when the probability of decay is constant over time. For humans, this is not the case. Why not?

Challenge 28, page 48: There are no non-physical processes: anything that can be observed is a physical process. Consciousness is due to processes in the brain, thus inside matter; thus it is a quantum process. At body temperature, coherence has lifetimes much smaller than the typical though processes.

Challenge 29, page 49: Radioactive dating methods can be said to be based on the nuclear interactions, even though the detection is again electromagnetic.

Challenge 30, page 49: All detectors of light can be called relativistic, as light moves with maximal speed. Touch sensors are not relativistic following the usual sense of the word, as the speeds involved are too small. The energies are small compared to the rest energies; this is the case even if the signal energies are attributed to electrons only.

Challenge 31, page 50: The noise is due to the photoacoustic effect; the periodic light periodically heats the air in the jam glass at the blackened surface and thus produces sound. See M. EULER, *Kann man Licht hören?*, Physik in unserer Zeit 32, pp. 180–182, 2001.

Challenge 32, page 52: It implies that neither resurrection nor reincarnation nor eternal life are possible.

Challenge 35, page 61: The ethanol disrupts the hydrogen bonds between the water molecules so that, on average, they can get closer together. A video of the experiment is found at www.youtube.com/watch?v=LUW7a7H-KuY.

Challenge 38, page 62: You get an intense yellow colour due to the formation of lead iodide (PbI_2).

Challenge 42, page 75: With a combination of the methods of Table 5, this is indeed possible. In fact, using cosmic rays to search for unknown chambers in the pyramids has been already done in the 1960s. The result was that no additional chambers exist. Ref. 291

Challenge 44, page 78: For example, a heavy mountain will push down the Earth's crust into the mantle, makes it melt on the bottom side, and thus lowers the position of the top.

Challenge 45, page 78: These developments are just starting; the results are still far from the original one is trying to copy, as they have to fulfil a second condition, in addition to being a 'copy' of original feathers or of latex: the copy has to be cheaper than the original. That is often a much tougher request than the first.

Challenge 46, page 79: About 0.2 m.

Challenge 48, page 79: Since the height of the potential is always finite, walls can always be overcome by tunnelling.

Challenge 49, page 79: The lid of a box can never be at rest, as is required for a tight closure, but is always in motion, due to the quantum of action.

Challenge 50, page 79: The unit of thermal conductance is $T\pi^2 k^2/3\hbar$, where T is temperature and k is the Boltzmann constant.

Challenge 51, page 82: The concentrations and can be measured from polar ice caps, by measuring how the isotope concentration changes over depth. Both in evaporation and in condensation of water, the isotope ratio depends on the temperature. The measurements in Antarctica and in Greenland coincide, which is a good sign of their trustworthiness.

Challenge 52, page 84: In the summer, tarmac is soft.

Challenge 55, page 99: The one somebody else has thrown away. Energy costs about 10 cents/kWh. For new lamps, the fluorescent lamp is the best for the environment, even though it is the least friendly to the eye and the brain, due to its flickering.

Challenge 56, page 104: This old dream depends on the precise conditions. How flexible does the display have to be? What lifetime should it have? The newspaper like display is many years away and maybe not even possible.

Challenge 57, page 104: The challenge here is to find a cheap way to deflect laser beams in a controlled way. Cheap lasers are already available.

Challenge 58, page 104: There is only speculation on the answer; the tendency of most researchers is to say no.

Challenge 59, page 104: No, as it is impossible because of momentum conservation and because of the no-cloning theorem.

Challenge 60, page 104: There are companies trying to sell systems based on quantum cryptology; but despite the technical interest, the commercial success is questionable.

Challenge 61, page 105: I predicted since the year 2000 that mass-produced goods using this technology (at least 1 million pieces sold) will not be available before 2025.

Challenge 62, page 105: Maybe, but for extremely high prices.

Challenge 63, page 110: For example, you could change gravity between two mirrors.

Challenge 64, page 110: As usual in such statements, either group or phase velocity is cited, but not the corresponding energy velocity, which is always below c.

Challenge 66, page 113: Echoes do not work once the speed of sound is reached and do not work well when it is approached. Both the speed of light and that of sound have a finite value. Moving with a mirror still gives a mirror image. This means that the speed of light cannot be reached. If it cannot be reached, it must be the same for all observers.

Challenge 67, page 114: Mirrors do not usually work for matter; in addition, if they did, matter, because of its rest energy, would require much higher acceleration values.

Challenge 70, page 116: The classical radius of the electron, which is the size at which the field energy would make up the hole electron mass, is about 137 times smaller, thus much smaller, than the Compton wavelength of the electron.

Challenge 71, page 117: The overhang can have any value whatsoever. There is no limit. Taking the indeterminacy principle into account introduces a limit as the last brick or card must not allow the centre of gravity, through its indeterminacy, to be over the edge of the table.

Challenge 72, page 117: A larger charge would lead to a field that spontaneously generates electron positron pairs, the electron would fall into the nucleus and reduce its charge by one unit.

Challenge 74, page 118: The Hall effect results from the deviation of electrons in a metal due to an applied magnetic field. Therefore it depends on their speed. One gets values around 1 mm. Inside atoms, one can use Bohr's atomic model as approximation.

Challenge 39, page 64: The usual way to pack oranges on a table is the densest way to pack spheres.

Challenge 40, page 65: Just use a paper drawing. Draw a polygon and draw it again at latter times, taking into account how the sides grow over time. You will see by yourself how the faster growing sides disappear over time.

Challenge 75, page 118: The steps are due to the particle nature of electricity and all other moving entities.

Challenge 76, page 118: If we could apply the Banach–Tarski paradox to vacuum, it seems that we could split, without any problem, one ball of vacuum into *two* balls of vacuum, each with the same volume as the original. In other words, one ball with vacuum energy E could not be distinguished from two balls of vacuum energy $2E$. Vol. I, page 56

We used the Banach–Tarski paradox in this way to show that chocolate (or any other matter) possesses an intrinsic length. But it is *not* clear that we can now deduce that the vacuum has an intrinsic length. Indeed, the paradox cannot be applied to vacuum for *two* reasons. First, there indeed is a maximum energy and minimum length in nature. Secondly, there is no place in nature without vacuum energy; so there is no place were we could put the second ball. We thus do not know why the Banach–Tarski paradox for vacuum cannot be applied, and thus cannot use it to deduce the existence of a minimum length in vacuum. Vol. I, page 288

It is better to argue in the following way for a minimum length in vacuum. If there were no intrinsic length cut-off, the vacuum energy would be infinite. Experiments however, show that it is finite.

Challenge 77, page 118: Mud is a suspension of sand; sand is not transparent, even if made of clear quartz, because of the scattering of light at the irregular surface of its grains. A suspension cannot be transparent if the index of refraction of the liquid and the suspended particles is different. It is never transparent if the particles, as in most sand types, are themselves not transparent.

Challenge 78, page 119: The first answer is probably no, as composed systems cannot be smaller than their own compton wavelength; only elementary systems can. However, the universe is not a system, as it has no environment. As such, its length is not a precisely defined concept, as an environment is needed to measure and to define it. (In addition, gravity must be taken into account in those domains.) Thus the answer is: in those domains, the question makes no sense.

Challenge 79, page 119: Methods to move on perfect ice from mechanics:

— if the ice is perfectly flat, rest is possible only in one point – otherwise you oscillate around that point, as shown in challenge 24;
— do nothing, just wait that the higher centrifugal acceleration at body height pulls you away;
— to rotate yourself, just rotate your arm above your head;
— throw a shoe or any other object away;
— breathe in vertically, breathing out (or talking) horizontally (or vice versa);
— wait to be moved by the centrifugal acceleration due to the rotation of the Earth (and its oblateness);
— jump vertically repeatedly: the Coriolis acceleration will lead to horizontal motion;
— wait to be moved by the Sun or the Moon, like the tides are;
— 'swim' in the air using hands and feet;
— wait to be hit by a bird, a flying wasp, inclined rain, wind, lava, earthquake, plate tectonics, or any other macroscopic object (all objects pushing count only as one solution);
— wait to be moved by the change in gravity due to convection in Earth's mantle;
— wait to be moved by the gravitation of some comet passing by;
— counts only for kids: spit, sneeze, cough, fart, pee; or move your ears and use them as wings.

Note that gluing your tongue is not possible on perfect ice.

Challenge 80, page 119: Methods to move on perfect ice using thermodynamics and electrodynamics:

— use the radio/TV stations nearby to push you around;
— use your portable phone and a mirror;

— switch on a pocket lamp, letting the light push you;
— wait to be pushed around by Brownian motion in air;
— heat up one side of your body: black body radiation will push you;
— heat up one side of your body, e.g. by muscle work: the changing airflow or the evaporation will push you;
— wait for one part of the body to be cooler than the other and for the corresponding black body radiation effects;
— wait for the magnetic field of the Earth to pull on some ferromagnetic or paramagnetic metal piece in your clothing or in your body;
— wait to be pushed by the light pressure, i.e. by the photons, from the Sun or from the stars, maybe using a pocket mirror to increase the efficiency;
— rub some polymer object to charge it electrically and then move it in circles, thus creating a magnetic field that interacts with the one of the Earth.

Note that perfect frictionless surfaces do not melt.

Challenge 81, page 119: Methods to move on perfect ice using general relativity:

— move an arm to emit gravitational radiation;
— deviate the cosmic background radiation with a pocket mirror;
— wait to be pushed by gravitational radiation from star collapses;
— wait for the universe to contract.

Challenge 82, page 119: Methods to move on perfect ice using quantum effects:

— wait for your wave function to spread out and collapse at the end of the ice surface;
— wait for the pieces of metal in the clothing to attract to the metal in the surrounding through the Casimir effect;
— wait to be pushed around by radioactive decays in your body.

Challenge 83, page 120: Methods to move on perfect ice using materials science, geophysics and astrophysics:

— be pushed by the radio waves emitted by thunderstorms and absorbed in painful human joints;
— wait to be pushed around by cosmic rays;
— wait to be pushed around by the solar wind;
— wait to be pushed around by solar neutrinos;
— wait to be pushed by the transformation of the Sun into a red giant;
— wait to be hit by a meteorite.

Challenge 84, page 120: A method to move on perfect ice using self-organization, chaos theory, and biophysics:

— wait that the currents in the brain interact with the magnetic field of the Earth by controlling your thoughts.

Challenge 85, page 120: Methods to move on perfect ice using quantum gravity:

— accelerate your pocket mirror with your hand;
— deviate the Unruh radiation of the Earth with a pocket mirror;
— wait for proton decay to push you through the recoil.

Challenge 87, page 126: This is a trick question: if you can say why, you can directly move to the last volume of this adventure and check your answer. The gravitational potential changes the phase of a wave function, like any other potential does; but the reason *why* this is the case will only become clear in the last volume of this series.

Challenge 92, page 128: No. Bound states of massless particles are always unstable.

Challenge 93, page 130: This is easy only if the black hole size is inserted into the entropy bound by Bekenstein. A simple deduction of the black hole entropy that includes the factor 1/4 is not yet at hand; more on this in the last volume.

Challenge 94, page 130: An entropy limit implies an information limit; only a given information can be present in a given region of nature. This results in a memory limit.

Challenge 95, page 130: In natural units, the exact expression for entropy is $S = 0.25A$. If each Planck area carried one bit (degree of freedom), the entropy would be $S = \ln W = \ln(2^A) = A \ln 2 = 0.693A$. This close to the exact value.

Challenge 99, page 136: The universe has about 10^{22} stars; the Sun has a luminosity of about 10^{26} W; the total luminosity of the visible matter in the universe is thus about 10^{48} W. A γ ray burster emits up to $3 \cdot 10^{47}$ W.

Challenge 105, page 138: They are carried away by the gravitational radiation.

Challenge 111, page 142: No system is known in nature which emits or absorbs only one graviton at a time. This is another point speaking against the existence of gravitons.

Challenge 115, page 150: Two stacked foils show the same effect as one foil of the same total thickness. Thus the surface plays no role.

Challenge 117, page 155: The electron is held back by the positive charge of the nucleus, if the number of protons in the nucleus is sufficient, as is the case for those nuclei we are made of.

Challenge 119, page 164: The half-time $t_{1/2}$ is related to the life-time τ by $t_{1/2} = \tau \ln 2$.

Challenge 120, page 165: The number is small compared with the number of cells. However, it is possible that the decays are related to human ageing.

Challenge 122, page 167: By counting decays and counting atoms to sufficient precision.

Challenge 124, page 170: The radioactivity necessary to keep the Earth warm is low; lava is only slightly more radioactive than usual soil.

Challenge 125, page 180: There is way to conserve both energy and momentum in such a decay.

Challenge 126, page 180: The combination of high intensity X-rays and UV rays led to this effect.

Challenge 129, page 189: The nuclei of nitrogen and carbon have a high electric charge which strongly repels the protons.

Challenge 131, page 197: See the paper by C.J. Hogan mentioned in Ref. 257.

Challenge 132, page 200: Touching something requires getting near it; getting near means a small time and position indeterminacy; this implies a small wavelength of the probe that is used for touching; this implies a large energy.

Challenge 135, page 207: The processes are electromagnetic in nature, thus electric charges give the frequency with which they occur.

Challenge 136, page 216: Designing a nuclear weapon is not difficult. University students can do it, and even have done so a few times. The first students who did so were two physics graduates in 1964, as told on www.guardian.co.uk/world/2003/jun/24/usa.science. It is not hard to conceive a design and even to build it. By far the hardest problem is getting or making the nuclear material. That requires either an extensive criminal activity or a vast technical effort, with numerous large factories, extensive development, and coordination of many technological activities. Most importantly, such a project requires a large financial investment, which poor countries cannot afford without great sacrifices for all the population. The problems are thus not technical, but financial.

Challenge 140, page 246: In 2008, an estimated 98% of all physicists agreed. Time will tell whether they are right.

Challenge 142, page 258: A mass of 100 kg and a speed of 8 m/s require 43 m^2 of wing surface.

Challenge 144, page 263: The issue is a red herring. The world has three dimensions.

Challenge 145, page 265: The largest rotation angle $\Delta\varphi$ that can be achieved in one stroke C is found by maximizing the integral

$$\Delta\varphi = -\int_C \frac{a^2}{a^2 + b^2} \, d\theta \tag{155}$$

Since the path C in shape space is closed, we can use Stokes' theorem to transform the line integral to a surface integral over the surface S enclosed by C in shape space:

$$\Delta\varphi = \int_S \frac{2ab^2}{(a^2 + b^2)^2} \, da \, d\theta \,. \tag{156}$$

The maximum angle is found by noting that θ can vary at most between 0 and π, and that a can vary at most between 0 and ∞. This yields

$$\Delta\varphi_{\text{max}} = \int_{\theta=0}^{\pi} \int_{a=0}^{\infty} \frac{2ab^2}{(a^2 + b^2)^2} \, da \, d\theta = \pi \,. \tag{157}$$

Challenge 147, page 269: A cloud is kept afloat and compact by convection currents. Clouds without convection can often be seen in the summer: they diffuse and disappear. The details of the internal and external air currents depend on the cloud type and are a research field on its own.

Challenge 152, page 275: Lattices are not isotropic, lattices are not Lorentz invariant.

Challenge 154, page 278: The infinite sum is not defined for numbers; however, it is defined for a knotted string.

Challenge 155, page 280: The research race for the solution is ongoing, but the goal is still far.

Challenge 156, page 281: This is a simple but hard question. Find out!

Challenge 159, page 283: Large raindrops are pancakes with a massive border bulge. When the size increases, e.g. when a large drop falls through vapour, the drop splits, as the central membrane is then torn apart.

Challenge 160, page 283: It is a drawing; if it is interpreted as an image of a three-dimensional object, it either does not exist, or is not closed, or is an optical illusion of a torus.

Challenge 161, page 283: See T. FINK & Y. MAO, *The 85 Ways to Tie a Tie*, Broadway Books, 2000.

Challenge 162, page 284: See T. CLARKE, *Laces high*, Nature Science Update 5th of December, 2002, or www.nature.com/nsu/021202/021202-4.html.

Challenge 165, page 285: In fact, nobody has even tried to do so yet. It may also be that the problem makes no sense.

Challenge 167, page 291: Most macroscopic matter properties fall in this class, such as the change of water density with temperature.

Challenge 169, page 296: Before the speculation can be fully tested, the relation between particles and black holes has to be clarified first.

Challenge 170, page 297: Never expect a correct solution for personal choices. Do what you yourself think and feel is correct.

Challenge 175, page 305: Planck limits can be exceeded for extensive observables for which many particle systems can exceed single particle limits, such as mass, momentum, energy or electrical resistance.

Challenge 177, page 307: Do not forget the relativistic time dilation.

Challenge 178, page 307: The formula with $n - 1$ is a better fit. Why?

Challenge 184, page 314: The slowdown goes *quadratically* with time, because every new slowdown adds to the old one!

Challenge 185, page 315: No, only properties of parts of the universe are listed. The universe itself has no properties, as shown in the last volume. Vol. VI, page 104

Challenge 186, page 317: The gauge coupling constants, via the Planck length, determine the size of atoms, the strength of chemical bonds and thus the size of all things.

Challenge 187, page 332: Covalent bonds tend to produce full shells; this is a smaller change on the right side of the periodic table.

Challenge 189, page 334: The metric is regular, positive definite and obeys the triangle inequality.

Challenge 193, page 336: The solution is the set of all two by two matrices, as each two by two matrix specifies a linear transformation, if one defines a transformed point as the product of the point and this matrix. (Only multiplication with a fixed matrix can give a linear transformation.) Can you recognize from a matrix whether it is a rotation, a reflection, a dilation, a shear, or a stretch along two axes? What are the remaining possibilities?

Challenge 196, page 337: The (simplest) product of two functions is taken by point-by-point multiplication.

Challenge 197, page 337: The norm $\|f\|$ of a real function f is defined as the supremum of its absolute value:

$$\|f\| = \sup_{x \in \mathbb{R}} |f(x)| \ . \tag{158}$$

In simple terms: the maximum value taken by the absolute of the function is its norm. It is also called 'sup'-norm. Since it contains a supremum, this norm is only defined on the subspace of *bounded* continuous functions on a space X, or, if X is compact, on the space of all continuous functions (because a continuous function on a compact space must be bounded).

Challenge 200, page 340: Take out your head, then pull one side of your pullover over the corresponding arm, continue pulling it over the over arm; then pull the other side, under the first, to the other arm as well. Put your head back in. Your pullover (or your trousers) will be inside out.

Challenge 201, page 340: Both can be untied.

Challenge 205, page 345: The transformation from one manifold to another with different topology can be done with a tiny change, at a so-called *singular point*. Since nature shows a minimum action, such a tiny change cannot be avoided.

Challenge 206, page 347: The product $M^\dagger M$ is Hermitean, and has positive eigenvalues. Thus H is uniquely defined and Hermitean. U is unitary because $U^\dagger U$ is the unit matrix.

BIBLIOGRAPHY

> " Gedanken sind nicht stets parat. Man schreibt
> auch, wenn man keine hat.*
> Wilhelm Busch, *Aphorismen und Reime.*

1   The use of radioactivity for breeding of new sorts of wheat, rice, cotton, roses, pineapple and many more is described by B. S. AHLOOWALIA & M. MALUSZYNSKI, *Induced mutations – a new paradigm in plant breeding*, Euphytica 11, pp. 167–173, 2004. Cited on page 19.

2   The motorized screw used by viruses was described by A. A. SIMPSON & al., *Structure of the bacteriophage phi29 DNA packaging motor*, Nature 408, pp. 745–750, 2000. Cited on page 21.

3   S. M. BLOCK, *Real engines of creation*, Nature 386, pp. 217–219, 1997. Cited on page 21.

4   Early results and ideas on molecular motors are summarised by B. GOSS LEVI, *Measured steps advance the understanding of molecular motors*, Physics Today pp. 17–19, April 1995. Newer results are described in R. D. ASTUMIAN, *Making molecules into motors*, Scientific American pp. 57–64, July 2001. Cited on page 22.

5   R. BARTUSSEK & P. HÄNGGI, *Brownsche Motoren*, Physikalische Blätter 51, pp. 506–507, 1995. See also R. ALT-HADDOU & W. HERZOG, *Force and motion generation of myosin motors: muscle contraction*, Journal of Electromyography and kinesiology 12, pp. 435–445, 2002. Cited on page 23.

6   N. HIROKAWA, S. NIWA & Y. TANAKA, *Molecular motors in neurons: transport mechanisms and roles in brain function, development, and disease*, Neuron 68, pp. 610–638, 2010. Cited on page 25.

7   J. WEBER & A. E. SENIOR, *ATP synthesis driven by proton transport in F1Fo-ATP synthase*, FEBS letters 545, pp. 61–70, 2003. Cited on page 26.

8   This truly fascinating research result, worth a Nobel Prize, is summarized in N. HIROKAWA, Y. TANAKA & Y. OKADA, *Left-right determination: involvement of molecular motor KIF3, cilia, and nodal flow*, Cold Spring Harbor Perspectives in Biology 1, p. a000802, 2009, also available at www.cshperspectives.org. The web site also links to numerous captivating films of the involved microscopic processes, found at beta.cshperspectives.cshlp.org. Cited on page 26.

9   R. J. CANO & M. K. BORUCKI, *Revival and identification of bacterial spores in 25- to 40-million-year-old Dominican amber*, Science 26, pp. 1060–1064, 1995. Cited on page 30.

10   The first papers on bacteria from salt deposits were V. R. OTT & H. J. DOMBRWOSKI, *Mikrofossilien in den Mineralquellen zu Bad Nauheim*, Notizblatt des Hessischen Landesamtes für Bodenforschung 87, pp. 415–416, 1959, H. J. DOMBROWSKI, *Bacteria from*

---

* 'Thoughts are not always available. Many write even without them.'

Paleozoic salt deposits, Annals of the New York Academy of Sciences **108**, pp. 453–460, 1963. A recent confirmation is R. H. VREELAND, W. D. ROSENZWEIG & D. W. POWERS, *Isolation of a 250 million-year-old halotolerant bacterium from a primary salt crystal*, Nature **407**, pp. 897–899, 2000. Cited on page 30.

11 This is explained in D. GRAUR & T. PUPKO, *The permian bacterium that isn't*, Molecular Biology and Evolution **18**, pp. 1143–1146, 2001, and also in M. B. HEBSGAARD, M. J. PHILLIPS & E. WILLERSLEV, *Geologically ancient DNA: fact or artefact?*, Trends in Microbiology **13**, pp. 212–220, 2005. Cited on page 30.

12 GABRIELE WALKER, *Snowball Earth – The Story of the Great Global Catastrophe That Spawned Life as We Know It*, Crown Publishing, 2003. No citations.

13 The table and the evolutionary tree are taken from J. O. MCINERNEY, M. MULLARKEY, M. E. WERNECKE & R. POWELL, *Bacteria and Archaea: Molecular techniques reveal astonishing diversity*, Biodiversity **3**, pp. 3–10, 2002. The tree might still change a little in the coming years. Cited on page 31.

14 E. K. COSTELLO, C. L. LAUBER, M. HAMADY, N. FIERER, J. I. GORDON & R. KNIGHT, *Bacterial community variation in human body habitats across space and time*, Science Express 5 November 2009. Cited on page 32.

15 This is taken from the delightful children text HANS J. PRESS, *Spiel das Wissen schafft*, Ravensburger Buchverlag 1964, 2004. Cited on page 35.

16 The discovery of a specific taste for fat was published by F. LAUGERETTE, P. PASSILLY-DEGRACE, B. PATRIS, I. NIOT, M. FEBBRAIO, J. P. MONTMAYEUR & P. BESNARD, *CD36 involvement in orosensory detection of dietary lipids, spontaneous fat preference, and digestive secretions*, Journal of Clinical Investigation **115**, pp. 3177–3184, 2005. Cited on page 37.

17 There is no standard procedure to learn to enjoy life to the maximum. A good foundation can be found in those books which teach the ability to those which have lost it.

The best experts are those who help others to overcome traumas. PETER A. LEVINE & ANN FREDERICK, *Waking the Tiger – Healing Trauma – The Innate Capacity to Transform Overwhelming Experiences*, North Atlantic Books, 1997. GEOFF GRAHAM, *How to Become the Parent You Never Had - a Treatment for Extremes of Fear, Anger and Guilt*, Real Options Press, 1986. A good complement to these texts is the approach presented by BERT HELLINGER, *Zweierlei Glück*, Carl Auer Verlag, 1997. Some of his books are also available in English. The author presents a simple and efficient technique for reducing entanglement with one's family past. Another good book is PHIL STUTZ & BARRY MICHELS, *The Tools – Transform Your Problems into Courage, Confidence, and Creativity*, Random House, 2012.

The next step, namely full mastery in the enjoyment of life, can be found in any book written by somebody who has achieved mastery in any one topic. The topic itself is not important, only the passion is. A few examples:

A. DE LA GARANDERIE, *Le dialogue pédagogique avec l'élève*, Centurion, 1984, A. DE LA GARANDERIE, *Pour une pédagogie de l'intelligence*, Centurion, 1990, A. DE LA GARANDERIE, *Réussir ça s'apprend*, Bayard, 1994. De la Garanderie explains how the results of teaching and learning depend in particular on the importance of evocation, imagination and motivation.

PLATO, *Phaedrus*, Athens, 380 BCE.

FRANÇOISE DOLTO, *La cause des enfants*, Laffont, 1985, and her other books. Dolto (b. 1908 Paris, d. 1988 Paris), a child psychiatrist, is one of the world experts on the growth of the child; her main theme was that growth is only possible by giving the highest possible responsibility to every child during its development.

In the domain of art, many had the passion to achieve full pleasure. A good piece of music, a beautiful painting, an expressive statue or a good film can show it. On a smaller scale, the art to typeset beautiful books, so different from what many computer programs do by default, the best introduction are the works by Jan Tschichold (b. 1902 Leipzig, d. 1974 Locarno), the undisputed master of the field. Among the many books he designed are the beautiful Penguin books of the late 1940s; he also was a type designer, e.g. of the *Sabon* typeface. A beautiful summary of his views is the short but condensed text JAN TSCHICHOLD, *Ausgewählte Aufsätze über Fragen der Gestalt des Buches und der Typographie*, Birkhäuser Verlag, Basel, 1993. An extensive and beautiful textbook on the topic is HANS PETER WILLBERG & FRIEDRICH FORSSMAN, *Lesetypographie*, Verlag Hermann Schmidt, Mainz, 1997. See also ROBERT BRINGHURST, *The Elements of Typographic Style*, Hartley & Marks, 2004.

Many scientists passionately enjoyed their occupation. Any biography of Charles Darwin will purvey his fascination for biology, of Friedrich Bessel for astronomy, of Albert Einstein for physics and of Linus Pauling for chemistry. Cited on page 38.

18 The group of John Wearden in Manchester has shown by experiments with humans that the accuracy of a few per cent is possible for any action with a duration between a tenth of a second and a minute. See J. MCCRONE, *When a second lasts forever*, New Scientist pp. 53–56, 1 November 1997. Cited on page 39.

19 The chemical clocks in our body are described in JOHN D. PALMER, *The Living Clock*, Oxford University Press, 2002, or in A. AHLGREN & F. HALBERG, *Cycles of Nature: An Introduction to Biological Rhythms*, National Science Teachers Association, 1990. See also the www.msi.umn.edu/~halberg/introd website. Cited on page 39.

20 D.J MORRÉ & al., *Biochemical basis for the biological clock*, Biochemistry 41, pp. 11941–11945, 2002. Cited on page 39.

21 An introduction to the sense of time as result of clocks in the brain is found in R. B. IVRY & R. SPENCER, *The neural representation of time*, Current Opinion in Neurobiology 14, pp. 225–232, 2004. The interval timer is explain in simple words in K. WRIGHT, *Times in our lives*, Scientific American pp. 40–47, September 2002. The MRI research used is S. M. RAO, A. R. MAYER & D. L. HARRINGTON, *The evolution of brain activation during temporal processing*, Nature Neuroscience 4, pp. 317–323, 2001. Cited on page 41.

22 See, for example, JAN HILGEVOORD, *Time in quantum mechanics*, American Journal of Physics 70, pp. 301–306, 2002. Cited on page 41.

23 E. J. ZIMMERMAN, *The macroscopic nature of space-time*, American Journal of Physics 30, pp. 97–105, 1962. Cited on page 42.

24 See P.D. PEŞIĆ, *The smallest clock*, European Journal of Physics 14, pp. 90–92, 1993. Cited on page 43.

25 The possibilities for precision timing using single-ion clocks are shown in W.H. OSKAY & al., *Single-atom clock with high accuracy*, Physical Review Letters 97, p. 020801, 2006. Cited on page 43.

26 A pretty example of a quantum mechanical system showing exponential behaviour at all times is given by H. NAKAZATO, M. NAMIKI & S. PASCAZIO, *Exponential behaviour of a quantum system in a macroscopic medium*, Physical Review Letters 73, pp. 1063–1066, 1994. Cited on page 45.

27 See the delightful book about the topic by PAOLO FACCHI & SAVERIO PASCAZIO, *La regola d'oro di Fermi*, Bibliopolis, 1999. An experiment observing deviations at short times is S. R. WILKINSON, C. F. BHARUCHA, M. C. FISCHER, K. W. MADISON, P. R. MORROW, Q. NIU, B. SUNDARAM & M. G. RAIZEN, Nature **387**, p. 575, 1997. Cited on page 45.

28 See, for example, R. EFRON, *The duration of the present*, Annals of the New York Academy of Sciences **138**, pp. 713–729, 1967. Cited on page 45.

29 W. M. ITANO, D. J. HEINZEN, J. J. BOLLINGER & D. J. WINELAND, *Quantum Zeno effect*, Physical Review A **41**, pp. 2295–2300, 1990. M. C. FISCHER, B. GUTIÉRREZ-MEDINA & M. G. RAIZEN, *Observation of the Quantum Zeno and Anti-Zeno effects in an unstable system*, Physical Review Letters **87**, p. 040402, 2001, also www-arxiv.org/abs/quant-ph/0104035. Cited on page 47.

30 See P. FACCHI, Z. HRADIL, G. KRENN, S. PASCAZIO & J. ŘEHÁČEK *Quantum Zeno tomography*, Physical Review A **66**, p. 012110, 2002. Cited on page 47.

31 See P. FACCHI, H. NAKAZATO & S. PASCAZIO *From the quantum Zeno to the inverse quantum Zeno effect*, Physical Review Letters **86**, pp. 2699–2702, 2001. Cited on page 47.

32 H. KOBAYASHI & S. KOHSHIMA, *Unique morphology of the human eye*, Nature **387**, pp. 767–768, 1997. No citations.

33 JAMES W. PRESCOTT, *Body pleasure and the origins of violence*, The Futurist Bethseda, 1975, also available at www.violence.de/prescott/bullettin/article.html. Cited on page 50.

34 FRANCES ASHCROFT, *The Spark of Life: Electricity in the Human Body*, Allen Lane, 2012. Cited on page 50.

35 FELIX TRETTER & MARGOT ALBUS, *Einführung in die Psychopharmakotherapie – Grundlagen, Praxis, Anwendungen*, Thieme, 2004. Cited on page 51.

36 See the talk on these experiments by Helen Fisher on www.ted.org and the information on helenfisher.com. Cited on page 51.

37 See for example P. PYYKKÖ, *Relativity, gold, closed-shell interactions, and CsAu.NH3*, Angewandte Chemie, International Edition **41**, pp. 3573–3578, 2002, or L. J. NORRBY, *Why is mercury liquid? Or, why do relativistic effects not get into chemistry textbooks?*, Journal of Chemical Education **68**, pp. 110–113, 1991. Cited on page 54.

38 On the internet, the 'spherical' periodic table is credited to Timothy Stowe; but there is no reference for that claim, except an obscure calendar from a small chemical company. The original table (containing a number errors) used to be found at the chemlab.pc.maricopa.edu/periodic/stowetable.html website; it is now best found by searching for images called 'stowetable' with any internet search engine. Cited on page 55.

39 For good figures of atomic orbitals, take any modern chemistry text. Or go to csi.chemie.tu-darmstadt.de/ak/immel/. Cited on page 56.

40 For experimentally determined pictures of the orbitals of dangling bonds, see for example F. GIESSIBL & al., *Subatomic features on the silicon (111)-(7x7) surface observed by atomic force microscopy*, Science **289**, pp. 422–425, 2000. Cited on page 56.

41 L. GAGLIARDI & B. O. ROOS, *Quantum chemical calculations show that the uranium molecule* U_2 *has a quintuple bond*, Nature **433**, pp. 848–851, 2005. B. O. ROOS, A. C. BORIN & L. GAGLIARDI, *Reaching the maximum multiplicity of the covalent chemical bond*, Angewandte Chemie, International Edition **46**, pp. 1469–1472, 2007. Cited on page 58.

42 H. -W. FINK & C. ESCHER, *Zupfen am Lebensfaden – Experimente mit einzelnen DNS-Molekülen*, Physik in unserer Zeit **38**, pp. 190–196, 2007. Cited on page 59.

43 L. Bindi, P. J. Steinhardt, N. Yai & P. J. Lu, *Natural Quasicrystal*, Science **324**, pp. 1306–1309, 2009. Cited on page 72.

44 The present attempts to build cheap THz wave sensors – which allow seeing through clothes – is described by D. Clery, *Brainstorming their way to an imaging revolution*, Science **297**, pp. 761–763, 2002. Cited on page 75.

45 The switchable mirror effect was discovered by J. N. Huiberts, R. Griessen, J. H. Rector, R. J. Wijngarden, J. P. Dekker, D. G. de Groot & N. J. Koeman, *Yttrium and lanthanum hydride films with switchable optical properties*, Nature **380**, pp. 231–234, 1996. A good introduction is R. Griessen, *Schaltbare Spiegel aus Metallhydriden*, Physikalische Blätter **53**, pp. 1207–1209, 1997. Cited on page 76.

46 V. F. Weisskopf, *Of atoms, mountains and stars: a study in qualitative physics*, Science **187**, p. 605, 1975. Cited on page 78.

47 K. Schwab, E. A. Henriksen, J. M. Worlock & M. L. Roukes, *Measurement of the quantum of thermal conductance*, Nature **404**, pp. 974–977, 2000. Cited on page 79.

48 K. Autumn, Y. Liang, T. Hsich, W. Zwaxh, W. P. Chan, T. Kenny, R. Fearing & R. J. Full, *Adhesive force of a single gecko foot-hair*, Nature **405**, pp. 681–685, 2000. Cited on page 80.

49 A. B. Kesel, A. Martin & T. Seidl, *Getting a grip on spider attachment: an AFM approach to microstructure adhesion in arthropods*, Smart Materials and Structures **13**, pp. 512–518, 2004. Cited on page 80.

50 R. Ritchie, M. J. Buehler & P. Hansma, *Plasticity and toughness in bones*, Physics Today **82**, pp. 41–47, 2009. This is a beautiful article. Cited on page 80.

51 H. Inaba, T. Saitou, K. -I. Tozaki & H. Hayashi, *Effect of the magnetic field on the melting transition of H2O and D2O measured by a high resolution and supersensitive differential scanning calorimeter*, Journal of Applied Physics **96**, pp. 6127–6132, 2004. Cited on page 81.

52 The graph for temperature is taken from J. Jouzel & al., *Orbital and millennial Antarctic climate variability over the past 800,000 years*, Science **317**, pp. 793–796, 2007. The graph for CO_2 is combined from six publications: D. M. Etheridge & al., *Natural and anthropogenic changes in atmospheric CO_2 over the last 1000 years from air in Antarctic ice and firm*, Journal of Geophysical Research **101**, pp. 4115–4128, 1996, J. R. Petit & al., *Climate and atmospheric history of the past 420,000 years from the Vostok ice core, Antarctica*, Nature **399**, pp. 429–436, 1999, A. Indermühle & al., *Atmospheric CO_2 concentration from 60 to 20 kyr BP from the Taylor Dome ic cores, Antarctica*, Geophysical Research Letters **27**, pp. 735–738, 2000, E. Monnin & al., *Atmospheric CO_2 concentration over the last glacial termination*, Science **291**, pp. 112–114, 2001, U. Siegenthaler & al., *Stable carbon cycle-climate relationship during the late pleistocene*, Science **310**, pp. 1313–1317, 2005, D. Lüthi & al., *High-resolution carbon dioxide concentration record 650,000-800,000 years before present*, Nature **453**, pp. 379–382, 2008. Cited on page 82.

53 H. Oerlemans, *De afkoeling van de aarde*, Nederlands tijdschrift voor natuurkunde pp. 230–234, 2004. This excellent paper by one of the experts of the field also explains that the Earth has two states, namely today's *tropical* state and the *snowball* state, when it is covered with ice. And there is hysteresis for the switch between the two states. Cited on page 82.

54 R. Hemley & al., presented these results at the International Conference on New Diamond Science in Tsukuba (Japan) 2005. See figures at www.carnegieinstitution.org/diamond-13may2005. Cited on page 84.

55 INGO DIERKING, *Textures of Liquid Crystals*, Wiley-VCH, 2003. Cited on page 84.

56 The *photon hall effect* was discovered by GEERT RIKKEN, BART VAN TIGGELEN & ANJA SPARENBERG, *Lichtverstrooiing in een magneetveld*, Nederlands tijdschrift voor natuurkunde **63**, pp. 67–70, maart 1998. Cited on page 85.

57 C. STROHM, G. L. J. A. RIKKEN & P. WYDER, *Phenomenological evidence for the phonon Hall effect*, Physical Review Letters **95**, p. 155901, 2005. Cited on page 85.

58 E. CHIBOWSKI, L. HOŁYSZ, K. TERPIŁOWSKI, *Effect of magnetic field on deposition and adhesion of calcium carbonate particles on different substrates*, Journal of Adhesion Science and Technology **17**, pp. 2005–2021, 2003, and references therein. Cited on page 85.

59 K. S. NOVOSELOV, D. JIANG, F. SCHEDIN, T. BOOTH, V. V. KHOTKEVICH, S. V. MOROZOV & A. K. GEIM, *Two Dimensional Atomic Crystals*, Proceedings of the National Academy of Sciences **102**, pp. 10451–10453, 2005. See also the talk by ANDRE GEIM, *QED in a pencil trace*, to be downloaded at online.kitp.ucsb.edu/online/graphene_m07/ geim/ or A. C. NETO, F. GUINEA & N. M. PERES, *Drawing conclusions from graphene*, Physics World November 2006. Cited on pages 85, 86, and 390.

60 R. R. NAIR & al., *Fine structure constant defines visual transparency of graphene*, Science **320**, p. 1308, 2008. Cited on page 87.

61 The first paper was J. BARDEEN, L. N. COOPER & J. R. SCHRIEFFER, *Microscopic Theory of Superconductivity*, Physical Review **106**, pp. 162–164, 1957. The full, so-called BCS theory, is then given in the masterly paper J. BARDEEN, L. N. COOPER & J. R. SCHRIEFFER, *Theory of Superconductivity*, Physical Review **108**, pp. 1175–1204, 1957. Cited on page 90.

62 A. P. FINNE, T. ARAKI, R. BLAAUWGEERS, V. B. ELTSOV, N. B. KOPNIN, M. KRUSIUS, L. SKRBEK, M. TSUBOTA & G. E. VOLOVIK, *An intrinsic velocity-independent criterion for superfluid turbulence*, Nature **424**, pp. 1022–1025, 2003. Cited on page 91.

63 The original prediction, which earned Laughlin a Nobel Prize in Physics, is in R. B. LAUGHLIN, *Anomalous quantum Hall effect: an incompressible quantum fluid with fractionally charged excitations*, Physical Review Letters **50**, pp. 1395–1398, 1983. Cited on pages 92 and 93.

64 The first experimental evidence is by V. J. GOLDMAN & B. SU, *Resonance tunnelling in the fractional quantum Hall regime: measurement of fractional charge*, Science **267**, pp. 1010–1012, 1995. The first unambiguous measurements are by R. DE PICCIOTTO & al., *Direct observation of a fractional charge*, Nature **389**, pp. 162–164, 1997, and L. SAMINADAYAR, D. C. GLATTLI, Y. JIN & B. ETIENNE, *Observation of the e,3 fractionally charged Laughlin quasiparticle/* Physical Review Letters **79**, pp. 2526–2529, 1997. or arxiv.org/abs/cond-mat/9706307. Cited on page 92.

65 The discovery of the original quantum Hall effect, which earned von Klitzing the Nobel Prize in Physics, was published as K. VON KLITZING, G. DORDA & M. PEPPER, *New method for high-accuracy determination of the fine-structure constant based on quantised Hall resistance*, Physical Review Letters **45**, pp. 494–497, 1980. Cited on page 92.

66 The fractional quantum Hall effect was discovered by D. C. TSUI, H. L. STORMER & A. C. GOSSARD, *Two-dimensional magnetotransport in the extreme quantum limit*, Physical Review Letters **48**, pp. 1559–1562, 1982. Tsui and Stormer won the Nobel Prize in Physics with this paper. Cited on page 93.

67 K. S. NOVOSELOV, Z. JIANG, Y. ZHANG, S. V. MOROZOV, H. L. STORMER, U. ZEITLER, J. C. MANN, G. S. BOEBINGER, P. KIM & A. K. GEIM, *Room-*

temperature quantum Hall effect in graphene, Science 315, p. 1379, 2007. Cited on page 93.

68 See for example the overview by M. BORN & T. JÜSTEL, *Umweltfreundliche Lichtquellen*, Physik Journal 2, pp. 43–49, 2003. Cited on page 94.

69 K. AN, J. J. CHILDS, R. R. DESARI & M. S. FIELD, *Microlaser: a laser with one atom in an optical resonator*, Physical Review Letters 73, pp. 3375–3378, 19 December 1994. Cited on page 100.

70 J. KASPARIAN, *Les lasers femtosecondes : applications atmosph ériques*, La Recherche pp. RE14 1–7, February 2004/ See also the www.teramobile.org website. Cited on page 101.

71 A. TAKITA, H. YAMAMOTO, Y. HAYASAKI, N. NISHIDA & H. MISAWA, *Three-dimensional optical memory using a human fingernail*, Optics Express 13, pp. 4560–4567, 2005. They used a Ti:sapphire oscillator and a Ti:sapphire amplifier. Cited on page 101.

72 The author predicted laser umbrellas in 2011. Cited on page 102.

73 W. GUO & H. MARIS, *Observations of the motion of single electrons in liquid helium*, Journal of Low Temperature Physics 148, pp. 199–206, 2007. Cited on page 103.

74 The story is found on many places in the internet. Cited on page 104.

75 S. L. BOERSMA, *Aantrekkende kracht tussen schepen*, Nederlands tidschrift voor natuurkunde 67, pp. 244–246, Augustus 2001. See also his paper S. L. BOERSMA, *A maritime analogy of the Casimir effect*, American Journal of Physics 64, pp. 539–541, 1996. Cited on page 107.

76 P. BALL, *Popular physics myth is all at sea – does the ghostly Casimir effect really cause ships to attract each other?*, online at www.nature.com/news/2006/060501/full/news060501-7.html. Cited on pages 107 and 110.

77 A. LARRAZA & B. DENARDO, *An acoustic Casimir effect*, Physics Letters A 248, pp. 151–155, 1998. See also A. LARRAZA, *A demonstration apparatus for an acoustic analog of the Casimir effect*, American Journal of Physics 67, pp. 1028–1030, 1999. Cited on page 109.

78 H. B. G. CASIMIR, *On the attraction between two perfectly conducting bodies*, Proceedings of the Koninklijke Nederlandse Akademie van Wetenschappen 51, pp. 793–795, 1948. Cited on page 109.

79 H. B. G. CASIMIR & D. POLDER, *The influence of retardation on the London–Van der Waals forces*, Physical Review 73, pp. 360–372, 1948. Cited on page 109.

80 B. W. DERJAGUIN, I. I. ABRIKOSOVA & E. M. LIFSHITZ, *Direct measurement of molecular attraction between solids separated by a narrow gap*, Quaterly Review of the Chemical Society (London) 10, pp. 295–329, 1956. Cited on page 109.

81 M. J. SPARNAAY, *Attractive forces between flat plates*, Nature 180, pp. 334–335, 1957. M. J. SPARNAAY, *Measurements of attractive forces between flat plates*, Physica (Utrecht) 24, pp. 751–764, 1958. Cited on page 109.

82 S. K. LAMOREAUX, *Demonstration of the Casimir force in the 0.6 to 6 μm range*, Physical Review Letters 78, pp. 5–8, 1997. U. MOHIDEEN & A. ROY, *Precision measurement of the Casimir force from 0.1 to 0.9 μm*, Physical Review Letters 81, pp. 4549–4552, 1998. Cited on page 109.

83 A. LAMBRECHT, *Das Vakuum kommt zu Kräften*, Physik in unserer Zeit 36, pp. 85–91, 2005. Cited on page 109.

84 T. H. BOYER, *Van der Waals forces and zero-point energy for dielectric and permeable materials*, Physical Review A 9, pp. 2078–2084, 1974. This was taken up again in O. KENNETH,

I. KLICH, A. MANN & M. REVZEN, *Repulsive Casimir forces*, Physical Review Letters **89**, p. 033001, 2002. However, none of these effects has been verified yet. Cited on page 110.

85 The effect was discovered by K. SCHARNHORST, *On propagation of light in the vacuum between plates*, Physics Letters B **236**, pp. 354–359, 1990. See also G. BARTON, *Faster-than-c light between parallel mirrors: the Scharnhorst effect rederived*, Physics Letters B **237**, pp. 559–562, 1990, and P. W. MILONNI & K. SVOZIL, *Impossibility of measuring faster-than-c signalling by the Scharnhorst effect*, Physics Letters B **248**, pp. 437–438, 1990. The latter corrects an error in the first paper. Cited on page 110.

86 D.L. BURKE & al., *Positron production in multiphoton light-by-light scattering*, Physical Review Letters **79**, pp. 1626–1629, 1997. A simple summary is given in BERTRAM SCHWARZSCHILD, *Gamma rays create matter just by plowing into laser light*, Physics Today **51**, pp. 17–18, February 1998. Cited on page 115.

87 M. KARDAR & R. GOLESTANIAN, *The 'friction' of the vacuum, and other fluctuation-induced forces*, Reviews of Modern Physics **71**, pp. 1233–1245, 1999. Cited on page 116.

88 G. GOUR & L. SRIRAMKUMAR, *Will small particles exhibit Brownian motion in the quantum vacuum?*, arxiv.org/abs/quant/phys/9808032 Cited on page 116.

89 P. A. M. DIRAC, *The requirements of fundamental physical theory*, European Journal of Physics **5**, pp. 65–67, 1984. This article transcribes a speech by Paul Dirac just before his death. Disturbed by the infinities of quantum electrodynamics, he rejects as wrong the theory which he himself found, even though it correctly describes all experiments. The humility of a great man who is so dismissive of his main and great achievement was impressive when hearing him, and still is impressive when reading the talk. Cited on page 117.

90 H. EULER & B. KOCKEL, Naturwissenschaften **23**, p. 246, 1935, H. EULER, Annalen der Physik **26**, p. 398, 1936, W. HEISENBERG & H. EULER, *Folgerung aus der Diracschen Theorie des Electrons*, Zeitschrift für Physik **98**, pp. 714–722, 1936. Cited on page 117.

91 TH. STOEHLKER & al., *The 1s Lamb shift in hydrogenlike uranium measured on cooled, decelerated ions beams*, Physical Review Letters **85**, p. 3109, 2000. Cited on page 118.

92 M. ROBERTS & al., *Observation of an electric octupole transition in a single ion*, Physical Review Letters **78**, pp. 1876–1879, 1997. A lifetime of 3700 days was determined. Cited on page 118.

93 See the website by Tom Hales on www.math.lsa.umich.edu/~hales/countdown. An earlier book on the issue is the delightful text by MAX LEPPMAIER, *Kugelpackungen von Kepler bis heute*, Vieweg Verlag, 1997. Cited on page 64.

94 For a short overview, also covering binary systems, see C. BECHINGER, H.-H. VON GRÜNBERG & P. LEIDERER, *Entropische Kräfte - warum sich repulsiv wechselwirkende Teilchen anziehen können*, Physikalische Blätter **55**, pp. 53–56, 1999. Cited on page 64.

95 H. -W. FINK & G. EHRLICH, *Direct observation of three-body interactions in adsorbed layers: Re on W(110)*, Physical Review Letters **52**, pp. 1532–1534, 1984, and H. -W. FINK & G. EHRLICH, *Pair and trio interactions between adatoms: Re on W(110)*, Journal of Chemical Physics **61**, pp. 4657–4665, 1984. Cited on pages 64 and 65.

96 See for example the textbook by A. PIMPINELLI & J. VILLAIN, *Physics of Crystal Growth*, Cambridge University Press, 1998. Cited on page 65.

97 Y. FURUKAWA, *Faszination der Schneekristalle - wie ihre bezaubernden Formen entstehen*, Chemie in unserer Zeit **31**, pp. 58–65, 1997. His www.lowtem.hokudai.ac.jp/~frkw/index_e. html website gives more information. Cited on page 65.

98 P.-G. DE GENNES, *Soft matter*, Reviews of Modern Physics **63**, p. 645, 1992. No citations.

99 The fluid helium based quantum gyroscope was first used to measure the rotation of the Earth by the French research group O. AVENEL & E. VAROQUAUX, *Detection of the Earth rotation with a superfluid double-hole resonator*, Czechoslovak Journal of Physics (Supplement S6) **48**, pp. 3319–3320, 1996. The description of the set-up is found in YU. MUKHARSKY, O. AVENEL & E. VAROQUAUX, *Observation of half-quantum defects in superfluid* ^{3}He – B, Physical Review Letters **92**, p. 210402, 2004. The gyro experiment was later repeated by the Berkeley group, K. SCHWAB, N. BRUCKNER & R. E. PACKARD, *Detection of the Earth's rotation using superfluid phase coherence*, Nature **386**, pp. 585–587, 1997. For an review of this topic, see O. AVENEL, YU. MUKHARSKY & E. VAROQUAUX, *Superfluid gyrometers*, Journal of Low Temperature Physics **135**, p. 745, 2004. Cited on page 105.

100 K. SCHWAB, E. A. HENRIKSEN, J. M. WORLOCK & M. L. ROUKES, *Measurement of the quantum of thermal conductance*, Nature **404**, p. 974, 2000. For the optical analog of the experiment, see the beautiful paper by E. A. MONTIE, E. C. COSMAN, G. W. 'T HOOFT, M. B. VAN DER MARK & C. W. J. BEENAKKER, *Observation of the optical analogue of quantized conductance of a point contact*, Nature **350**, pp. 594–595, 18 April 1991, and the longer version E. A. MONTIE, E. C. COSMAN, G. W. 'T HOOFT, M. B. VAN DER MARK & C. W. J. BEENAKKER, *Observation of the optical analogue of the quantised conductance of a point contact*, Physica B **175**, pp. 149–152, 1991. For more details on this experiment, see the volume *Light, Charges and Brains* of the Motion Mountain series. Cited on page 118.

101 The present knowledge on how humans move is summarized by … Cited on page 122.

102 See, e.g., D. A. BRYANT, *Electron acceleration in the aurora*, Contemporary Physics **35**, pp. 165–179, 1994. For a recent development, see D. L. CHANDLER, *Mysterious electron acceleration explained*, at web.mit.edu/newsoffice/2012/plasma-phenomenon-explained-0227.html which explains the results of the paper by J. EGEDAL, W. DAUGHTON & A. LE, *Large-scale acceleration by parallel electric fields during magnetic reconnection*, Nature Physics 2012. Cited on page 123.

103 The discharges above clouds are described, e.g., by … Cited on page 123.

104 About Coulomb explosions and their importance for material processing with lasers, see for example D. MÜLLER, *Picosecond lasers for high-quality industrial micromachining*, Photonics Spectra pp. 46–47, November 2009. Cited on page 123.

105 See, e.g., L. O'C. DRURY, *Acceleration of cosmic rays*, Contemporary Physics **35**, pp. 231–242, 1994. Cited on page 123.

106 See DIRK KREIMER, *New mathematical structures in renormalizable quantum field theories*, arxiv.org/abs/hep-th/0211136 or Annals of Physics **303**, pp. 179–202, 2003, and the erratum ibid., **305**, p. 79, 2003. Cited on page 123.

107 See, for example, the paper by M. URBAN, *A particle mechanism for the index of refraction*, preprint at arxiv.org/abs/0709.1550. Cited on page 123.

108 S. FRAY, C. ALVAREZ DIEZ, T. W. HÄNSCH & M. WEITZ, *Atomic interferometer with amplitude gratings of light and its applications to atom based tests of the equivalence principle*, Physical Review Letters **93**, p. 240404, 2004, preprint at arxiv.org/abs/physics/0411052. Cited on page 124.

109 The original papers on neutron behaviour in the gravitational field are all worth reading. The original feat was reported in V. V. NESVIZHEVSKY, H. G. BOERNER, A. K. PETOUKHOV, H. ABELE, S. BAESSLER, F. RUESS, TH. STOEFERLE,

A. WESTPHAL, A. M. GAGARSKI, G. A. PETROV & A. V. STRELKOV, *Quantum states of neutrons in the Earth's gravitational field*, Nature 415, pp. 297–299, 17 January 2002. Longer explanations are found in V. V. NESVIZHEVSKY, H. G. BOERNER, A. M. GAGARSKY, A. K. PETOUKHOV, G. A. PETROV, H. ABELE, S. BAESSLER, G. DIVKOVIC, F. J. RUESS, TH. STOEFERLE, A. WESTPHAL, A. V. STRELKOV, K. V. PROTASOV & A. YU. VORONIN, *Measurement of quantum states of neutrons in the Earth's gravitational field*, Physical Review D 67, p. 102002, 2003, preprint at arxiv.org/abs/hep-ph/0306198, as well as V. V. NESVIZHEVSKY, A. K. PETUKHOV, H. G. BOERNER, T. A. BARANOVA, A. M. GAGARSKI, G. A. PETROV, K. V. PROTASOV, A. YU. VORONIN, S. BAESSLER, H. ABELE, A. WESTPHAL & L. LUCOVAC, *Study of the neutron quantum states in the gravity field*, European Physical Journal C 40, pp. 479–491, 2005, preprint at arxiv.org/abs/hep-ph/0502081 and A. YU. VORONIN, H. ABELE, S. BAESSLER, V. V. NESVIZHEVSKY, A. K. PETUKHOV, K. V. PROTASOV & A. WESTPHAL, *Quantum motion of a neutron in a wave-guide in the gravitational field*, Physical Review D 73, p. 044029, 2006, preprint at arxiv.org/abs/quant-ph/0512129. Cited on page 125.

110 H. RAUCH, W. TREIMER & U. BONSE, *Test of a single crystal neutron interferometer*, Physics Letters A 47, pp. 369–371, 1974. Cited on page 126.

111 R. COLELLA, A. W. OVERHAUSER & S. A. WERNER, *Observation of gravitationally induced quantum mechanics*, Physical Review Letters 34, pp. 1472–1475, 1975. This experiment is usually called the COW experiment. Cited on page 126.

112 CH. J. BORDÉ & C. LÄMMERZAHL, *Atominterferometrie und Gravitation*, Physikalische Blätter 52, pp. 238–240, 1996. See also A. PETERS, K. Y. CHUNG & S. CHU, *Observation of gravitational acceleration by dropping atoms*, Nature 400, pp. 849–852, 26 August 1999. Cited on page 126.

113 J. D. BEKENSTEIN, *Black holes and entropy*, Physical Review D 7, pp. 2333–2346, 1973. Cited on page 129.

114 R. BOUSSO, *The holographic principle*, Review of Modern Physics 74, pp. 825–874, 2002, also available as arxiv.org/abs/hep-th/0203101. The paper is an excellent review; however, it has some argumentation errors, as explained on page 130. Cited on page 130.

115 This is clearly argued by S. CARLIP, *Black hole entropy from horizon conformal field theory*, Nuclear Physics B Proceedings Supplement 88, pp. 10–16, 2000. Cited on page 130.

116 S. A. FULLING, *Nonuniqueness of canonical field quantization in Riemannian space-time*, Physical Review D 7, pp. 2850–2862, 1973, P. DAVIES, *Scalar particle production in Schwarzchild and Rindler metrics*, Journal of Physics A: General Physics 8, pp. 609–616, 1975, W. G. UNRUH, *Notes on black hole evaporation*, Physical Review D 14, pp. 870–892, 1976. Cited on page 130.

117 About the possibility to measure Fulling–Davies–Unruh radiation directly, see for example the paper by H. ROSU, *On the estimates to measure Hawking effect und Unruh effect in the laboratory*, International Journal of Modern Physics D3 p. 545, 1994. arxiv.org/abs/gr-qc/9605032 or the paper P. CHEN & T. TAJIMA, *Testing Unruh radiation with ultra-intense lasers*, Physical Review Letters 83, pp. 256–259, 1999. Cited on page 131.

118 E. T. AKHMEDOV & D. SINGLETON, *On the relation between Unruh and Sokolov–Ternov effects*, preprint at www.arxiv.org/abs/hep-ph/0610394; see also E. T. AKHMEDOV & D. SINGLETON, *On the physical meaning of the Unruh effect*, preprint at www.arxiv.org/abs/0705.2525. Unfortunately, as Bell and Leinaas mention, the Sokolov–Ternov effect cannot be used to confirm the expression of the black hole entropy, because the radiation is not thermal, and so a temperature of the radiation is hard to define. Nevertheless, C. Schiller predicts (October 2009) that careful analysis of the spectrum could be used (1) to check the proportionality of entropy and area in black holes, and (2) to check the transformation of temperature in special relativity. Cited on page 131.

119 W. G. UNRUH & R. M. WALD, *Acceleration radiation and the generalised second law of thermodynamics*, Physical Review D 25, pp. 942–958 , 1982. Cited on page 132.

120 R. M. WALD, *The thermodynamics of black holes*, Living Reviews of Relativity 2001, www-livingreviews.org/lrr-2001-6. Cited on page 134.

121 For example, if neutrinos were massless, they would be emitted by black holes more frequently than photons. For a delightful popular account from a black hole expert, see IGOR NOVIKOV, *Black Holes and the Universe*, Cambridge University Press 1990. Cited on pages 134 and 138.

122 The original paper is W. G. UNRUH, *Experimental black hole evaporation?*, Physical Review Letters 46, pp. 1351–1353, 1981. A good explanation with good literature overview is the one by MATT VISSER, *Acoustical black holes: horizons, ergospheres and Hawking radiation*, arxiv.org/abs/gr-qc/9712010. Cited on page 135.

123 Optical black holes are explored in W. G. UNRUH & R. SCHÜTZHOLD, *On Slow Light as a Black Hole Analogue*, Physical Review D 68, p. 024008, 2003, preprint at arxiv.org/abs/gr-qc/0303028. Cited on page 135.

124 T. DAMOUR & R. RUFFINI, *Quantum electrodynamical effects in Kerr–Newman geometries*, Physical Review Letters 35, pp. 463–466, 1975. Cited on page 137.

125 These were the Vela satellites; their existence and results were announced officially only in 1974, even though they were working already for many years. Cited on page 135.

126 An excellent general introduction into the topic of gamma ray bursts is S. KLOSE, J. GREINER & D. HARTMANN, *Kosmische Gammastrahlenausbrüche – Beobachtungen und Modelle*, Teil I und II, Sterne und Weltraum March and April 2001. Cited on pages 136 and 137.

127 When the gamma ray burst encounters the matter around the black hole, it is broadened. The larger the amount of matter, the broader the pulse is. See G. PREPARATA, R. RUFFINI & S. -S. XUE, *The dyadosphere of black holes and gamma-ray bursts*, Astronomy and Astrophysics 338, pp. L87–L90, 1998, R. RUFFINI, J. D. SALMONSON, J. R. WILSON & S. -S. XUE, *On the pair electromagnetic pulse of a black hole with electromagnetic structure*, Astronomy and Astrophysics 350, pp. 334–343, 1999, R. RUFFINI, J. D. SALMONSON, J. R. WILSON & S. -S. XUE, *On the pair electromagnetic pulse from an electromagnetic black hole surrounded by a baryonic remnant*, Astronomy and Astrophysics 359, pp. 855–864, 2000, and C. L. BIANCO, R. RUFFINI & S. -S. XUE, *The elementary spike produced by a pure e^+e^- pair-electromagnetic pulse from a black hole: the PEM pulse*, Astronomy and Astrophysics 368, pp. 377–390, 2001. For a very personal account by Ruffini on his involvement in gamma ray bursts, see his paper *Black hole formation and gamma ray bursts*, arxiv.org/abs/astro-ph/0001425. Cited on page 137.

128 arxiv.org/abs/astro-ph/0301262, arxiv.org/abs/astro-ph/0303539, D. W. FOX & al., *Early optical emission from the γ-ray burst of 4 October 2002*, Nature 422, pp. 284–286, 2003. Cited on page 136.

129 Negative heat capacity has also been found in atom clusters and in nuclei. See, e.g., M. SCHMIDT & al., *Negative heat capacity for a cluster of 147 sodium atoms*, Physical Review Letters 86, pp. 1191–1194, 2001. Cited on page 138.

130 H. -P. NOLLERT, *Quasinormal modes: the characteristic 'sound' of black holes and neutron stars*, Classical and Quantum Gravity 16, pp. R159–R216, 1999. Cited on page 138.

131 About the membrane description of black holes, see KIP S. THORNE, RICHARD H. PRICE & DOUGLAS A. MACDONALD, editors, *Black Holes: the Membrane Paradigm*, Yale University Press, 1986. Cited on page 138.

132 Page wrote a series of papers on the topic; a beautiful summary is DON N. PAGE, *How fast does a black hole radiate information?*, International Journal of Modern Physics 3, pp. 93–106, 1994, which is based on his earlier papers, such as *Information in black hole radiation*, Physical Review Letters 71, pp. 3743–3746, 1993. See also his preprint at arxiv.org/abs/hep-th/9305040. Cited on page 139.

133 See DON N. PAGE, *Average entropy of a subsystem*, Physical Review Letters 71, pp. 1291–1294, 1993. The entropy formula of this paper, used above, was proven by S. K. FOONG & S. KANNO, *Proof of Page's conjecture on the average entropy of a subsystem*, Physical Review Letters 72, pp. 1148–1151, 1994. Cited on page 140.

134 R. LAFRANCE & R. C. MYERS, *Gravity's rainbow: limits for the applicability of the equivalence principle*, Physical Review D 51, pp. 2584–2590, 1995, arxiv.org/abs/hep-th/9411018. Cited on page 141.

135 M. YU. KUCHIEV & V. V. FLAMBAUM, *Scattering of scalar particles by a black hole*, arxiv.org/abs/gr-qc/0312065. See also M. YU. KUCHIEV & V. V. FLAMBAUM, *Reflection on event horizon and escape of particles from confinement inside black holes*, arxiv.org/abs/gr-qc/0407077. Cited on page 141.

136 See the widely cited but wrong paper by G. C. GHIRARDI, A. RIMINI & T. WEBER, *Unified dynamics for microscopic and macroscopic systems*, Physical Review D 34, pp. 470–491, 1986. Cited on page 127.

137 I speculate that this version of the coincidences could be original; I have not found it in the literature. Cited on page 128.

138 STEVEN WEINBERG, *Gravitation and Cosmology*, Wiley, 1972. See equation 16.4.3 on page 619 and also page 620. Cited on page 128.

139 It could be that knot theory provides a relation between a local knot invariant, related to particles, and a global one. Cited on page 128.

140 This point was made repeatedly by STEVEN WEINBERG, namely in *Derivation of gauge invariance and the equivalence principle from Lorentz invariance of the S-matrix*, Physics Letters 9, pp. 357–359, 1964, in *Photons and gravitons in S-matrix theory: derivation of charge conservation and equality of gravitational and inertial mass*, Physical Review B 135, pp. 1049–1056, 1964, and in *Photons and gravitons in perturbation theory: derivation of Maxwell's and Einstein's equations*, Physical Review B 138, pp. 988–1002, 1965. Cited on page 142.

141 E. JOOS, *Why do we observe a classical space-time?*, Physics Letters A 116, pp. 6–8, 1986. Cited on page 143.

142 L. H. FORD & T. A. ROMAN, *Quantum field theory constrains traversable wormhole geometries*, arxiv.org/abs/gr-qc/9510071 or Physical Review D 53, pp. 5496–5507, 1996. Cited on page 143.

143 B. S. KAY, M. RADZIKOWSKI & R. M. WALD, *Quantum field theory on spacetimes with a compactly generated Cauchy horizon*, arxiv.org/abs/gr-qc/9603012 or Communications in Mathematical Physics 183 pp. 533–556, 1997. Cited on page 143.

144 M. J. PFENNING & L. H. FORD, *The unphysical nature of 'warp drive'*, Classical and Quantum Gravity 14, pp. 1743–1751, 1997. Cited on page 143.

145 A excellent technical introduction to nuclear physics is BOGDAN POVH, KLAUS RITH, CHRISTOPH SCHOLZ & FRANK ZETSCHE, *Teilchen und Kerne*, Springer, 5th edition, 1999. It is also available in English translation. Cited on page 146.

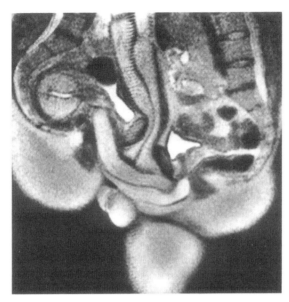

FIGURE 175 The wonderful origin of human life (© W.C.M. Weijmar Schultz).

146 For magnetic resonance imaging films of the heart beat, search for "cardiac MRI" on the internet. See for example www.youtube.com/watch?v=58l6oFhfZU. Cited on page 148.

147 C. Bamberg, G. Rademacher, F. Güttler, U. Teichgräber, M. Cremer, C. Bührer, C. Spies, L. Hinkson, W. Henrich, K. D. Kalache & J. W. Dudenhausen, *Human birth observed in real-time open magnetic resonance imaging*, American Journal of Obstetrics & Gynecology **206**, p. 505e1-505e6, 2012. Cited on page 148.

148 W.C.M. Weijmar Schultz & al., *Magnetic resonance imaging of male and female genitals during coitus and female sexual arousal*, British Medical Journal **319**, pp. 1596–1600, December 18, 1999, available online as www.bmj.com/cgi/content/full/319/7225/1596. Cited on page 148.

149 M. Chantell, T. C. Weekes, X. Sarazin & M. Urban, *Antimatter and the moon*, Nature **367**, p. 25, 1994. M. Amenomori & al., *Cosmic ray shadow by the moon observed with the Tibet air shower array*, Proceedings of the 23rd International Cosmic Ray Conference, Calgary **4**, pp. 351–354, 1993. M. Urban, P. Fleury, R. Lestienne & F. Plouin, *Can we detect antimatter from other galaxies by the use of the Earth's magnetic field and the Moon as an absorber?*, Nuclear Physics, Proceedings Supplement **14B**, pp. 223–236, 1990. Cited on page 161.

150 Joseph Magill & Jean Galy, *Radioactivity Radionuclides Radiation*, Springer, 2005. This dry but dense book contains most data about the topic, including much data on all the known nuclides. Cited on pages 165 and 181.

151 An good summary on radiometric dating is by R. Wiens, *Radiometric dating – a christian perpective*, www.asa3.org/ASA/resources/Wiens.html. The absurd title is due to the habit in many religious circles to put into question radiometric dating results. Putting apart the few religious statements in the review, the content is well explained. Cited on page 167.

152 See for example the excellent and free lecture notes by Heinz-Günther Stosch, *Einführung in die Isotopengeochemie*, 2004, available on the internet. Cited on page 167.

153 G. B. Dalrymple, *The age of the Earth in the twentieth century: a problem (mostly) solved,*

Special Publications, Geological Society of London **190**, pp. 205–221, 2001. Cited on page 169.

154 A good overview is given by A. N. HALLIDAY, *Radioactivity, the discovery of time and the earliest history of the Earth*, Contemporary Physics **38**, pp. 103–114, 1997. Cited on page 169.

155 J. DUDEK, A. GOD, N. SCHUNCK & M. MIKIEWICZ, *Nuclear tetrahedral symmetry: possibly present throughout the periodic table*, Physical Review Letters **88**, p. 252502, 24 June 2002. Cited on page 171.

156 A good introduction is R. CLARK & B. WODSWORTH, *A new spin on nuclei*, Physics World pp. 25–28, July 1998. Cited on page 171.

157 JOHN HORGAN, *The End of Science – Facing the Limits of Knowledge in the Twilight of the Scientific Age*, Broadway Books, 1997, chapter 3, note 1. Cited on page 175.

158 G. CHARPAK & R. L. GARWIN, *The DARI*, Europhysics News **33**, pp. 14–17, January/February 2002. Cited on page 177.

159 M. BRUNETTI, S. CECCHINI, M. GALLI, G. GIOVANNINI & A. PAGLIARIN, *Gamma-ray bursts of atmospheric origin in the MeV energy range*, Geophysical Research Letters **27**, p. 1599, 2000. Cited on page 179.

160 A book with nuclear explosion photographs is MICHAEL LIGHT, *100 Suns*, Jonathan Cape, 2003. Cited on page 180.

161 *A conversation with Peter Mansfield*, Europhysics Letters **37**, p. 26, 2006. Cited on page 177.

162 An older but still fascinating summary of solar physics is R. KIPPENHAHN, *Hundert Milliarden Sonnen*, Piper, 1980. It was a famous bestseller and is available also in English translation. Cited on page 183.

163 H. BETHE, *On the formation of deuterons by proton combination*, Physical Review **54**, pp. 862–862, 1938, and H. BETHE, *Energy production in stars*, Physical Review **55**, pp. 434–456, 1939. Cited on page 183.

164 M. NAUENBERG & V. F. WEISSKOPF, *Why does the sun shine?*, American Journal of Physics **46**, pp. 23–31, 1978. Cited on page 184.

165 The slowness of the speed of light inside the Sun is due to the frequent scattering of photons by solar matter. The most serious estimate is by R. MITALAS & K. R. SILLS, *On the photon diffusion time scale for the Sun*, The Astrophysical Journal **401**, pp. 759–760, 1992. They give a photon escape time of 0.17 Ma, an average photon free mean path of 0.9 mm and an average speed of 0.97 cm/s. See also the interesting paper by M. STIX, *On the time scale of energy transport in the sun*, Solar Physics **212**, pp. 3–6, 2003, which comes to the conclusion that the speed of energy transport is 30 Ma, two orders of magnitude higher than the photon diffusion time. Cited on page 185.

166 See the freely downloadable book by JOHN WESSON, *The Science of JET - The Achievements of the Scientists and Engineers Who Worked on the Joint European Torus 1973-1999*, JET Joint Undertaking, 2000, available at www.jet.edfa.org/documents/wesson/wesson.html. Cited on page 190.

167 J. D. LAWSON, *Some criteria for a power producing thermonuclear reactor*, Proceedings of the Physical Society, London B **70**, pp. 6–10, 1957. The paper had been kept secret for two years. However, the result was already known, before Lawson, to all Russian nuclear physicists several years earlier. Cited on page 191.

168 The classic paper is R. A. ALPHER, H. BETHE & G. GAMOW, *The Origin of Chemical Elements*, Physical Review **73**, pp. 803–804, 1948. Cited on page 193.

169 The standard reference is E. ANDERS & N. GREVESSE, *Abundances of the elements - meteoritic and solar*, Geochimica et Cosmochimica Acta 53, pp. 197–214, 1989. Cited on page 195.

170 S. GORIELY, A. BAUSWEIN & H. T. JANKA, *R-process nucleosynthesis in dynamically ejected matter of neutron star mergers*, Astrophysical Journal 738, p. L38, 2011, preprint at arxiv.org/abs/1107.0899. Cited on pages 193 and 196.

171 Kendall, Friedman and Taylor received the 1990 Nobel Prize for Physics for a series of experiments they conducted in the years 1967 to 1973. The story is told in the three Nobel lectures R. E. TAYLOR, *Deep inelastic scattering: the early years*, Review of Modern Physics 63, pp. 573–596, 1991, H. W. KENDALL, *Deep inelastic scattering: Experiments on the proton and the observation of scaling*, Review of Modern Physics 63, pp. 597–614, 1991, and J. I. FRIEDMAN, *Deep inelastic scattering: Comparisons with the quark model*, Review of Modern Physics 63, pp. 615–620, 1991. Cited on page 198.

172 G. ZWEIG, *An SU3 model for strong interaction symmetry and its breaking II*, CERN Report No. 8419TH. 412, February 21, 1964. Cited on page 198.

173 About the strange genius of Gell-Mann, see the beautiful book by GEORGE JOHNSON, *Murray Gell-Mann and the Revolution in Twentieth-Century Physics*, Knopf, 1999. Cited on page 198.

174 The best introduction might be the wonderfully clear text by DONALD H. PERKINS, *Introduction to High Energy Physics*, Cambridge University Press, fourth edition, 2008. Also beautiful, with more emphasis on the history and more detail, is KURT GOTTFRIED & VICTOR F. WEISSKOPF, *Concepts of Particle Physics*, Clarendon Press, Oxford, 1984. Victor Weisskopf was one of the heroes of the field, both in theoretical research and in the management of CERN, the European organization for particle research. Cited on pages 200, 210, and 377.

175 The official reference for all particle data, worth a look for every physicist, is the massive collection of information compiled by the Particle Data Group, with the website pdg.web.cern.ch containing the most recent information. A printed review is published about every two years in one of the major journals on elementary particle physics. See for example C. AMSLER & al., *The Review of Particle Physics*, Physics Letters B 667, p. 1, 2008. For some measured properties of these particles, the official reference is the set of so-called CODATA values given in reference Ref. 273. Cited on pages 201, 211, 212, 213, 214, 223, 229, 233, and 240.

176 H. FRITSCH, M. GELL-MANN & H. LEUTWYLER, *Advantages of the color octet picture*, Physics Letters B 47, pp. 365–368, 1973. Cited on page 202.

177 Quantum chromodynamics can be explored in books of several levels. For the first, popular level, see the clear longseller by one of its founders, HARALD FRITZSCH, *Quarks – Urstoff unserer Welt*, Piper Verlag, 2006, or, in English language, *Quarks: the Stuff of Matter*, Penguin Books, 1983. At the second level, the field can be explored in texts on high energy physics, as those of Ref. 174. The third level contains books like KERSON HUANG, *Quarks, Leptons and Gauge Fields*, World Scientific, 1992, or FELIX J. YNDURÁIN, *The Theory of Quark and Gluon Interactions*, Springer Verlag, 1992, or WALTER GREINER & ANDREAS SCHÄFER, *Quantum Chromodynamics*, Springer Verlag, 1995, or the modern and detailed text by STEPHAN NARISON, *QCD as a Theory of Hadrons*, Cambridge University Press, 2004. As always, a student has to discover by himself or herself which text is most valuable. Cited on pages 203, 209, and 211.

178 S. DÜRR & al., *Ab initio determination of the light hadron masses*, Science 322, pp. 1224–1227, 2008. Cited on page 207.

179 See for example C. BERNARD & al., *Light hadron spectrum with Kogut–Susskind quarks*, Nuclear Physics, Proceedings Supplement 73, p. 198, 1999, and references therein. Cited on page 207.

180 R. BRANDELIK & al., *Evidence for planar events in e^+e^- annihilation at high energies*, Physics Letters B 86, pp. 243–249, 1979. Cited on page 215.

181 For a pedagogical introduction to lattice QCD calculations, see R. GUPTA, *Introduction to Lattice QCD*, preprint at arxiv.org/abs/hep-lat/9807028, or the clear introduction by MICHAEL CREUTZ, *Quarks, Gluons and Lattices*, Cambridge University Press, 1983. Cited on page 210.

182 S. BETHKE, *Experimental tests of asymptotic freedom*, Progress in Particle and Nuclear Physics 58, pp. 351–368, 2007, preprint at arxiv.org/abs/hep-ex/0606035. Cited on page 211.

183 F. ABE & al., *Measurement of dijet angular distributions by the collider detector at Fermilab*, Physical Review Letters 77, pp. 5336–5341, 1996. Cited on page 212.

184 The approximation of QCD with zero mass quarks is described by F. WILCZEK, *Getting its from bits*, Nature 397, pp. 303–306, 1999. It is also explained in F. WILCZEK, *Asymptotic freedom*, Nobel lecture 2004. The proton's mass is set by the energy scale at which the strong coupling, coming from its value at Planck energy, becomes of order unity. Cited on pages 212, 213, and 236.

185 A. J. BUCHMANN & E. M. HENLEY, *Intrinsic quadrupole moment of the nucleon*, Physical Review C 63, p. 015202, 2000. Alfons Buchmann also predicts that the quadrupole moment of the other, strange $J = 1/2$ octet baryons is positive, and predicts a prolate structure for all of them (private communication). For the decuplet baryons, with $J = 3/2$, the quadrupole moment can often be measured spectroscopically, and is always negative. The four Δ baryons are thus predicted to have a negative intrinsic quadrupole moment and thus an oblate shape. This explained in A. J. BUCHMANN & E. M. HENLEY, *Quadrupole moments of baryons*, Physical Review D 65, p. 073017, 2002. For recent updates, see A. J. BUCHMANN, *Charge form factors and nucleon shape*, pp. 110–125, in the Shape of Hadrons Workshop Conference, Athens Greece, 27-29 April 2006, AIP Conference Proceedings 904, Eds. C.N. Papanicolas, Aron Bernstein. For updates on other baryons, see A. J. BUCHMANN, *Structure of strange baryons*, Hyp 2006, International Conference on Hypernuclear and Strange Particle Physics, Oct.10-14, Mainz, Germany, to be published in European Physics Journal A 2007. The topic is an active field of research; for example, results on magnetic octupole moments are expected soon. Cited on page 213.

186 S. STRAUCH & al., *Polarization transfer in the ^{4}He $(e,e'p)$ ^{3}H reaction up to $Q^2 = 2.6(GeV/c)^2$*, Physical Review Letters 91, p. 052301, 2003. Cited on page 213.

187 F. CLOSE, *Glueballs and hybrids: new states of matter*, Contemporary Physics 38, pp. 1–12, 1997. See also the next reference. Cited on page 216.

188 A tetraquark is thought to be the best explanation for the $f_0(980)$ resonance at 980 MeV. The original proposal of this explanation is due to R. L. JAFFE, *Multiquark hadrons I: phenomenology of $Q^2\overline{Q}^2$ mesons*, Physical Review D 15, pp. 267–280, 1977, R. L. JAFFE, *Multiquark hadrons II: methods*, Physical Review D 15, pp. 281–289, 1977, and R. L. JAFFE, Physical Review D $Q^2\overline{Q}^2$ *resonances in the baryon-antibaryon system*, 17, pp. 1444–1458, 1978. For a clear and detailed modern summary, see the excellent review by E. KLEMPT & A. ZAITSEV, *Glueballs, hybrids, multiquarks: experimental facts versus QCD inspired concepts*, Physics Reports 454, pp. 1–202, 2007, preprint at arxiv.org/abs/0708.4016. See also F. GIACOSA, *Light scalars as tetraquarks*, preprint at arxiv.org/abs/0711.3126, and

V. CREDE & C. A. MEYER, *The experimental status of glueballs*, preprint at arxiv.org/abs/ 0812.0600. Cited on page 216.

189 Pentaquarks were first predicted by Maxim Polyakov, Dmitri Diakonov, and Victor Petrov in 1997. Two experimental groups in 2003 claimed to confirm their existence, with a mass of 1540 MeV; see K. HICKS, *An experimental review of the Θ^+ pentaquark*, arxiv.org/abs/ hep-ex/0412048. Results from 2005 and later, however, ruled out that the 1540 MeV particle is a pentaquark. Cited on page 216.

190 See, for example, YA. B. ZEL'DOVICH & V. S. POPOV, *Electronic structure of superheavy atoms*, Soviet Physics Uspekhi 17, pp. 673–694, 2004 in the English translation. Cited on page 118.

191 J. TRAN THANH VAN, editor, *CP violation in Particle Physics and Astrophysics*, Proc. Conf. Chateau de Bois, France, May 1989, Editions Frontières, 1990. Cited on page 223.

192 P.L. ANTHONY & al., *Observation of parity nonconservation in Møller scattering*, Physical Review Letters **92**, p. 181602, 2004. Cited on page 224.

193 M. A. BOUCHIAT & C. C. BOUCHIAT, *Weak neutral currents in atomic physics*, Physics Letters B 48, pp. 111–114, 1974. U. AMALDI, A. BÖHM, L. S. DURKIN, P. LANGACKER, A. K. MANN, W. J. MARCIANO, A. SIRLIN & H. H. WILLIAMS, *Comprehensive analysis of data pertaining to the weak neutral current and the intermediate-vector-boson masses*, Physical Review D 36, pp. 1385–1407, 1987. Cited on page 224.

194 M. C. NOECKER, B. P. MASTERSON & C. E. WIEMANN, *Precision measurement of parity nonconservation in atomic cesium: a low-energy test of electroweak theory*, Physical Review Letters **61**, pp. 310–313, 1988. See also D.M. MEEKHOF & al., *High-precision measurement of parity nonconserving optical rotation in atomic lead*, Physical Review Letters 71, pp. 3442–3445, 1993. Cited on page 225.

195 S. C. BENNET & C. E. WIEMANN, *Measurement of the 6S – 7S transition polarizability in atomic cesium and an improved test of the standard model*, Physical Review Letters **82**, pp. 2484–2487, 1999. The group has also measured the spatial distribution of the weak charge, the so-called the *anapole moment*; see C.S. WOOD & al., *Measurement of parity nonconservation and an anapole moment in cesium*, Science 275, pp. 1759–1763, 1997. Cited on page 225.

196 C. JARLSKOG, *Commutator of the quark mass matrices in the standard electroweak model and a measure of maximal CP nonconservation*, Physical Review Letters 55, pp. 1039–1042, 1985. Cited on page 228.

197 The correct list of citations is a topic of intense debate. It surely includes Y. NAMBU & G. JONA-LASINIO, *Dynamical model of the elementary particles based on an analogy with superconductivity - I*, Physical Review **122**, p. 345-358, 1961, P. W. ANDERSON, *Plasmons, gauge invariance, and mass*, Physical Review 130, pp. 439–442, 1963, The list then continues with reference Ref. 199. Cited on pages 230 and 380.

198 K. GROTZ & H. V. KLAPDOR, *Die schwache Wechselwirkung in Kern-, Teilchen- und Astrophysik*, Teubner Verlag, Stuttgart, 1989. Also available in English and in several other languages. Cited on page 233.

199 P. W. HIGGS, *Broken symmetries, massless particles and gauge fields*, Physics Letters 12, pp. 132–133, 1964, P. W. HIGGS, *Broken symmetries and the masses of the gauge bosons*, Physics Letters 13, pp. 508–509, 1964. He then expanded the story in P. W. HIGGS, *Spontaneous symmetry breakdown without massless bosons*, Physical Review 145, pp. 1156–1163, 1966. Higgs gives most credit to Anderson, instead of to himself; he also mentions Brout and Englert, Guralnik, Hagen, Kibble and 't Hooft. These papers are F. ENGLERT & R. BROUT,

Broken symmetry and the mass of the gauge vector mesons, Physics Review Letters 13, pp. 321–323, 1964, G. S. GURALNIK, C. R. HAGEN & T. W. B. KIBBLE, *Global conservations laws and massless particles*, Physical Review Letters 13, pp. 585–587, 1964, T. W. B. KIBBLE, *Symmetry breaking in non-Abelian gauge theories*, Physical Review 155, pp. 1554–1561, 1967. For the ideas that inspired all these publications, see Ref. 197. Cited on pages 235 and 379.

200 D. TREILLE, *Particle physics from the Earth and from the sky: Part II*, Europhysics News 35, no. 4, 2004. Cited on page 234.

201 Rumination is studied in P. JORDAN & DE LAER KRONIG, in Nature 120, p. 807, 1927. Cited on page 236.

202 K. W. D. LEDINGHAM & al., *Photonuclear physics when a multiterawatt laser pulse interacts with solid targets*, Physical Review Letters 84, pp. 899–902, 2000. K. W. D. LEDINGHAM & al., *Laser-driven photo-transmutation of Iodine-129 – a long lived nuclear waste product*, Journal of Physics D: Applied Physics 36, pp. L79–L82, 2003. R. P. SINGHAL, K. W. D. LEDINGHAM & P. MCKENNA, *Nuclear physics with ultra-intense lasers – present status and future prospects*, Recent Research Developments in Nuclear Physics 1, pp. 147–169, 2004. Cited on page 237.

203 The electron radius limit is deduced from the $g - 2$ measurements, as explained in the Nobel Prize talk by HANS DEHMELT, *Experiments with an isolated subatomic particle at rest*, Reviews of Modern Physics 62, pp. 525–530, 1990, or in HANS DEHMELT, *Is the electron a composite particle?*, Hyperfine Interactions 81, pp. 1–3, 1993. Cited on page 242.

204 G. GABRIELSE, H. DEHMELT & W. KELLS, *Observation of a relativistic, bistable hysteresis in the cyclotron motion of a single electron*, Physical Review Letters 54, pp. 537–540, 1985. Cited on page 242.

205 For the bibliographic details of the latest print version of the *Review of Particle Physics*, see Appendix B. The online version can be found at pdg.web.cern.ch. The present status on grand unification can also be found in the respective section of the overview. Cited on pages 246, 247, and 248.

206 Slides of a very personal review talk by H. GEORGI, *The future of grand unification*, can be found at www2.yukawa.kyoto-u.ac.jp/~yt100sym/files/yt100sym_georgi.pdf. A modern research approach is S. RABY, *Grand unified theories*, arxiv.org/abs/hep-ph/0608183. Cited on page 247.

207 H. JEON & M. LONGO, *Search for magnetic monopoles trapped in matter*, Physical Review Letters 75, pp. 1443–1447, 1995. Cited on page 247.

208 The quantization of charge as a consequence of magnetic monopoles is explained, e.g., in … Cited on page 247.

209 On proton decay rates, see the latest data of the Particle Data Group, at pdg.web.cern.ch. Cited on page 248.

210 U. AMALDI, DE BOER & H. FÜRSTENAU, *Comparison of grand unified theories with electroweak and strong coupling constants measured at LEP*, Physics Letters 260, pp. 447–455, 1991. This widely cited paper is the standard reference for this issue. An update is found in DE BOER & C. SANDER, *Global electroweak fits and gauge coupling unification*, Physics Letters B 585 pp. 276–286, 2004, or preprint at arxiv.org/abs/hep-ph/0307049. The figure is taken with permission from the home page of Wim de Boer www-ekp.physik.uni-karlsruhe.de/~deboer/html/Forschung/forschung.html. Cited on page 248.

211 PETER G. O. FREUND, *Introduction to Supersymmetry*, Cambridge 1988. JULIUS WESS & JONATHAN BAGGER, *Supersymmetry and Supergravity*, Princeton University Press, 1992. This widely cited book contains a lot of mathematics but little physics. Cited on page 249.

212 S. COLEMAN & J. MANDULA, *All possible symmetries of the S matrix*, Physical Review 159, pp. 1251–1256, 1967. Cited on page 250.

213 P. C. ARGYRES, *Dualities in supersymmetric field theories*, Nuclear Physics Proceedings Supplement **61A**, pp. 149–157, 1998, preprint available at arxiv.org/abs/hep-th/9705076. Cited on page 252.

214 MICHAEL STONE editor, *Bosonization*, World Scientific, 1994. R. RAJARAMAN, *Solitons and Instantons*, North Holland, 1987. Cited on page 253.

215 In 1997, the smallest human-made flying object was the helicopter built by a group of the Institut für Mikrotechnik in Mainz, in Germany. A picture is available at their web page, to be found at www.imm-mainz.de/English/billboard/f_hubi.html. The helicopter is 24 mm long, weighs 400 mg and flies (though not freely) using two built-in electric motors driving two rotors, running at between 40 000 and 100 000 revolutions per minute. See also the helicopter from Stanford University at www-rpl.stanford.edu/RPL/htmls/mesoscopic/mesicopter/mesicopter.html, with an explanation of its battery problems. Cited on page 257.

216 HENK TENNEKES, *De wetten van de vliegkunst – over stijgen, dalen, vliegen en zweven*, Aramith Uitgevers, 1992. This clear and interesting text is also available in English. Cited on page 257.

217 The most recent computational models of lift still describe only two-dimensional wing motion, e.g., Z. J. WANG, *Two dimensional mechanism for insect hovering*, Physical Review Letters 85 pp. 2216–2219, 2000. A first example of a mechanical bird has been constructed by Wolfgang Send; it can be studied on the www.aniprop.de website. See also W. SEND, *Physik des Fliegens*, Physikalische Blätter 57, pp. 51–58, June 2001. Cited on page 258.

218 R. B. SRYGLEY & A. L. R. THOMAS, *Unconventional lift-generating mechanisms in free-flying butterflies*, Nature 420, pp. 660–664, 2002. Cited on page 258.

219 The book by JOHN BRACKENBURY, *Insects in Flight*, 1992. is a wonderful introduction into the biomechanics of insects, combining interesting science and beautiful photographs. Cited on page 260.

220 The simulation of insect flight using enlarged wing models flapping in oil instead of air is described for example in www.dickinson.caltech.edu/research_robofly.html. The higher viscosity of oil allows achieving the same Reynolds number with larger sizes and lower frequencies than in air. See also the 2013 video of the talk on insect flight at www.ted.com/talks/michael_dickinson_how_a_fly_flies.html. Cited on page 260.

221 E. PURCELL, *Life at low Reynolds number*, American Journal of Physics 45, p. 3, 1977. Cited on page 261.

222 Most bacteria are flattened, ellipsoidal sacks kept in shape by the membrane enclosing the cytoplasma. But there are exceptions; in salt water, quadratic and triangular bacteria have been found. More is told in the corresponding section in the interesting book by BERNARD DIXON, *Power Unseen – How Microbes Rule the World*, W.H. Freeman, New York, 1994. Cited on page 262.

223 S. PITNICK, G. SPICER & T. A. MARKOW, *How long is a giant sperm?*, Nature 375, p. 109, 1995. Cited on page 262.

224 M. KAWAMURA, A. SUGAMOTO & S. NOJIRI, *Swimming of microorganisms viewed from string and membrane theories*, Modern Journal of Physics Letters A 9, pp. 1159–1174, 1994. Also available as arxiv.org/abs/hep-th/9312200. Cited on page 262.

225 W. NUTSCH & U. RÜFFER, *Die Orientierung freibeweglicher Organismen zum Licht, dargestellt am Beispiel des Flagellaten Chlamydomonas reinhardtii*, Naturwissenschaften 81, pp. 164–174, 1994. Cited on page 262.

226 They are also called *prokaryote flagella*. See for example S. C. Schuster & S. Khan, *The bacterial flagellar motor*, Annual Review of Biophysics and Biomolecular Structure 23, pp. 509–539, 1994, or S. R. Caplan & M. Kara-Ivanov, *The bacterial flagellar motor*, International Review of Cytology 147, pp. 97–164, 1993. See also the information on the topic that can be found on the website www.id.ucsb.edu:16080/fscf/library/origins/graphics-captions/flagellum.html. Cited on page 262.

227 For an overview of the construction and the motion of coli bacteria, see H. C. Berg, *Motile behavior of bacteria*, Physics Today 53, pp. 24–29, January 2000. Cited on page 262.

228 J. W. Shaevitz, J. Y. Lee & D. A. Fletcher, *Spiroplasma swim by a processive change in body helicity*, Cell 122, pp. 941–945, 2005. Cited on page 263.

229 This is from the book by David Dusenbery, *Life at a Small Scale*, Scientific American Library, 1996. Cited on page 263.

230 F. Wilczek & A. Zee, *Appearance of gauge structures in simple dynamical systems*, Physical Review Letters 52, pp. 2111–2114, 1984, A. Shapere & F. Wilczek, *Self-propulsion at low Reynold number*, Physical Review Letters 58, pp. 2051–2054, 1987, A. Shapere & F. Wilczek, *Gauge kinematics of deformable bodies*, American Journal of Physics 57, pp. 514–518, 1989, A. Shapere & F. Wilczek, *Geometry of self-propulsion at low Reynolds number*, Journal of Fluid Mechanics 198, pp. 557–585, 1989, A. Shapere & F. Wilczek, *Efficiencies of self-propulsion at low Reynolds number*, Journal of Fluid Mechanics 198, pp. 587–599, 1989. See also R. Montgomery, *Gauge theory of the falling cat*, Field Institute Communications 1, pp. 75–111, 1993. Cited on page 264.

231 E. Putterman & O. Raz, *The square cat*, American Journal of Physics 76, pp. 1040–1045, 2008. Cited on page 264.

232 J. Wisdom, *Swimming in spacetime: motion by cyclic changes in body shape*, Science 299, pp. 1865–1869, 21st of March, 2003. A similar effect was discovered later on by E. Guéron & R. A. Mosna, *The relativistic glider*, Physical Review D 75, p. 081501(R), 2007, preprint at arxiv.org/abs/gr-qc/0612131. Cited on page 266.

233 S. Smale, *A classification of immersions of the two-sphere*, Transactions of the American Mathematical Society 90, pp. 281–290, 1958. Cited on page 267.

234 G. K. Francis & B. Morin, *Arnold Shapiro's Eversion of the Sphere*, Mathematical Intelligencer pp. 200–203, 1979. See also the unique manual for drawing manifolds by George Francis, *The Topological Picturebook*, Springer Verlag, 1987. It also contains a chapter on sphere eversion. Cited on page 267.

235 B. Morin & J.-P. Petit, *Le retournement de la sphere*, Pour la Science 15, pp. 34–41, 1979. See also the clear article by A. Phillips, *Turning a surface inside out*, Scientific American pp. 112–120, May 1966. Cited on page 267.

236 S. Levy, D. Maxwell & T. Munzner, *Making Waves – a Guide to the Ideas Behind Outside In*, Peters, 1995. Cited on page 267.

237 George K. Batchelor, *An Introduction to Fluid Mechanics*, Cambridge University Press, 1967, and H. Hashimoto, *A soliton on a vortex filament*, Journal of Fluid Mechanics 51, pp. 477–485, 1972. A summary is found in H. Zhou, *On the motion of slender vortex filaments*, Physics of Fluids 9, p. 970-981, 1997. Cited on pages 269 and 271.

238 V. P. Dmitriyev, *Helical waves on a vortex filament*, American Journal of Physics 73, pp. 563–565, 2005, and V. P. Dmitriyev, *Mechanical analogy for the wave-particle: helix on a vortex filament*, arxiv.org/abs/quant-ph/0012008. Cited on page 271.

239 T. Jacobson, *Thermodynamics of spacetime: the Einstein equation of state*, Physical Review Letters 75, pp. 1260–1263, 1995, or arxiv.org/abs/gr-qc/9504004. Cited on page 273.

240 J. Frenkel & T. Kontorowa, *Über die Theorie der plastischen Verformung*, Physikalische Zeitschrift der Sowietunion 13, pp. 1–10, 1938. F.C. Frank, *On the equations of motion of crystal dislocations*, Proceedings of the Physical Society A 62, pp. 131–134, 1949, J. Eshelby, *Uniformly moving dislocations*, Proceedings of the Physical Society A 62, pp. 307–314, 1949. See also G. Leibfried & H. Dietze, Zeitschrift für Physik 126, p. 790, 1949. A general introduction can be found in A. Seeger & P. Schiller, *Kinks in dislocations lines and their effects in internal friction in crystals*, Physical Acoustics 3A, W. P. Mason, ed., Academic Press, 1966. See also the textbooks by Frank R. N. Nabarro, *Theory of Crystal Dislocations*, Oxford University Press, 1967, or J. P. Hirth & J. Lothe, *Theory of Dislocations*, McGraw Hills Book Company, 1968. Cited on page 275.

241 Enthusiastic introductions into the theoretical aspects of polymers are Alexander Yu. Grosberg & Alexei R. Khokhlov, *Statistical Physics of Macromolecules*, AIP, 1994, and Pierre-Gilles de Gennes, *Scaling Concepts in Polymer Physics*, Cornell University Press, 1979. Cited on page 275.

242 The master of combining research and enjoyment in mathematics is Louis Kauffman. An example is his art is the beautiful text Louis H. Kauffman, *Knots and Physics*, World Scientific, third edition, 2001. It gives a clear introduction to the mathematics of knots and their applications. Cited on page 277.

243 A good introduction to knot tabulation is the paper by J. Hoste, M. Thistlethwaite & J. Weeks, *The first 1,701,936 knots*, The Mathematical Intelligencer 20, pp. 33–47, 1998. Cited on page 277.

244 I. Stewart, *Game, Set and Math*, Penguin Books, 1989, pp. 58–67. Cited on page 278.

245 P. Pieranski, S. Przybyl & A. Stasiak, *Tight open knots*, arxiv.org/abs/physics/0103016. No citations.

246 T. Ashton, J. Cantarella, M. Piatek & E. Rawdon, *Self-contact sets for 50 tightly knotted and linked tubes*, arxiv.org/abs/math/0508248. Cited on page 280.

247 J. Cantarella, J. H. G. Fu, R. Kusner, J. M. Sullivan & N. C. Wrinkle, *Criticality for the Gehring link problem*, Geometry and Topology 10, pp. 2055–2116, 2006, preprint at arxiv.org/abs/math/0402212. Cited on page 280.

248 W. R. Taylor, *A deeply knotted protein structure and how it might fold*, Nature 406, pp. 916–919, 2000. Cited on page 282.

249 Alexei Sossinsky, *Nœuds – histoire d'une théorie mathématique*, Editions du Seuil, 1999. D. Jensen, *Le poisson noué*, Pour la science, dossier hors série, pp. 14–15, April 1997. Cited on page 282.

250 D. M. Raymer & D. E. Smith, *Spontaneous knotting of an agitated string*, Proceedings of the National Academy of Sciences (USA) 104, pp. 16432–16437, 2007, or www.pnas.org/cgi/doi/10.1073/pnas.0611320104. This work won the humorous Ignobel Prize in Physics in 2008; seeimprobable.com/ig. Cited on page 282.

251 A. C. Hirshfeld, *Knots and physics: Old wine in new bottles*, American Journal of Physics 66, pp. 1060–1066, 1998. Cited on page 282.

252 For some modern knot research, see P. Holdin, R. B. Kusner & A. Stasiak, *Quantization of energy and writhe in self-repelling knots*, New Journal of Physics 4, pp. 20.1–20.11, 2002. Cited on page 283.

253 H. R. PRUPPACHER & J. D. KLETT, *Microphysics of Clouds and Precipitation*, Reidel, 1978, pp. 316–319. Falling drops are flattened and look like a pill, due to the interplay between surface tension and air flow. See also U. THIELE, *Weine nicht, wenn der Regen zerfällt*, Physik Journal 8, pp. 16–17, 2009. Cited on page 283.

254 J. J. SOCHA, *Becoming airborne without legs: the kinematics of take-off in a flying snake, Chrysopelea paradisi*, Journal of Experimental Biology 209, pp. 3358–3369, 2006, J. J. SOCHA, T. O'DEMPSEY & M. LABARBERA, *A three-dimensional kinematic analysis of gliding in a flying snake, Chrysopelea paradisi*, Journal of Experimental Biology 208, pp. 1817–1833, 2005, J. J. SOCHA & M. LABARBERA, *Effects of size and behavior on aerial performance of two species of flying snakes (Chrysopelea)*, Journal of Experimental Biology 208, pp. 1835–1847, 2005. A full literature list on flying snakes can be found on the website www.flyingsnake.org. Cited on page 284.

255 An informative account of the world of psychokinesis and the paranormal is given by the famous professional magician JAMES RANDI, *Flim-flam!*, Prometheus Books, Buffalo 1987, as well as in several of his other books. See also the www.randi.org website. No citations.

256 JAMES CLERK MAXWELL, *Scientific Papers*, 2, p. 244, October 1871. Cited on page 294.

257 A good introduction is C. J. HOGAN, *Why the universe is just so*, Reviews of Modern Physics 72, pp. 1149–1161, 2000. Most of the material of Table 25 is from the mighty book by JOHN D. BARROW & FRANK J. TIPLER, *The Anthropic Cosmological Principle*, Oxford University Press, 1986. Discarding unrealistic options is also an interesting pastime. See for example the reasons why life can only be carbon-based, as explained in the essay by I. ASIMOV, *The one and only*, in his book *The Tragedy of the Moon*, Doubleday, Garden City, New York, 1973. Cited on pages 294 and 359.

258 L. SMOLIN, *The fate of black hole singularities and the parameters of the standard models of particle physics and cosmology*, arxiv.org/abs/gr-qc/9404011. Cited on page 296.

259 ARISTOTLE, *Treaty of the heaven*, III, II, 300 b 8. See JEAN-PAUL DUMONT, *Les écoles présocratiques*, Folio Essais, Gallimard, p. 392, 1991. Cited on page 299.

260 *Le Système International d'Unités*, Bureau International des Poids et Mesures, Pavillon de Breteuil, Parc de Saint Cloud, 92310 Sèvres, France. All new developments concerning SI units are published in the journal *Metrologia*, edited by the same body. Showing the slow pace of an old institution, the BIPM launched a website only in 1998; it is now reachable at www.bipm.fr. See also the www.utc.fr/~tthomass/Themes/Unites/index.html website; this includes the biographies of people who gave their names to various units. The site of its British equivalent, www.npl.co.uk/npl/reference, is much better; it provides many details as well as the English-language version of the SI unit definitions. Cited on page 300.

261 The bible in the field of time measurement is the two-volume work by J. VANIER & C. AUDOIN, *The Quantum Physics of Atomic Frequency Standards*, Adam Hilge, 1989. A popular account is TONY JONES, *Splitting the Second*, Institute of Physics Publishing, 2000.

  The site opdafl.obspm.fr/www/lexique.html gives a glossary of terms used in the field. For precision *length* measurements, the tools of choice are special lasers, such as mode-locked lasers and frequency combs. There is a huge literature on these topics. Equally large is the literature on precision *electric current* measurements; there is a race going on for the best way to do this: counting charges or measuring magnetic forces. The issue is still open. On *mass* and atomic mass measurements, see page 66 in volume II. On high-precision *temperature* measurements, see page 478 in volume I. Cited on page 301.

262 The unofficial prefixes were first proposed in the 1990s by Jeff K. Aronson of the University of Oxford, and might come into general usage in the future. Cited on page 302.

263 For more details on electromagnetic unit systems, see the standard text by JOHN DAVID JACKSON, *Classical Electrodynamics*, 3rd edition, Wiley, 1998. Cited on page 305.

264 D.J. BIRD & al., *Evidence for correlated changes in the spectrum and composition of cosmic rays at extremely high energies*, Physical Review Letters 71, pp. 3401–3404, 1993. Cited on page 306.

265 P.J. HAKONEN, R.T. VUORINEN & J.E. MARTIKAINEN, *Nuclear antiferromagnetism in rhodium metal at positive and negative nanokelvin temperatures*, Physical Review Letters 70, pp. 2818–2821, 1993. See also his article in Scientific American, January 1994. Cited on page 306.

266 A. ZEILINGER, *The Planck stroll*, American Journal of Physics 58, p. 103, 1990. Can you find another similar example? Cited on page 306.

267 An overview of this fascinating work is given by J.H. TAYLOR, *Pulsar timing and relativistic gravity*, Philosophical Transactions of the Royal Society, London A 341, pp. 117–134, 1992. Cited on page 306.

268 The most precise clock built in 2004, a caesium fountain clock, had a precision of one part in 10^{15}. Higher precision has been predicted to be possible soon, among others by M. TAKAMOTO, F.-L. HONG, R. HIGASHI & H. KATORI, *An optical lattice clock*, Nature 435, pp. 321–324, 2005. Cited on page 306.

269 J. BERGQUIST, ed., *Proceedings of the Fifth Symposium on Frequency Standards and Metrology*, World Scientific, 1997. Cited on page 306.

270 See the information on $D_s^{\pm}$ mesons from the particle data group at pdg.web.cern.ch/pdg. Cited on page 307.

271 About the long life of tantalum 180, see D. BELIC & al., *Photoactivation of $^{180}Ta^m$ and its implications for the nucleosynthesis of nature's rarest naturally occurring isotope*, Physical Review Letters 83, pp. 5242–5245, 20 December 1999. Cited on page 307.

272 The various concepts are even the topic of a separate international standard, ISO 5725, with the title *Accuracy and precision of measurement methods and results*. A good introduction is JOHN R. TAYLOR, *An Introduction to Error Analysis: the Study of Uncertainties in Physical Measurements*, 2nd edition, University Science Books, Sausalito, 1997. Cited on page 307.

273 P.J. MOHR, B.N. TAYLOR & D.B. NEWELL, *CODATA recommended values of the fundamental physical constants: 2010*, preprint at arxiv.org/abs/1203.5425. This is the set of constants resulting from an international adjustment and recommended for international use by the Committee on Data for Science and Technology (CODATA), a body in the International Council of Scientific Unions, which brings together the International Union of Pure and Applied Physics (IUPAP), the International Union of Pure and Applied Chemistry (IUPAC) and other organizations. The website of IUPAC is www.iupac.org. Cited on pages 309 and 377.

274 Some of the stories can be found in the text by N.W. WISE, *The Values of Precision*, Princeton University Press, 1994. The field of high-precision measurements, from which the results on these pages stem, is a world on its own. A beautiful introduction to it is J.D. FAIRBANKS, B.S. DEAVER, C.W. EVERITT & P.F. MICHAELSON, eds., *Near Zero: Frontiers of Physics*, Freeman, 1988. Cited on page 309.

275 The details are given in the well-known astronomical reference, P. KENNETH SEIDELMANN, *Explanatory Supplement to the Astronomical Almanac*, 1992. Cited on page 314.

276 For information about the number π, and about some other mathematical constants, the website oldweb.cecm.sfu.ca/pi/pi.html provides the most extensive information and refer-

Challenge 208 e

ences. It also has a link to the many other sites on the topic, including the overview at mathworld.wolfram.com/Pi.html. Simple formulae for π are

$$\pi + 3 = \sum_{n=1}^{\infty} \frac{n\, 2^n}{\binom{2n}{n}} \tag{159}$$

or the beautiful formula discovered in 1996 by Bailey, Borwein and Plouffe

$$\pi = \sum_{n=0}^{\infty} \frac{1}{16^n} \left(\frac{4}{8n+1} - \frac{2}{8n+4} - \frac{1}{8n+5} - \frac{1}{8n+6} \right) . \tag{160}$$

The mentioned site also explains the newly discovered methods for calculating specific binary digits of π without having to calculate all the preceding ones. The known digits of π pass all tests of randomness, as the mathworld.wolfram.com/PiDigits.html website explains. However, this property, called *normality*, has never been proven; it is the biggest open question about π. It is possible that the theory of chaotic dynamics will lead to a solution of this puzzle in the coming years.

Another method to calculate π and other constants was discovered and published by D. V. Chudnovsky & G. V. Chudnovsky, *The computation of classical constants*, Proceedings of the National Academy of Sciences (USA) **86**, pp. 8178–8182, 1989. The Chudnowsky brothers have built a supercomputer in Gregory's apartment for about 70 000 euros, and for many years held the record for calculating the largest number of digits of π. They have battled for decades with Kanada Yasumasa, who held the record in 2000, calculated on an industrial supercomputer. However, the record number of (consecutive) digits in 2010 was calculated in 123 days on a simple desktop PC by Fabrice Bellard, using a Chudnovsky formula. Bellard calculated over 2.7 million million digits, as told on bellard.org. New formulae to calculate π are still occasionally discovered.

For the calculation of Euler's constant γ see also D. W. DeTemple, *A quicker convergence to Euler's constant*, The Mathematical Intelligencer, pp. 468–470, May 1993. Cited on page 316.

277 The proton charge radius was determined by measuring the frequency of light emitted by hydrogen atoms to high precision by T. Udem, A. Huber, B. Gross, J. Reichert, M. Prevedelli, M. Weitz & T. W. Hausch, *Phase-coherent measurement of the hydrogen 1S–2S transition frequency with an optical frequency interval divider chain*, Physical Review Letters **79**, pp. 2646–2649, 1997. Cited on page 317.

278 For a full list of isotopes, see R. B. Firestone, *Table of Isotopes, Eighth Edition, 1999 Update*, with CD-ROM, John Wiley & Sons, 1999. For a list of isotopes on the web, see the Corean website by J. Chang, atom.kaeri.re.kr. For a list of precise isotope masses, see the csnwww.in2p3.fr website. Cited on pages 318, 321, and 332.

279 The ground state of bismuth 209 was thought to be stable until early 2003. It was then discovered that it was radioactive, though with a record lifetime, as reported by P. de Marcillac, N. Coron, G. Dambier, J. Leblanc & J.-P. Moalic, *Experimental detection of α-particles from the radioactive decay of natural bismuth*, Nature **422**, pp. 876–878, 2003. By coincidence, the excited state 83 MeV above the ground state of the same bismuth 209 nucleus is the shortest known radioactive nuclear state. Cited on page 318.

280 For information on the long life of tantalum 180, see D. Belic & al., *Photoactivation of ^{180}Ta^m and its implications for the nucleosynthesis of nature's rarest naturally occurring isotope*, Physical Review Letters **83**, pp. 5242–5245, 20 December 1999. Cited on page 318.

281 STEPHEN J. GOULD, *The Panda's Thumb*, W.W. Norton & Co., 1980. This is one of several interesting and informative books on evolutionary biology by the best writer in the field. Cited on page 319.

282 F. MARQUES & al., *Detection of neutron clusters*, Physical Review C **65**, p. 044006, 2002. Opposite results have been obtained by B. M. SHERRILL & C. A. BERTULANI, *Proton-tetraneutron elastic scattering*, Physical Review C **69**, p. 027601, 2004, and D.V. ALEKSANDROV & al., *Search for resonances in the three- and four-neutron systems in the 7Li(7Li, 11C)3n and 7Li(7Li, 10C)4n reactions*, JETP Letters **81**, p. 43, 2005. No citations.

283 For a good review, see the article by P. T. GREENLAND, *Antimatter*, Contemporary Physics **38**, pp. 181–203, 1997. Cited on page 319.

284 Almost everything known about each element and its chemistry can be found in the encyclopaedic GMELIN, *Handbuch der anorganischen Chemie*, published from 1817 onwards. There are over 500 volumes, now all published in English under the title *Handbook of Inorganic and Organometallic Chemistry*, with at least one volume dedicated to each chemical element. On the same topic, an incredibly expensive book with an equally bad layout is PER ENHAG, *Encyclopedia of the Elements*, Wiley–VCH, 2004. Cited on page 321.

285 The atomic masses, as given by IUPAC, can be found in Pure and Applied Chemistry **73**, pp. 667–683, 2001, or on the www.iupac.org website. For an isotope mass list, see the csnwww.in2p3.fr website. Cited on pages 321, 331, and 332.

286 The metallic, covalent and Van der Waals radii are from NATHANIEL W. ALCOCK, *Bonding and Structure*, Ellis Horwood, 1999. This text also explains in detail how the radii are defined and measured. Cited on page 331.

287 M. FLATO, P. SALLY & G. ZUCKERMAN (editors), *Applications of Group Theory in Physics and Mathematical Physics*, Lectures in applied mathematics, volume 21, American Mathematical Society, 1985. This interesting book was written before the superstring revolution, so that the topic is missing from the otherwise excellent presentation. Cited on page 340.

288 For more puzzles, see the excellent book JAMES TANTON, *Solve This – Math Activities for Students and Clubs*, Mathematical Association of America, 2001. Cited on page 340.

289 For an introduction to topology, see for example MIKIO NAKAHARA, *Geometry, Topology and Physics*, IOP Publishing, 1990. Cited on page 341.

290 An introduction to the classification theorem is R. SOLOMON, *On finite simple groups and their classification*, Notices of the AMS **42**, pp. 231–239, 1995, also available on the web as www.ams.org/notices/199502/solomon.ps Cited on page 345.

291 A pedagogical explanation is given by C. G. WOHL, *Scientist as detective: Luis Alvarez and the pyramid burial chambers, the JFK assassination, and the end of the dinosaurs*, American Journal of Physics **75**, pp. 968–977, 2007. Cited on page 355.

CREDITS

Many people who have kept their gift of curiosity alive have helped to make this project come true. Most of all, Saverio Pascazio has been – present or not – a constant reference for this project. Fernand Mayné, Anna Koolen, Ata Masafumi, Roberto Crespi, Serge Pahaut, Luca Bombelli, Herman Elswijk, Marcel Krijn, Marc de Jong, Martin van der Mark, Kim Jalink, my parents Peter and Isabella Schiller, Mike van Wijk, Renate Georgi, Paul Tegelaar, Barbara and Edgar Augel, M. Jamil, Ron Murdock, Carol Pritchard, Richard Hoffman, Stephan Schiller and, most of all, my wife Britta have all provided valuable advice and encouragement.

Many people have helped with the project and the collection of material. Most useful was the help of Mikael Johansson, Bruno Barberi Gnecco, Lothar Beyer, the numerous improvements by Bert Sierra, the detailed suggestions by Claudio Farinati, the many improvements by Eric Sheldon, the detailed suggestions by Andrew Young, the continuous help and advice of Jonatan Kelu, the corrections of Elmar Bartel, and in particular the extensive, passionate and conscientious help of Adrian Kubala.

Important material was provided by Bert Peeters, Anna Wierzbicka, William Beaty, Jim Carr, John Merrit, John Baez, Frank DiFilippo, Jonathan Scott, Jon Thaler, Luca Bombelli, Douglas Singleton, George McQuarry, Tilman Hausherr, Brian Oberquell, Peer Zalm, Martin van der Mark, Vladimir Surdin, Julia Simon, Antonio Fermani, Don Page, Stephen Haley, Peter Mayr, Allan Hayes, Norbert Dragon, Igor Ivanov, Doug Renselle, Wim de Muynck, Steve Carlip, Tom Bruce, Ryan Budney, Gary Ruben, Chris Hillman, Olivier Glassey, Jochen Greiner, squark, Martin Hardcastle, Mark Biggar, Pavel Kuzin, Douglas Brebner, Luciano Lombardi, Franco Bagnoli, Lukas Fabian Moser, Dejan Corovic, Paul Vannoni, John Haber, Saverio Pascazio, Klaus Finkenzeller, Leo Volin, Jeff Aronson, Roggie Boone, Lawrence Tuppen, Quentin David Jones, Arnaldo Uguzzoni, Frans van Nieuwpoort, Alan Mahoney, Britta Schiller, Petr Danecek, Ingo Thies, Vitaliy Solomatin, Carl Offner, Nuno Proença, Elena Colazingari, Paula Henderson, Daniel Darre, Wolfgang Rankl, John Heumann, Joseph Kiss, Martha Weiss, Antonio González, Antonio Martos, André Slabber, Ferdinand Bautista, Zoltán Gácsi, Pat Furrie, Michael Reppisch, Enrico Pasi, Thomas Köppe, Martin Rivas, Herman Beeksma, Tom Helmond, John Brandes, Vlad Tarko, Nadia Murillo, Ciprian Dobra, Romano Perini, Harald van Lintel, Andrea Conti, François Belfort, Dirk Van de Moortel, Heinrich Neumaier, Jarosław Królikowski, John Dahlman, Fathi Namouni, Paul Townsend, Sergei Emelin, Freeman Dyson, S.R. Madhu Rao, David Parks, Jürgen Janek, Daniel Huber, Alfons Buchmann, William Purves, Pietro Redondi, Douglas Singleton, Emil Akhmedov Damoon Saghian, Zach Joseph Espiritu, plus a number of people who wanted to remain unnamed.

I also thank the lawmakers and the taxpayers in Germany, who, in contrast to most other countries in the world, allow residents to use the local university libraries.

The software tools were refined with extensive help on fonts and typesetting by Michael Zedler and Achim Blumensath and with the repeated and valuable support of Donald Arseneau; help came also from Ulrike Fischer, Piet van Oostrum, Gerben Wierda, Klaus Böhncke, Craig Upright, Herbert Voss, Andrew Trevorrow, Danie Els, Heiko Oberdiek, Sebastian Rahtz, Don Story, Vincent Darley, Johan Linde, Joseph Hertzlinger, Rick Zaccone, John Warkentin, Ulrich Diez, Uwe Siart, Will Robertson, Joseph Wright, Enrico Gregorio, Rolf Niepraschk, Alexander Grahn and Paul Townsend.

The typesetting and book design is due to the professional consulting of Ulrich Dirr. The typography was improved with the help of Johannes Küster. The design of the book and its website owe also much to the suggestions and support of my wife Britta.

 Since May 2007, the electronic edition and distribution of the Motion Mountain text is generously supported by the Klaus Tschira Foundation.

FILM CREDITS

The animation of the actin–myosin system on page 22 is copyright and courtesy by San Diego State University, Jeff Sale and Roger Sabbadini. It can be found on the website www.sci.sdsu.edu/movies/actin_myosin.html. The figure of the violin on page 102 is copyright Franz Aichinger and courtesy of EOS. The figure of floating display on page 102 is copyright and courtesy of Burton Inc. and found on www.burton-jp.com. The film on page 104, showing single electrons moving through liquid helium, is copyright and courtesy of Humphrey Maris; it can be found on his website physics.brown.edu/physics/researchpages/cme/bubble. The film of the spark chamber showing the cosmic rays on page 159 is courtesy and copyright Wolfgang Rueckner and found on isites.harvard.edu/icb. The film of the solar flare on page 186 was taken by NASA's TRACE satellite, whose website is trace.lmsal.com. The film of a flying snake on page 284 is copyright and courtesy by Jake Socha. It can be found on his website www.flyingsnake.org.

IMAGE CREDITS

The photograph of the east side of the Langtang Lirung peak in the Nepalese Himalayas, shown on the front cover, is courtesy and copyright by Dave Thompson and used to be on his website www.daveontrek.co.uk. The photograph of a soap bubble on page 14 is courtesy and copyright of Jason Tozer for Creative Review/Sony; its making is described on creativereview.co.uk/crblog. The whale photograph on page 16 is courtesy by the NOAA in the USA. The photograph of a baobab with elephant is courtesy and copyright by Ferdinand Reus. The picture of the orchid on page 19 is courtesy and copyright of Elmar Bartel. The sea urchin photograph on page 23 was taken by Kristina Yu, and is copyright and courtesy of the Exploratorium, found at www.exploratorium.edu. The images of embronic nodes on page 27 are courtesy and copyright by Hirokawa Nobutaka and found in the cited paper. The illustrations of DNA on page 59 are copyright and courtesy by David Deerfield and dedicated to his memory. They are taken from his website www.psc.edu/~deerfiel/Nucleic_Acid-SciVis.html. The images of DNA molecules on page 60 are copyright and courtesy by Hans-Werner Fink and used with permission of Wiley VCH. The mineral photographs on page 67 onwards are copyright and courtesy of Rob Lavinsky at irocks.com, and taken from his large and beautiful collection there and at www.mindat.org/photo-49529.html. The corundum photograph on page 67, the malachite photograph on page 161 and the magnetite photograph on page page 163 are copyright and courtesy of Stephan Wolfsried and taken from his beautiful collection of microcrystals at www.mindat.org/user-1664.html. The images of crystal growth techniques on page 83 are copyright of the Ivan Golota and used with his permission and that of Andrey Vesselovsky. The photograph of synthetic corundum on page 68 is copyright of the Morion Company at www.motioncompany.com and courtesy of Uriah Prichard. The photograph of Paraiba tourmalines on page page 68 is copyright and courtesy of Manfred Fuchs and found on his website at www.tourmalinavitalis.de. The synthetic crystals photographs on page 69 are copyright of Northrop Grumman and courtesy of Scott Griffin.

The alexandrite photographs on page 69 are copyright and courtesy of John Veevaert, Trinity Mineral Co., trinityminerals.com. The photograph of PZT – lead zirconium titanate – products on page 70 is copyright and courtesy of Ceramtec. The photographs of the silicon crystal growing machines and of the resulting silicon crystals are copyright of PVATePla at www.PVATePla.com, and are used by kind permission of Gert Fisahn. The shark teeth photograph on page 73 is courtesy and copyright of Peter Doe and found at www.flickr.com/photos/peteredin. The climate graph on page 82 is copyright and courtesy of Dieter Lüthi. The photograph and drawings of a switchable mirror on page 76 are courtesy and copyright Ronald Griessen and found on his website www.nat.vu.nl/~griessen. The photograph of a bystander with light and terahertz waves on page 81 are courtesy of the Jefferson Lab. The photograph of a Hall probe on page 84 is courtesy and copyright Metrolab, and found on their website at www.metrolab.com. The collection of images of single atom sheets on page 86 is taken from Ref. 59, copyright of the National Academy of Sciences, and are courtesy of the authors and of Tiffany Millerd. The graphene photograph on page 86 is copyright and courtesy of Andre Geim. The rhenium photographs on page 65 are copyright and courtesy of Hans-Werner Fink, APS and AIP and reprinted with permission from the cited references. The pictures of snow flakes on page 65 are courtesy and copyright of Furukawa Yoshinori. The pictures of rainbows on page 122 are courtesy and copyright of ed g2s, found on wikimedia commons, and of Christophe Afonso, found on his beautiful site www.flickr.com/photos/chrisafonso21. The picture of neutron interferometers on page 126 is courtesy and copyright of Helmut Rauch and Erwin Seidl, and found on the website of the Atominstitut in Vienna at www.ati.ac.at. The simulated view of a black hole on page 129 is copyright and courtesy of Ute Kraus and can be found on her splendid website www.tempolimit-lichtgeschwindigkeit.de. The map with the γ ray bursts on page 136 is courtesy of NASA and taken from the BATSE website at www.batse.msfc.nasa.gov/batse. The MRI images of the head and the spine on page 147 are courtesy and copyright of Joseph Hornak and taken from his website www.cis.rit.edu/htbooks/mri. The MRI image on page 375 of humans making love is courtesy and copyright of Willibrord Weijmar Schultz. The nuclide chart on page 156 is used with permission of Alejandro Sonzogni of the National Nuclear Data Centre, using data extracted from the NuDat database. The electroscope photograph on page 157 is courtesy and copyright of Harald Chmela and taken from his website www.hcrs.at. The photograph of an aurora on page 162 is courtesy and copyright of Jan Curtis and taken from his website climate.gi.alaska.edu/Curtis/curtis.html. The photograph of the Geigerpod on page 158 is courtesy and copyright of Joseph Reinhardt, and found on his website flickr.com/photos/javamoose. The figure of the Erta Ale volcano on page 169 is courtesy and copyright of Marco Fulle and part of the wonderful photograph collection at www.stromboli.net. The drawing of the JET reactor on page 191 is courtesy and copyright of EFDA-JET. The drawing of the strong coupling constant on page page 211 is courtesy and copyright of Siegfried Bethke. The photograph of the watch on page 219 is courtesy and copyright of Traser, found at www.traser.com. The photograph of the Sudbury Neutrino Observatory on page 234 is courtesy and copyright SNO, and found on their website at http://www.sno.phy.queensu.ca/sno/images. The unification graph on page 249 is courtesy and copyright of W. de Boer and taken from his home page at www-ekp.physik.uni-karlsruhe.de/~deboer. The picture of the butterfly in a wind tunnel on page 257 is courtesy and copyright of Robert Srygley and Adrian Thomas. The picture of the vulture on page 259 is courtesy and copyright of S.L. Brown from the website SLBrownPhoto.com. The picture of the hummingbird on page 259 is courtesy and copyright of the Pennsylvania Game Commission and Joe Kosack and taken from the website www.pgc.state.pa.us. The picture of the dragonfly on page 259 is courtesy and copyright of nobodythere and found on his website xavier.favre2.free.fr/coppermine. The pictures of feathers insects on page 260 are material used with kind permission of HortNET, a product of The Horticulture and Food Research Institute of New Zealand. The picture of the scallop on page 261 is courtesy and copyright of Dave Colwell.

NAME INDEX

IRR

G

ASCAZIO

Z

ZEL'DOVICH

SUBJECT INDEX

C

CORE

ARTH

ADRON

K

KILOGRAM

M

AGNETIC

M
MOTORS

P

─────

PREDICTIONS

S

SNAKES

V

VACUUM

7195207R00224

Made in the USA
San Bernardino, CA
26 December 2013